Lecture Notes in Bioinformatics 8492

Subseries of Lecture Notes in Computer Science

T0212823

Mitra Basu Yi Pan Jianxin Wang (Eds.)

Bioinformatics Research and Applications

10th International Symposium, ISBRA 2014
Zhangjiajie, China, June 28-30, 2014
Proceedings

 Springer

Volume Editors

Mitra Basu
Johns Hopkins University
Computer Science Department
Baltimore, MD 21218, USA
and National Science Foundation, CCF
Arlington, VA 22230, USA
E-mail: mbasu@nsf.gov

Yi Pan
Georgia State University
Department of Computer Science
Atlanta, GA 30303, USA
E-mail: yipan@gsu.edu

Jianxin Wang
Central South University
School of Information Science and Engineering
Changsha, 410083, China
E-mail: jxwang@mail.csu.edu.cn

ISSN 0302-9743   e-ISSN 1611-3349
ISBN 978-3-319-08170-0   e-ISBN 978-3-319-08171-7
DOI 10.1007/978-3-319-08171-7
Springer Cham Heidelberg New York Dordrecht London

Library of Congress Control Number: 2014940855

LNCS Sublibrary: SL 8 – Bioinformatics

Typesetting: Camera-ready by author, data conversion by Scientific Publishing Services, Chennai, India

Printed on acid-free paper

Springer is part of Springer Science+Business Media (www.springer.com)

Preface

The International Symposium on Bioinformatics Research and Applications (IS-BRA) provides a forum for the exchange of ideas and results among researchers, developers, and practitioners working on all aspects of bioinformatics and computational biology and their applications. Submissions presenting original research are solicited in all areas of bioinformatics and computational biology, including the development of experimental or commercial systems.

The 10*th* edition of the International Symposium on Bioinformatics Research and Applications (ISBRA 2014) was held during June 28–30, 2014, in Zhangjiajie, China. ISBRA 2014 received 119 full paper submissions. Every paper went through a very rigorous review process. Each paper was reviewed by two to five Program Committee members. After careful consideration, 33 papers were accepted as regular papers in Track 1 (28.57% acceptance rate) and 32 papers were accepted as one-page abstracts in Track 2 (26.05% acceptance rate). Additionally, the symposium included three invited keynote talks by distinguished speakers: Prof. John E. Hopcroft from Cornell University, USA, Prof. Ming Li from University of Waterloo, Canada, and Prof. Ying Xu from University of Georgia, USA.

We would like to thank the Program Committee members and external reviewers for volunteering their time to review and discuss the symposium papers. We would like to extend special thanks to the steering and general chairs of the symposium for their leadership, and to the finance, publication, publicity, and local organization chairs for their hard work in making ISBRA 2014 a successful event. Last but not least, we would like to thank all authors for presenting their work at the symposium.

June 2014

Mitra Basu
Yi Pan
Jianxin Wang

Symposium Organization

Steering Chairs

Alex Zelikovsky   Georgia State University, USA
Dan Gusfield   University of California, Davis, USA
Ion Mandoiu   University of Connecticut, USA
Marie-France Sagot   Inria, France
Yi Pan   Georgia State University, USA
Ying Xu   University of Georgia, USA

General Chairs

Albert Zomaya   University of Sydney, Australia
Ming Li   University of Waterloo, Canada

Program Chairs

Mitra Basu   Johns Hopkins University, National Science
         Foundation, USA
Yi Pan   Georgia State University, USA
Jianxin Wang   Central South University, China

Publication Chair

Min Li   Central South University, China

Local Organizing Chairs

Jianxin Wang   Central South University, China
Qingping Zhou   Jishou University, China

Local Organizing Committee

Min Li   Central South University, China
Mingxing Zeng   Jishou University, China
Yu Sheng   Central South University, China
Guihua Duan   Central South University, China
Li Wang   Jishou University, China
Yanping Yang   Jishou University, China

Program Committee

Srinivas Aluru	IIT Bombay/Iowa State University, India/USA
Mitra Basu	National Science Foundation, USA
Robert Beiko	Dalhousie University, Canada
Paola Bonizzoni	Università di Milano-Bicocca, Italy
Zhipeng Cai	Georgia State University, USA
Doina Caragea	Kansas State University, USA
Tien-Hao Chang	National Cheng Kung University
Ovidiu Daescu	University of Texas at Dallas, USA
Bhaskar Dasgupta	University of Illinois at Chicago, USA
Amitava Datta	University of Western Australia
Oliver Eulenstein	Iowa State University, USA
Guillaume Fertin	LINA, UMR CNRS 6241, University of Nantes, France
Lin Gao	Xidian University, China
Katia Guimaraes	UFPE, Brazil
Jiong Guo	Universität des Saarlandes, Germany
Jieyue He	Southeast University, China
Matthew He	Nova Southeastern University, USA
Steffen Heber	NCSU, USA
Wei Hu	Houghton College, USA
Xiaohua Tony Hu	Drexel University, USA
Jinling Huang	East Carolina University, USA
Lars Kaderali	University of Technology Dresden, Germany
Iyad Kanj	DePaul University, USA
Ming-Yang Kao	Northwestern University, USA
Yury. Khudyakov	Centers for Disease Control and Prevention, USA
Wooyoung Kim	University of Washington Bothell, USA
Danny Krizanc	Wesleyan University, USA
Guojun Li	Shandong University, China
Jing Li	Case Western Reserve University, USA
Min Li	Central South University, China
Shuaicheng Li	City University of Hong Kong, SAR China
Yanchun Liang	Jilin University, China
Zhiyong Liu	Institute of Computing Technology, Chinese Academy of Science
Ion Mandoiu	University of Connecticut, USA
Fenglou Mao	University of Georgia, USA
Osamu Maruyama	Kyushu University, Japan
Giri Narasimhan	Florida International University, USA
Yi Pan	Georgia State University, USA

Andrei Paun	University of Bucharest, Romania
Nadia Pisanti	Università di Pisa, Italy
Teresa Przytycka	NIH, USA
Sven Rahmann	University of Duisburg-Essen, Germany
David Sankoff	University of Ottawa, Canada
Daniel Schwartz	University of Connecticut, USA
Russell Schwartz	Carnegie Mellon University, USA
Joao Setubal	University of São Paulo, Brazil
Xinghua Shi	University of North Carolina at Charlotte, USA
Ileana Streinu	Smith College, Northampton, USA
Zhengchang Su	University of North Carolina at Charlotte, USA
Raj Sunderraman	Georgia State University, USA
Wing-Kin Sung	National University of Singapore
Sing-Hoi Sze	Texas A&M University, USA
Gabriel Valiente	Technical University of Catalonia, Spain
Stéphane Vialette	Université Paris-Est LIGM UMR CNRS 8049, France
Jianxin Wang	Central South University, China
Li-San Wang	University of Pennsylvania, USA
Lusheng Wang	City University of Hong Kong, SAR China
Peng Wang	Chinese Academy of Sciences
Fangxiang Wu	University of Saskatchewan, Canada
Yufeng Wu	University of Connecticut, USA
Minzhu Xie	Hunan Normal University, China
Dechang Xu	Harbin Institute of Technology, China
Zhenyu Xuan	University of Texas at Dallas, USA
Zuguo Yu	Queensland University of Technology, Australia
Alex Zelikovsky	GSU, USA
Fa Zhang	Institute of Computing Technology, China
Fengfeng Zhou	Chinese Academy of Sciences
Leming Zhou	University of Pittsburgh, USA

Additional Reviewers

Alonso Alemany, Daniel
Anghelache, Andreea
Beissbarth, Tim
Beißer, Daniela
Bingbo, Wang
Campo, David S.
Caravagna, Giulio
Cardona, Gabriel
Cho, Dongyeon
Chowdhury, Salim

Cliquet, Freddy
Curé, Olivier
Dao, Phuong
Dondi, Riccardo
Du, Xiangjun
Falca, Elena-Bianca
Guo, Xingli
Hayes, Matthew
Herrmann, Carl
Hoinka, Jan

Hwang, Yih-Chii
Jiang, Ruhua
Jiang, Xingpeng
Kim, Yooah
Knapp, Bettina
Komusiewicz, Christian
Köster, Johannes
Lara, James
Lauber, Chris
Li, Fan
Li, Xiaojie
Llabrés, Mercè
Luo, Junwei
Menconi, Giulia
Mirzaei, Sajad
Pan, Yi
Pei, Jingwen
Peng, Wei
Peng, Xiaoqing

Peterlongo, Pierre
Pirola, Yuri
Rizzi, Raffaella
Rocha, Jairo
Roman, Theodore
Sun, Peng Gang
Valentini, Giorgio
Venturini, Rossano
Wan, Xiaohua
Wang, Weichao
Wang, Wenhui
Wang, Yan
Wohlers, Inken
Wójtowicz, Damian
Zhang, Fa
Zheng, Qi
Zhong, Jiancheng
Zhou, Chan

Elucidation of Key Drivers and Facilitators of Cancer Initiation and Metastasis: A Data-Mining Approach

(Invited Talk)

Ying Xu, Ph.D.

Department of Biochemistry and Molecular Biology, and Institute of
Bioinformatics,
University of Georgia, USA, and College of Computer Science and Technology,
Jilin University, China

Numerous theories and hypotheses have been proposed in the past 100 years regarding what drive a cancer to initiate, progress and metastasize, including (1) the now popular view of cancer as a result of genomic mutations; (2) cancer being induced by viral or bacterial infection; and (3) cancer resulted from malfunctioning mitochondria. I will present our recent work on (i) key drivers of cancer initiation and (ii) drivers of post-metastatic cancer's explosive growth, based on comparative and integrative analyses of very large scale of multiple type of omic data collected on cancer tissues. On (i), our starting point is a speculation made by Nobel Laureate Otto Warburg in the 1960s: "Cancer ... has countless secondary causes. But ... there is only one prime cause, [which] is the replacement of respiration of oxygen in normal body cells by a fermentation of sugar." While increasingly more cancer researchers tend to agree with Warburg, the link between the observed reprogramming of energy metabolism and cell proliferation is unknown. We have recently discovered that hyaluronic acid may be the missing link through statistical analyses of omic data of different types of cancer; and developed a detailed model in linking energy metabolism reprogramming and cell proliferation. On (ii), metastatic cancer is responsible for 90% of cancer-related mortalities, and has been considered as a terminal illness, mainly based on past experience in largely unsuccessful treatment of metastatic cancers using drugs designed for primary cancer. We have recently discovered that fundamentally different from primary cancer, metastatic cancer is predominantly driven by a different force, i.e., oxidized cholesterols and their steroidogenic metabolites. A detailed model is proposed regarding (a) why metastatic cancer tends to have increased cholesterol influx and (b) how oxidized cholesterol products drive metastatic cancers. Both studies suggest fundamentally different ways to view and possibly treat cancer.

Table of Contents

Full Papers

Abstracts

Predicting Disease Risks Using Feature Selection Based on Random Forest and Support Vector Machine

Jing Yang[1], Dengju Yao[1,2], Xiaojuan Zhan[3], and Xiaorong Zhan[4]

[1] College of Computer Science and Technology, Harbin Engineering University, Harbin, China
yangjing@hrbeu.edu.cn
[2] School of Software, Harbin University of Science and Technology, Harbin, China
ydkvictory@163.com
[3] College of Computer Science and Technology,
Heilongjiang Institute of Technology, Harbin China
xiaojuanzhan@gmail.com
[4] Department of Endocrinology, First Affiliated Hospital,
Harbin Medical University, Harbin China
xiaorongzhan@sina.com

Abstract. Disease risk prediction is an important task in biomedicine and bioinformatics. To resolve the problem of high-dimensional features space and highly feature redundancy and to improve the intelligibility of data mining results, a new wrapper method of feature selection based on random forest variables importance measures and support vector machine was proposed. The proposed method combined sequence backward searching approach and sequence forward searching approach. Feature selection starts with the entire set of features in the dataset. At every iteration, two feature subsets are gained. One feature subset removes those most unimportant features and the most important feature at the same time, which is used to train random forest and to compute feature importance for next feature selection. Another feature subset removes only those most unimportant features while remains the most important feature, which is used as the optimal feature subset to train SVM classifier. Finally, the feature subset with the highest SVM classification accuracy was regarded as optimal feature subset. The experimental results on 11 UCI datasets, a real clinical data sets and a gene expression dataset show that the proposed algorithm can generate the smaller feature subset while improve the classification accuracy.

Keywords: Disease risk prediction, Feature selection, High dimensional data, Random forest, Support vector machine.

1   Introduction

Disease risk prediction is an important issue in biomedical and bioinformatics. High-dimensional and redundant features in medical and biological data have created an urgent need for feature selection techniques [1]. In general, feature selection algorithms can be divided into Filter methods and Wrapper methods by the adopted

M. Basu, Y. Pan, and J. Wang (Eds.): ISBRA 2014, LNBI 8492, pp. 1–11, 2014.

feature selection strategy [2]. Filter methods are independent to machine learning algorithms and can quickly remove out noise features and narrows searching range of the optimal feature subset, but it does not guarantee find out a smaller optimized feature subset. Conversely, Wrapper methods use the selected feature subset directly to train classifiers in the feature selection process and evaluate the quality of the feature subsets according to the performance of the classifier in the test set. Wrapper methods are computationally less efficient than Filter methods, but these methods can result smaller optimal feature subset than Filter methods [3].

Random forest (RF henceforth) [4] is a popular ensemble machine learning algorithm, which provides a unique combination of prediction accuracy and model interpretability among popular machine learning method [1]. RF uses Bootstrap [16] to sample samples randomly from original samples with replacement and train the decision trees in each Bootstrap sampling. In the process of node splitting of each tree, a feature is randomly selects as splitting attribute from a feature subset [5, 6, 7]. Finally, the class of a new sample is decided by voting of multiple decision trees. Currently, RF has been widely used in various classifications, prediction, the variables importance, feature selection, and outlier detection issues [8, 9, 10, 11]. Especially in the biomedicine and bioinformatics, random forest is favored because it can efficiently identify complex interaction among multiple predictors. Diaz-Uriarte et al [12] investigated the use of random forest for classification of microarray data and proposed a method for gene selection in classification problems based on random forest. Their experimental results showed that random forest has comparable performance to other classification methods, including DLDA, KNN, and SVM, and the proposed gene selection procedure yielded very small sets of genes while preserving predictive accuracy. However, this approach made the decision as to the number of genes to retain arbitrarily, and it is not the most appropriate if the objective is to obtain the smaller possible sets of genes that will allow good predictive performance. Herbert Pang et al [13] developed an iterative feature elimination method based on the random survival forests to identify a set of prognostic genes. Indeed, it is an extension of the method proposed by Diaz-Uriarte in survival outcomes prediction. This approach ordered the genes by variable importance in descending order and removed genes of the bottom 20 percent (default), where 20 percent is also the default chosen by Diaz-Uriarte. Dessì et al [14] proposed a pre-filtering feature selection method based on random forests for microarray data classification. They examined random forests from an experimental perspective and evaluated the effects of a filtering process which preceded the actual construction of the random forest. However, within this approach, a first critical issue is the choice of a threshold value denoting the cut-off point of the list of ranked features. Ali Anaissi et al [15] introduced a balanced iterative random forest (BIRF) algorithm to select the most relevant genes for a disease from imbalanced high-throughput gene expression microarray data. The experimental results showed the BIRF approach outperformed these state-of-the-art methods, such as Support Vector Machine-Recursive Feature Elimination (SVM-RFE), Multi-class SVM-RFE (MSVM-RFE), Random Forest (RF) and Naive Bayes (NB) classifiers, especially in the case of imbalanced datasets. However, BIRF algorithm has a limitation that random forest will not be able to get global correlation due to the splitting of the dataset.

In all these methods mentioned above, random forest was directly used for classifier to evaluate the quality of feature subsets in the process of feature selection, but the applicability of the random forest and the comparison with other classification algorithms were not been systematically researched. This paper studied the performance of random forest used as feature subset evaluating function and compared it with k-nearest neighbor (KNN) and support vector machine (SVM) classification algorithms. The experimental results on acute lymphoblastic leukemia (ALL) dataset showed the SVM is similar to RF but superior to KNN with respect to classification performance when they were used as feature subset evaluating function. On this basis, we proposed a new method of feature selection based on random forest, called RF&SVMFS, which is a wrapper feature selector that combined the random forest with support vector machine. RF&SVMFS also combined sequence backward searching approach and sequence forward searching approach. The base learning algorithm is random forest, which is used to compute variable importance for each feature and to determine what features are removed or selected at each step. The SVM algorithm is used for evaluating the quality of feature subsets. Feature selection starts with the entire set of features in the dataset. At every iteration, two feature subsets are gained. One feature subset removes those most unimportant features and the most important feature at the same time, which is used to train random forest and to compute feature importance for next feature selection. Another feature subset removes only those most unimportant features while remains the most important feature, which is used as the optimal feature subset to train SVM classifier. The experimental results on 11 UCI datasets, a real clinical data sets and a gene expression dataset show that the proposed algorithm can generate the smaller feature subset while improve the classification accuracy.

2   Method

In this paper, we proposed a new feature selection method called RF&SVMFS based on random forest and support vector machine, which combined sequence backward searching approach and sequence forward searching approach. In the RF&SVMFS, RF was run firstly to compute importance score for each feature. Then, all features were sorted based on the importance scores. In order to ensure the stability and reliability of the result, RF was run 5 times and the average of 5 times running result was used as the basis of sorting features in every iteration. Next, the generalized sequence backward searching strategy and sequence forward searching strategy was used to generate feature subset. In detail, L most unimportant features (with minimal importance score) and the most important feature were removed from original dataset, and a new dataset was generated. Meanwhile, another dataset was generated by removing only the L most unimportant features. The first dataset was used to train random forest and to compute variable importance for next iteration. The second dataset was used to train support vector machine and to evaluate the quality of the feature subset. In order to ensure the stability of results, 10-fold cross-validation was used while calculating the classification accuracy. The above process was repeated iteratively until the number of features in the feature set meeting the requirements (only 5 features are left in the feature set in this research). Finally, feature set with highest classification accuracy of SVM in all iterations was selected as the optimal features set, and the variable importance scores are calculated for each feature at the same time. The proposed algorithm is designed as follows:

```
Input:  the original dataset S
        L value in generalized sequence backward search
Output: highest classification accuracy MaxAccuracy
        optimal feature subset OptFeatureSet
        importance scores of features FeatureScore
Steps:
1. Initialization: MaxAccuracy <- 0
                   OptFeatureSet <- S
                   TmpFeatureSet <- S
                   GloOptFeatureSet <- NUll
2. while ( the number of features in OptFeatureSet > 5)
2.1 RF is run on 5 times TmpFeatureSet, the average value
    of variable importance score of each feature is
    computed and stored as vector RFAverageScore;
2.2 Features in TmpFeatureSet are ordered according to
    RFAverageScore, the results are saved as
    SortedFeatureSet;
2.3 According to SortedFeatureSet, remove L features with
    the lowest variable importance scores from
    OptFeatureSet and get new dataset OptFeatureSet;
2.4 According to SortedFeatureSet, remove L features with
    the lowest variable importance scores and a most
    important feature from TmpFeatureSet and get new
    dataset TmpFeatureSet;
2.5 Randomly divided OptFeatureSet into 10 equal parts,
    on nine of which run SVM algorithm to train
    classifier, the remaining one part is used as a test
    dataset, then calculate the SVM classifier accuracy
    in test set, such iterations was executed 10 times,
    calculate the average classification accuracy and
    save as SVMAverageAccuracy;
2.6 if( MaxAccuracy <=  SVMAverageAccuracy)
        MaxAccuracy <-  SVMAverageAccuracy
        GloOptFeatureSet <- features in OptFeatureSet
End while
3. print(MaxAccuracy)
   print(GloOptFeatureSet)
```

3 Experiments and Discussion

3.1 Datasets

For validating the effectiveness of the proposed feature selection algorithm, the paper selects 11 UCI datasets frequently used in literatures [17], a real diabetes clinical dataset (DiabetesDB), and an acute lymphoblastic leukemia (ALL) dataset [18]. The detailed information about these datasets is shown in Table 1. The dimensions of the UCI datasets range from 6 to 61, and the data types include not only discrete data but also continuous data or discrete and continuous mixed data. Particularly, Diabetes clinical data were collected from a Level-three hospital in Heilongjiang Province of China in 2006-2012, which includes 955 records of patients with type II diabetes. Each original record has 72 features. According to advices of the endocrine experts, a portion of obviously irrelevant and redundant features were removed, and the final dataset includes 46 classification variables and one objective variable. ALL dataset consist of 12625 genes from 128 different individuals with acute lymphoblastic leukemia (ALL), in which there are 33 T cells acute lymphoblastic leukemia and 95 B cells acute lymphoblastic leukemia. The data have been normalized (using rma) and it is the jointly normalized data that are available here. The data are presented in the form of an exprSet object. In this paper, we focus on the analysis of B cells acute lymphoblastic leukemia, and gene mutation is our target class. So we selected 94 B cells acute lymphoblastic leukemia samples with 12625 genes and an objective variable representing the type of acute lymphoblastic leukemia as our dataset, marked as ALLb.

Table 1. Summary of Datasets

NO.	Dataset	Nominal attributes	Continuous attributes	Instance size	Number of Class
1	Breast	1	9	699	2
2	Chess	36	0	3196	2
3	Credit	9	6	690	2
4	Diabetes	0	8	768	2
5	Heart	7	6	270	2
6	Liver	0	6	345	2
7	wpbc	1	33	198	2
8	wdbc	1	31	569	2
9	German-org	15	7	1000	2
10	Sonar	1	60	208	2
11	Ionosphere	0	34	351	2
12	DiabetesDB	1	46	955	2
13	ALLb	0	12626	94	4

3.2   Experiments on UCI Datasets

Experimental results on UCI datasets are shown in Table 2 and Table 3, where the "Feature" represents the number of features in optimal feature subset, "Acc" represents the classification accuracy in test set, and "NA" indicates that there are no experimental results in the referenced literatures. Here, the L value in RF&SVMFS is set as 1. From Table 2 and Table 3, it can be seen, in dataset 1, 3 and 5, RF&SVMFS selected out the smaller or equal optimal feature subset than CBFS [19], AMGA [20], but the classification accuracy has been significantly improved. In dataset 7, 8, 9, 10 and 11, the number of the selected features selected by RF&SVMFS is similar with other algorithms, but the classification accuracy is obvious higher than ACAHFS [21] and GA-cull [21]. In addition, as one can see, the higher the dimensions of the dataset are, the better the performance of RF&SVMFS is. In a word, the proposed algorithm outperformed the existing methods in literature with respect to both the quality of feature subset and the classification accuracy.

Table 2. The Performance Comparison among RF&SVMFS, CBFS and AMGA

No.	CBFS		AMGA		RF&SVMFS	
	Feature	Acc	Feature	Acc	Feature	Acc
1	6	0.943	NA	NA	5	**0.985**
2	22	0.954	5	**0.991**	22	0.967
3	5	0.848	NA	NA	5	**0.885**
4	4	0.670	NA	NA	4	**0.79**
5	9	0.791	5	0.804	3	**0.963**
6	3	0.628	8	**0.915**	3	0.768

Table 3. The Performance Comparison among RFVIMFS, ACAHFS and GA-cull

No.	ACAHFS		GA-cull		RF&SVMFS	
	Feature	Acc	Feature	Acc	Feature	Acc
7	3	0.796	10	0.694	6	**0.854**
8	5	0.979	2	0.979	9	**0.983**
9	17	0.732	2	0.692	14	**0.793**
10	16	0.824	2	0.827	17	**0.918**
11	6	0.965	6	0.937	6	**0.972**

3.3   Experiments on Real Diabetes Dataset

For DiabetesDB dataset, we performed RF, SVM and RF&SVMFS algorithm respectively. In order to ensure the stability of the results, the paper adopted 10-fold cross-validation method for each algorithm, and the average value of classification accuracy in 10 test sets was computed. The results were shown in Table 4. One can see that the classification accuracy of the RF&SVMFS is significantly better than original RF and SVM algorithm. In addition, the method based on random forests can provide variable importance scores of each feature. The bigger the importance score is, the greater the impact of the feature to target variable is. This can help medical experts to understand the results of data mining.

Table 4. Performance of RF, SVM and RF&SVMFS on DiabetesDB Dataset

Algorithm	Accuracy	Features Set
RF	0.742	All Features
SVM	0.723	All Features
RFVIMFS	0.832	Age CH ALT INS30 INS60 AST LDL-C Cr CP60 FCP CP30 INS120

We also studied the risk factors of Peripheral Arterial Disease. The top 10 risk factors were shown in Figure 2. As shown, age is the primary risk factor to Peripheral Arterial Disease. Smoking history is ranked in second. These results are consistent with previous findings. ALT is the third risk factor. Recent studies have shown that ALT is a flag that one's liver has been damaged, which is related with atherosclerosis. Therefore, our findings are consistent with previous researches. We also have a new discovery that INS30 and INS60 are important risk factors for Peripheral Arterial Disease and their impacts are similar. According to medical knowledge, insulin is a potent growth factor, which can increase collagen synthesis and stimulate vascular smooth muscle cell proliferation. This is a process of atherosclerosis, and therefore insulin levels reflect the lower limbs of atherosclerosis in some extent. Overall, the results of this study are highly consistent with previous studies, and the proposed RFVIMFS algorithm is reasonable.

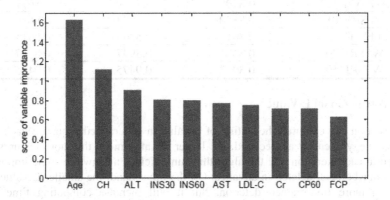

Fig. 1. Top-10 risk factors of Peripheral Arterial Disease

3.4 Experiments on Acute Lymphoblastic Leukemia Dataset

In this section, we studied the capability of different classifiers used as feature subset evaluating function and compared the performance of the proposed RF&SVMFS with some popular feature selection method using acute lymphoblastic leukemia dataset (ALLb) [18]. ALLb is a microarray gene express dataset, which include 94 B cell acute lymphoblastic leukemia samples with 12625 genes. The objective variable has four categories, including ALL/AF4, BCR/ABL, E2A/PBX1 and NEG. The dimensions of this dataset are very large, so feature selection was performed before the prediction model was trained. Firstly, we used interquartile range (IQR) to filter genes based on gene expression levels distribution, and all genes whose variability is less than 1/5 overall IQR are eliminated. After this process, the numbers of genes became 3970 from 12625. Next, we performed Factor Analysis (AVONA), filter feature selection based on random forest (FFSRF), filter feature selection based on combined feature clustering (FFSFC) [18] and our proposed method respectively. Incidentally, because KNN, randomForest and SVM were used as feature evaluating function respectively here, we represent the proposed feature selection method by WFSRF which is different from FFSRF. As a result, AVONA selected 752 genes; Both FFSRF and FFSFC selected top 30 genes, RF&SVMFS selected top 50 and top 20 genes. Finally, KNN, randomForest and SVM algorithm were performed on these feature subsets, and the classification accuracy of each case was shown as Table 5. As one can see from Table 5, the proposed WFSRF methods are overall superior to ANOVA, FFSRF and FFSFC with respect to classification accuracy. While KNN, randomForest and SVM were used as feature subset evaluating function respectively, randomForest and SVM are evenly matched, and both are superior to KNN. This proves the validity of our proposed method.

Table 5. Classification Accuracy of Different Methos

classifier feature selection	KNN	randomForest	SVM
ANOVA.752	0.8298	0.7979	0.8617
FFSRF.30	0.8830	0.8723	0.8617
FFSFC.30	0.8617	0.8298	0.8511
WFSRF.20	0.8421	0.8947	0.8947
WFSRF.50	0.8947	0.9475	0.9475

3.5   Discussion on L Value

In this section, we explored the setting of L value in generalized sequence backward search strategy. Readily appreciated, the larger L value mean the more features are removed at each iteration and the algorithm run quickly, meanwhile some important features maybe been deleted in advance. On the other hand, the smaller L value can provide a more fine-grained deletion, but it will increase computing time. We designed a set of experiments on 11 UCI datasets, and we run RF&SVMFS when L is set as 1, 2, 3, 4 and 5 respectively. The results are shown as Figure 1. As one can see, in the Sonar, when L is set as 3, RF&SVMFS obtains the best classification performance of 94.23%. Otherwise, in the other 10 data sets, with the increasing of L, the classification accuracy showed a downward trend, that is, when L is set as 1, RF&SVMFS has best results. This is maybe because that, when dimensions of the dataset is small, each time deleting one feature can effectively eliminate redundant features and irrelevant features, and larger L value will make some important features be removed together with non-related features. However, when the dimensions of the dataset are very large, such as ALLb, the larger L value makes it possible to remove redundant features and irrelevant features quickly and to improve the classification performance, as shown in Sonar dataset in this paper. As a rule of thumb, when the dimensions of dataset is very large, L should be set as $\sqrt{N}$, N is the number of features of dataset. For ALLb dataset in this paper, we adopted a combination method. When the dimension of the dataset is larger than 50, we set L as 50; when the dimension is smaller than 50, we set L as 5.

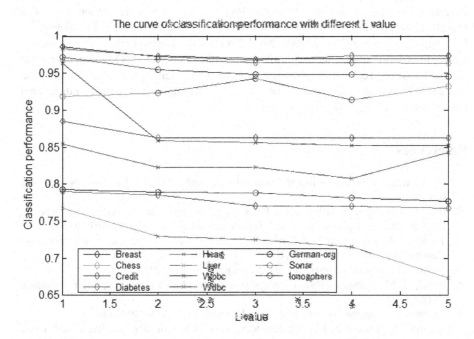

Fig. 2. Performance comparison of SVM classification while L adopts different values

4 Conclusions

Due to high-dimensional feature space and highly feature redundancy in biomedicine and bioinformatics dataset, the existing machine learning algorithms have been not competent data mining tasks in these field. Random forests algorithm has the capacity of analyzing complex interactions among features and can provide variable importance score which can be used as a convenient tool for the feature selection. The paper proposed a new Wrapper feature selection algorithms based on random forest variable importance measurement and support vector machine. The proposed method combined generalized sequence backward searching strategy and sequence forward sequence searching strategy for feature selection. Experimental results show that the proposed feature selection algorithm is responsible for finding the optimal feature subset and can effectively improve the classification accuracy. Simultaneously, the algorithm can give out the variable importance scores for each feature in the optimal feature subset, and enhance the comprehensibility of data mining results. In addition, we study the capability of different classification algorithms used as feature subset evaluating function, and experiment shows that SVM is evenly matched with random forest but superior to KNN in ALLb dataset. Experimental validation and deeper research on more datasets is the next direction of research.

Acknowledgements. This work is sponsored by the National Natural Science Foundation of China (No.61370083, No.61073043, and No.61073041), the National Research Foundation for the Doctoral Program of Higher Education of China (No.20112304110011, No.20122304110012), the Natural Science Foundation of Heilongjiang Province (No.F200901, No. F201313), the Harbin Outstanding Academic Leader Foundation of Heilongjiang Province of China (No.2011RFXXG015), the Harbin Special Funds for Technological Innovation Research of Heilongjiang Province of China (No.2013RFQXJ114), and the Foundation of Heilongjiang Province Educational Committee(No.12511233).

References

1. Qi, Y.: Random Forest for Bioinformatics. In: Ensemble Machine Learning, pp. 307–323 (2012)
2. Inza, I., Larranaga, P., Blanco, R.: Filter versus wrapper gene selection approaches in DNA microarray domains. Artificial Intelligence in Medicine 31(2), 91–103 (2008)
3. Tsymbal, A., Puuronen, S.: Ensemble feature selection with the simple Bayesian classification. Information Fusion 4(2), 87–100 (2010)
4. Breiman, L.: Random forests. Machine Learning 45, 5–32 (2001)
5. Bishop, C.M.: Bootstrap. Pattern Recognition and Machine Learning. Springer, Singapore (2006)
6. Breiman, L.: Bagging predictors. Machine Learning 24(2), 123–140 (1996)
7. Breiman, L., Friedman, J.H., Olshen, R.A., et al.: Classification and Regression Trees. Chapman&Hall (1993)
8. Strobl, C., Boulesteix, A.-L., Kneib, T., Augustin, T., Zeileis, A.: Conditional variable importance for random forests. BMC Bioinformatics 9, 307 (2008)
9. Verikas, A., Gelzinis, A., Bacauskiene, M.: Mining data with random forests: A survey and results of new tests. Pattern Recognition 44, 330–349 (2011)
10. Liu, H., Li, J.: A comparative study on feature selection and classification methods using gene expression profiles and proteomic patterns. Genome Informatics 13, 51–60 (2012)
11. Wang, A., Wan, G., Cheng, Z., et al.: Incremental Learning Extremely Random Forest Classifier for Online Learning. Journal of Software 22(9), 2059–2074 (2011)
12. Díaz-Uriarte, R., de Andrés, S.A.: Gene selection and classification of microarray data using random forest. BMC Bioinformatics 7, 3 (2006)
13. Pang, H., George, S.L., Hui, K., Tong, T.: Gene Selection Using Iterative Feature Elimination Random Forests for Survival Outcomes. IEEE/ACM Transactions on Computational Biology and Bioinformatics 9(5), 1422–1431 (2012)

14. Dessì, N., Milia, G., Pes, B.: Pre-filtering Features in Random Forests for Microarray Data Classification. In: New Frontiers in Mining Complex Patterns (NFMCP 2012). vol. 60 (2012)
15. Anaissi, A., Kennedy, P.J., Goyal, M., Catchpoole, D.R.: A balanced iterative random forest for gene selection from microarray data. BMC Bioinformatics 14, 261 (2013)
16. Yi, C., Li, J., Zhu, C.: A kind of feature selection based on classification accuracy of SVM. Journal of Shandong University 45(7), 119–124 (2010)
17. UC Irvine Machine Learning Repository, http://archive.ics.uci.edu/ml/
18. Torgo, L.: Data Mining with R: Learning with Case Studies. Luis Chapman & Hall/CRC (2010)
19. Jiang, S., Zheng, Q., Zhang, Q.: Clustering-Based Feature Selection. Acta Electronica Sinica 36(12), 157–160 (2008)
20. Liu, Y., Wang, G., Zhu, X.: Feature selection based on adaptive multi-population genetic algorithm. Journal of Jilin University 41(6), 1690–1693 (2011)
21. Zhang, J., He, Z., Wang, J.: Hybrid Feature Selection Algorithm Based on Adaptive Ant Colony Algorithm. Journal of System Simulation 21(6), 1605–1614 (2009)

Phylogenetic Bias in the Likelihood Method Caused by Missing Data Coupled with Among-Site Rate Variation: An Analytical Approach

Xuhua Xia

Department of Biology and Center for Advanced Research in Environmental Genomics,
University of Ottawa, 30 Marie Curie, P.O. Box 450, Station A,
Ottawa, Ontario, Canada, K1N 6N5
xxia@uottawa.ca

Abstract. More and more researchers in phylogenetics are concatenating gene sequences to produce supermatrices in the hope that larger data sets will lead to better phylogenetic resolution. Almost all of these supermatrices contain a high proportion of missing data which could potentially cause phylogenetic bias. Previous studies aiming to identify the missing-data-mediated bias in the maximum likelihood method have noted a bias associated with among-site rate variation. However, this finding is by sequence simulation and has been challenged by other simulation studies, with the controversy still unresolved. Here I illustrate analytically this bias caused by missing data coupled with among-site rate variation. This approach allows one to see how much the bias can contribute to likelihood differences among different topologies. The study highlights the point that, while supermatrices may lead to "robust" trees, such "robust" trees may be purchased with illegal phylogenetic currency.

Keywords: missing data, pruning algorithm, likelihood, phylogenetic bias, supermatrix.

1   Introduction

Many supermatrices have been compiled in recent years by concatenating sequences from many different genes [1-4]. Such concatenated genes typically have few shared sites among all included species. For example, while Regier et al. [3] claimed to have 41 kilobases of aligned DNA sequences, the actual number of sites that are completely unambiguous among all 80 species amounts to only 705 sites. Some genes are completely missing in nearly half of the 80 species. While the potential problems involving such "?"-laden supermatrices have been suspected before[5], specific biases associated with such missing data have not been well studied, especially not in the likelihood framework which has been the gold standard in phylogenetic reconstruction.

Previous studies [6-11] attempted to identify bias associated with missing data either by sequence simulation or by selectively eliminating sites in a real sequence alignment. While most publications suggest that phylogenetic reconstruction is not sensitive to missing data or that the benefit of including taxa with missing data

M. Basu, Y. Pan, and J. Wang (Eds.): ISBRA 2014, LNBI 8492, pp. 12–23, 2014.
© Springer International Publishing Switzerland 2014

out-weight the cost of their exclusion [6, 8-11], a recent study [7] suggested a significant bias associated with missing data and coupled with among-site rate variation. However, such simulation-based findings often cannot pin-point where the bias arises and consequently have been challenged by others on both empirical [6, 9, 11] and theoretical grounds [9], although these latter publications did not explicitly test the claimed bias [7] associated with among-site variation. Roure et al. [9] noted that, if sequences contain similar phylogenetic information, then phylogenetic reconstruction is not sensitive to missing data. However, they also noted that heterogeneous data could lead to phylogenetic bias based on extensive data analysis.

Here I demonstrate analytically the bias associated with the missing data coupled with among-site rate variation. The pruning algorithm [12, 13, 14, pp. 253-255] is briefly outlined, in conjunction with the conventional missing data handling by the likelihood method, so that the reader can verify the claimed bias introduced by missing data. I first illustrate the "bias" shown by Lemmon et al. [7] when branch lengths are not allowed to be zero, by using both JC69 [15] and F84 [16] models. Such a "bias" can be easily avoided by simply allow branches to be zero and should not be considered as estimation bias in the likelihood method. However, the bias due to the missing data associated with among-site rate variation [7] is real. This bias can lead to either increased tendency (and confidence) to group together OTUs (operational taxonomic units) that share the same stretches of missing sites or in the opposite direction. The results suggest that blindly concatenating sequence data to generate a supermatrix with many pieces of missing data will generate false confidence in phylogenetic resolution and should be avoided.

2   Missing Data Handling and the Pruning Algorithm

The likelihood approach features a convenient way to handle missing data, which is best illustrated with the pruning algorithm. Suppose we have four OTUs with sequence data in Fig. 1a, and with the last two sequences being entirely missing (represented by '?'). Obviously, we can only estimate the distance between S1 and S2 but not the evolutionary relationships involving OTUs S3 or S4. The maximum likelihood distance between S1 and S2, based on the JC69 model, is given by

$$L = \frac{8!}{4!4!} P_{ii}^4 P_{ij}^4 \qquad (1)$$

which, when maximized, leads to a distance of 0.8239592165.

Fig. 2 illustrates the computation of the likelihood by the pruning algorithm, given the first site of the aligned nucleotide sequence (Fig. 1a) and topology T_1 in Fig. 1. I included the numerical illustration here to facilitate the verification of subsequent claims that the maximum likelihood method does exhibit a true and identifiable bias in phylogenetic reconstruction involving missing data coupled with among-site rate variation.

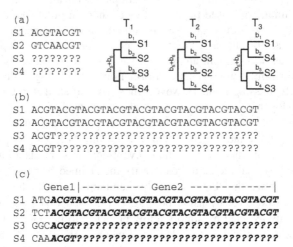

(a)

S1	ACGTACGT
S2	GTCAACGT
S3	????????
S4	????????

(b)

S1	ACGTACGTACGTACGTACGTACGTACGTACGTACGT
S2	ACGTACGTACGTACGTACGTACGTACGTACGTACGT
S3	ACGT???????????????????????????????
S4	ACGT???????????????????????????????

(c)

Gene1 |---------- Gene2 -------------|

S1	ATG**ACGTACGTACGTACGTACGTACGTACGTACGTACGT**
S2	TCT**ACGTACGTACGTACGTACGTACGTACGTACGTACGT**
S3	GGC**ACGT**???????????????????????????????
S4	CAA**ACGT**???????????????????????????????

Fig. 1. Three sets of sequences, (a), (b) and (c), for four OTUs (operational taxonomic units), and three alternative topologies (T1, T2 and T3) for illustrating phylogenetic bias introduced by missing data. Branch lengths are represented by bi. The sequences in bold italic in (c) are the same as those in (b). Note that the three variable sites at the 5' end could be diffused at different sites in the data instead of clumping together to be so easily recognized.

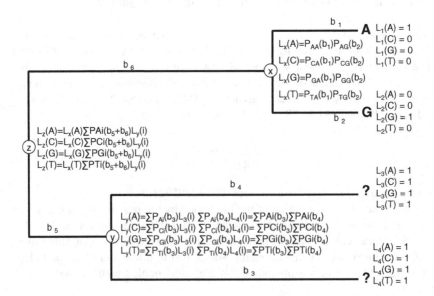

Fig. 2. Likelihood computation with the pruning algorithm [14, pp. 253-255]

We first define an array for each of the nodes including the leaf nodes. The array contains four elements for nucleotide sequences and 20 for amino acid sequences. For a leaf node i with a resolved nucleotide S, $L_i(S) = 1$, and $L_i(\text{not } S) = 0$. For an unknown or missing nucleotide, $L_i(1) = L_i(2) = L_i(3) = L_i(4) = 1$. For an internal node i with two offspring (o_1 and o_2), L_i is recursively defined as

$$L_i(s) = \left[\sum_{k=0}^{3} P_{sk}(b_{i,o_1}) L_{o_1}(k) \right] \left[\sum_{k=0}^{3} P_{sk}(b_{i,o_2}) L_{o_2}(k) \right] \qquad (2)$$

where $b_{i,o1}$ means the branch length between internal node i and its offspring o_1, and P_{sk} is the transition probability from state s to state k. For example, $b_{x,S1}$ (branch length between internal node x and its offspring S_1) is b_1 in Fig. 2. Internal node z is special in that we cannot estimate b_5 and b_6 separately because the resulting tree is unrooted. We simply move node z to the location of node y (or node x), so that either b_5 or b_6 is 0 and the other is then equal to (b_5+b_6). If b_5 is 0, then $P_{ii}(b_5) = 1$ and $P_{ij}(b_5) = 0$, i.e., no time for anything to change. This leads to the simplified equations for computing $L_z(i)$ in Fig. 2. The final likelihood is

$$L = \sum_{i=1}^{4} \pi_i L_z(i) \qquad (3)$$

where π_i is the frequency of nucleotide i.

Given the JC69 model, the sequences in Fig. 1a have two site patterns, with the first four sites sharing one site pattern and the last four sites sharing the other site pattern. Designating the likelihood of the two site patterns in Fig. 1a as $L_{a.pattern1}$ and $L_{a.pattern2}$, the log-likelihood (lnL) for all eight sites (Fig. 1a), given topology T_1 in Fig. 1), is

$$\ln L = 4\ln(L_{a.pattern1}) + 4\ln(L_{a.pattern2}) \qquad (4)$$

which, upon maximization, leads to $b_1 + b_2 = 0.8239592165$, and lnL = -21.02998149. This is perfectly consistent with the result from Eq. (1) as we would have expected. Terms including b_3, b_4 and b_5+b_6 all cancel out in Eq. (4), suggesting that the sequences in Fig. 1a have absolutely no information for estimating b_3, b_4 and b_5+b_6, which again is what we would have expected. Note that lnL would be greater if we treat the two site patterns as two separate partitions and estimate branch lengths separately. Assuming the JC69 model, the maximum likelihood is 0.25^2 for each site in the first partition (reached when b_1 and b_2 are infinitely large) and 0.25 for each site in the second partition (reached when $b_1 = b_2 = 0$), so lnL will then be

$$\ln L = 4\ln(0.25^2) + 4\ln(0.25) = -16.63553233 \qquad (5)$$

which indicates that maximizing lnL by partitioning the data may not be a good idea given the dramatically incompatible branch length estimates from the two partitions.

If we perform the computation again with topology T_2 in Fig. 1, we will have exactly the same lnL, but b_5+b_6 will be 0 and $b_1+b_3 = 0.8239592165$ (i.e., the distance between OTUs S1 and S2 is 0.8239592165 as before). This again is perfectly consistent with results from Eq. (1). Topology T_3 in Fig. 1 will lead to the same lnL and the same conclusion with distance between OTUs S1 and S2 being 0.8239592165.

We can also fit the F84 model to the data in Fig. 1a which now has three sites patterns, with the first two sites sharing the first site pattern, sites 3-4 sharing the second site pattern and sites 5-8 sharing the third site pattern. Because the nucleotide frequencies of the four sequences are all equal to 0.25, and because of equal number of transitions and transversions in the sequences so that $k = 1$, the F84 distance between sequences S1 and S2 is defined by the following likelihood function:

$$L = \frac{8!}{2!2!4!} P_s^2 P_v^2 P_{ii}^4$$

$$P_{ii} = \frac{1}{2} e^{-(1+k)D/c} + \frac{1}{4} e^{-D/c} + \frac{1}{4}$$

$$P_s = \frac{1}{4} e^{-D/c} - \frac{1}{2} e^{-(1+k)D/c} + \frac{1}{4} \tag{6}$$

$$P_v = -\frac{1}{4} e^{-D/c} + \frac{1}{4}$$

$$c = 2(\pi_T \pi_C (1+k/\pi_Y) + \pi_A \pi_G (1+k/\pi_R) + \pi_R \pi_Y) = 1.25$$

where D is the F84 distance, P_{ii}, P_s and P_v corresponding to transition probabilities for no change, transition and transversion, respectively. Maximizing L leads to $k = 1$ (which is expected because we observe the same number of transitions and transversions in the sequences in Fig. 1a) and $D = 0.8664339758$ which is slightly larger than the JC69 distance.

Applying the pruning algorithm to the four sequences (Fig. 1a) and topology T_1 in Fig. 1, we obtain a final likelihood function that includes only k and (b_1+b_2), i.e., there is no information to estimate b_3, b_4 and b_5+b_6 in topology T_1 in Fig. 1. Maximizing the likelihood function leads to the maximum lnL = -20.79441542, reached when $k = 1$ and $(b_1+b_2) = 0.8664339758$. This is exactly the same as the result derived from Eq. (6). If we perform the computation again with topology T_2 in Fig. 1, we will have exactly the same lnL, but b_5+b_6 will be 0 and $b_1+b_3 = 0.8664339758$ (i.e., the distance between OTUs S1 and S2 is 0.8664339758). This again is perfectly consistent with results from Eq. (6). We can use topology T_3 in Fig. 1 and will again obtain the same lnL and the same conclusion with distance between OTUs S1 and S2 being 0.8664339758.

Note that the application of the F84 model resulted in a small increase in lnL from -21.02998149 with the JC69 model to -20.79441542. This is expected from the sequence data in Fig. 1a which do not conform strictly to the JC69 model. S1 and S2 differ by two transitions and two transversions instead of the 1:2 ratio expected under the JC69 model, so F84 is a more appropriate substitution model than JC69.

The transition/transversion ratio for the DNAML program (R_{DNAML}) is defined [17. p. 18] as

$$R_{DNAML} = \frac{\pi_T \pi_C (1 + k / \pi_Y) + \pi_A \pi_G (1 + k / \pi_R)}{\pi_R \pi_Y} \qquad (7)$$

Given the equal nucleotide frequencies and $k = 1$, lnL from DNAML is maximized when $R_{DNAML} = 1.5$, and DNAML outputs lnL = -20.79442 which is the same as shown above. The lnL values are the same for all three topologies. BASEML outputs the same k and lnL. Of course, if one uses DNAML with the default R_{DNAML} of 2, then the three possible topologies will lead to different likelihood values. For this reason, one should not always use default values when running phylogenetic tools. However, misleading phylogenetic results due to misuse of default values should not be attributed to bias in phylogenetic methods.

3   A "Bias" That Is Not True Bias

Suppose we now have the sequence data in Fig. 1b. The four sequences are identical except that S3 and S4 have part of the sequences missing, so there are only two site patterns assuming the JC69 model (with the first shared by the first four sites and the second by the last 32 sites containing unknown nucleotides). These sequences again allow us to have two straightforward expectations. First, the three topologies should have the same lnL. Second, all branches should have length equal to 0 (i.e., $b_i = 0$). Third, the likelihood for each site is simply 0.25, so that the maximum lnL for the entire sequence alignment and for any of the three topologies is

$$\ln L = 4 \ln(0.25) + 32 \ln(0.25) = -49.906597 \qquad (8)$$

which is reached when $b_1 = b_2 = b_3 = b_4 = b_5 + b_6 = 0$. One could replace the JC69 model by the F84 model, but the results will be the same because the greater generality of the F84 model relative to the JC69 model is not necessary for the sequence data in Fig. 1b.

Both DNAML and BASEML produce results and conclusions quite different from our expectations when they are used to evaluate the three alternative topologies. First, topology T_1 in Fig. 1 has higher lnL than the other two alternative topologies, and is declared by both DNAML and BASEML as significantly better than the other two alternative topologies. Second, the b_i values listed in the output of DNAML and BASEML are greater than zero and their consequent lnL values are less than the maximum -49.906597 reached when b_i values are all zero.

This "bias" was analytically identified before [Supplemental Materials in 7], and it is not a true bias in the maximum likelihood method. The problem is caused by both DNAML and BASEML not allowing branch lengths to zero during their evaluation of the three alternative topologies. Most likelihood-based phylogenetic programs set a small constant as the lower bound for estimating branch lengths. If we force DNAML and BASEML to evaluate the four-taxon tree with zero branch lengths, they will find lnL to be equal to that in Eq. (8). As soon as we allow branch lengths to be greater than zero, topology T_1 in Fig. 1 will be favored against the other two alternative topologies by DNAML and BASEML.

The effect is easy to see if we simply set all branch lengths (b_i values) to a small constant C and write down the likelihood functions for the two site patterns (shared by the first four sites and the last 32 sites, respectively) in sequences in Fig. 1b for topologies T_1 and T_2. For T_1, the likelihood functions for the two site patterns ($L_{T1.pattern1}$, $L_{T1.pattern2}$), given the JC69 model, can be obtained by traversing the tree in Fig. 2 and expressed as

$$L_{T1.pattern1} = \frac{1}{4}b^2(b^3 + 3a^3) + \frac{3}{4}a^2(ab^2 + ba^2 + 2a^3)$$

$$L_{T1.pattern2} = \frac{1}{4}b^2 + \frac{3}{4}a^2$$

$$a = \frac{1}{4} - \frac{1}{4}e^{-4C/3}$$

$$b = \frac{1}{4} + \frac{3}{4}e^{-4C/3}$$

(9)

where all branch lengths are equal to C. Both $L_{T1.pattern1}$ and $L_{T1.Pattern2}$ reach the maximum 0.25 when C = 0 as one would expect.

For topology T_2, the likelihood function for the first site pattern shared by the first four sites is exactly the same as $L_{T1.pattern1}$ in Eq. (9). However, the likelihood function for the second site pattern shared by the 32 "?"-containing sites, is different between topologies T_2 and T_1. For topology T_2, the likelihood function for each of these 32 sites is equal to

$$L_{T2.pattern2} = \frac{1}{4}b(b^2 + 3a^2) + \frac{3}{4}a(2ab + 2a^2)$$

(10)

which reaches the maximum 0.25 when C = 0 as one would expect. With the increase in C, $L_{T2.pattern2}$ becomes smaller than $L_{T1.pattern2}$, leading to T_1 preferred over T_2 (or T_3). However, in practical data analysis, the difference should be negligible because the minimum branch length in software is usually set to a value in the order of 0.000001 or smaller. With such a small C, the lnL difference contributed by one site is in the order of 0.000001.

4   True Bias Involving Missing Data Coupled with Among-Site Rate Variation

Suppose now we have sequence data in Fig. 1c, with Gene1 being variable but Gene2, which is missing in S3 and S4, is so conservative as to be invariant. In practice, Gene1 and Gene2 could be different segments within the same gene, e.g., the conserved and variable domains in ribosomal RNAs with no clear boundary between them. I used this configuration because (1) it has been used before in simulations [7], and (2) it represents a recurring pattern in published supermatrices. Note that the three variable sites at the 5'-end could be diffused at different sites in the data instead of clumping together to be so easily recognizable in real data.

The sequences are intentionally made not to favor any one of the three possible topologies (Fig. 1). For Gene1, the four OTUs are exactly equally divergent from each other given the JC69 and F84 models, i.e., each pair of sequences differ in exactly one transition and two transversions so that no particular topology is favored over the other two. Gene2 is extremely conservative and no substitution has been observed, so it also should not favor any topology over the other two.

With the sequence data in Fig. 1c and topology T_1 in Fig. 1, we can apply the pruning algorithm and the JC69 model to compute the likelihood. There are only three different site patterns with the JC69 model, i.e., sites 1 to 3 share the first site pattern, sites 4 to 7 sharing the second and sites 8 to 39 sharing the third. Maximizing the likelihood will lead to $\ln L$ = -83.56464029 which is reached when $b_1 = b_2 = 0.04153005797$, $b_3 = b_4 = 0.3787544804$, and $(b_5+b_6) = 0.3511004094$.

The maximum $\ln L$ value for topology T_2 in Fig., 1 is -83.96663731, reached when $b_1 = b_3 = 0.04184900$, $b_2 = b_4 = 0.60765526$, and $(b_5+b_6) = 0.000947018$. The maximum $\ln L$ value for topology T_3 is the same as that for T_2 and both are significantly smaller ($p < 0.001$) than that for T_1 (Fig. 1) based on either the Kishino-Hasegawa test and RELL test [16] or Shimodaira & Hasegawa test [18]. DNAML reached exactly the same conclusion, so did BASEML with either the JC69 model or the F84 model. Note that, if the sequence alignment is 100 times as long (which is common in studies with supermatrices), the difference in $\ln L$ between T_1 and T_2 would be about 40, which is often greater than the difference between the best and the second best trees in a typical ML reconstruction.

This rejection of topologies T_2 and T_3 in favor of T_1 is not expected from the data in Fig. 1c because each pair of sequences differs by exactly one transition and two transversions. Why is topology T_1 strongly favored by the likelihood method over T_2 and T_3? We can find the answer by making a few observations below.

First, different sites require different branch lengths for maximizing its likelihood. For example, the maximum likelihood for each of the first three sites (Fig. 1c), given the JC69 model, is 0.00390625 ($=0.25^4$) reached when b_1 to b_4 are infinitely large. In contrast, the maximum likelihood for each site from site 4 to site 7 is 0.25 reached when b_1 to b_4 are all zero. Thus, the log-likelihood for the first seven sites ($\ln L_7$), if maximized separately, would be $3*\ln(0.25^4) + 4*\ln(0.25)$, i.e., -22.18070977 for topologies T_1, T_2 and T_3. However, as a compromise between the first three and the next four sites, $\ln L_7$ becomes -32.96754443, reached when $b_1 = b_2 = b_3 = b_4 = 0.3841581410$ and $(b_5+b_6) = 0.1220492271$. This result is applicable to all three topologies. Thus, among-site rate variation itself does not cause phylogenetic bias if it is not lineage-specific, although a previous study [19] suggested that it does based on simulation studies.

Second, the maximum log-likelihood for the 32 sites with missing values in Fig. 1c ($\ln L_{32}$) is also the same among the three topologies, being -44.36141955 when $b_1 = b_2 = 0$ for topology T_1 in Fig. 1a (all other branch lengths are irrelevant for computing $\ln L_{32}$ given T_1). For topology T_2 (Fig. 1a) to reach the same maximum $\ln L_{32}$, we need $b_1 = b_3 = (b_5+b_6) = 0$. Similarly, with T_3, we need $b_1 = b_4 = (b_5+b_6) = 0$. Thus, the branch lengths that maximize $\ln L_{32}$ (i.e., when all branch lengths are zero) are not the same as the branch lengths that maximize $\ln L_7$ (which is maximized when branch lengths are greater than zero, with optimal values specified above), and different topologies impose different constraints on maximizing likelihood.

Third, recall that $\ln L_{32}$ depends only on b_1 and b_2 for topology T_1, but on more branch lengths for T_2 and T_3. With T_1, b_1 and b_2 can be reduced to maximize $\ln L_{32}$ (although not to zero because of the first three variable sites in Fig. 1c). Other branch lengths such as b_3, b_4 and (b_5+b_6) can take optimal values to maximize $\ln L_7$ without affecting $\ln L_{32}$. Because b_1 and b_2 are reduced to maximize $\ln L_{32}$, and consequently deviated substantially from the optimal branch length (= 0.3841581410) for maximizing $\ln L_7$, (b_5+b_6) is increased to 0.3511004094 to compensate. In contrast, $\ln L_{32}$ for topology T_2 depends on b_1, b_3 and (b_5+b_6). Maximization of $\ln L_{32}$ for T_2 can be achieved by reducing b_1, b_3 and (b_5+b_6) and at the same time increasing b_2 and b_4 as a compensation to maximize $\ln L_7$. This explains why the final T_2 tree has relatively short b_1, b_3, both being 0.04184900, and a very small (b_5+b_6), being 0.000947018, but much larger b_2 and b_4, both being 0.60765526.

To recapitulate, maximizing $\ln L_7$ requires $b_1 = b_2 = b_3 = b_4 = 0.3841581410$ and $(b_5+b_6) = 0.1220492271$, and maximizing $\ln L_{32}$ requires $b_1 = b_2 = b_3 = b_4 = (b_5+b_6) = 0$. Obviously, conflicts in maximizing $\ln L_{32}$ and $\ln L_7$ is greater for topology T_2 than for topology T_1, leading to $\ln L$ greater for T_1 than for T_2. This result proves the finding by Lemmon et al. [7] reached through sequence simulation, i.e., missing data coupled with among-site rate variation could lead to phylogenetic bias. It should eliminate the doubt expressed on other empirical grounds [6, 11]. One way to eliminate the bias in favor one topology over others is to identify sites with different rates into different partitions. However, in real data, these variable sites may be diffused among conservative sites instead of clumping together as in Fig. 1c to be easily recognizable.

An alternative to partition the sequence alignment is to use a gamma distribution to accommodate rate variation among sites. Unfortunately, parameter estimation (e.g., the shape parameter of the gamma distribution) often depends on topology. Ideally, we should get the same shape parameter regardless of which topology we use, but this is almost never the case. When we get different shape parameters from different topologies, which shape parameter should we trust? If we know that topology T_1 is true, then we would give more credit to the shape parameter obtained with T_1. Alternatively, if we know the true shape parameter, we would trust more the topology that yields a shape parameter that is the same as the true parameter than other topologies that generate a shape parameter that is far from the true value. Such a chicken-egg problem lands us in an awkward dilemma.

Note that the first four sites in Fig. 1c are equivalent to a stretch of the alignment that has undergone substitution saturation. While phylogenetic information will be eroded by substitution saturation and tests have been developed to assess such substitution saturation [20, 21], it is perhaps the first time to link substitution saturation directly to phylogenetic bias in the context of missing data. Also note that, although sequences in Fig. 1c is biased in favor of grouping S_1 and S_2 together, one can easily envision scenarios in which S_1 and S_2 would repulse each other, e.g., when the last 32 sites in Fig. 1c are far more variable than the first seven sites. Thus, the direction of the bias cannot be predicted before data analysis.

Lemmon et al. [7] speculated that the bias they observed from simulated sequences with missing data may be associated with model misspecification. While there is possibility for such an association, the results I have presented show that the bias can be entirely independent of model misspecification.

The bias associated with missing data and rate heterogeneity among sites has been noted for a long time. For example, the 18S rRNA sequences contain the variable and conservative regions. Missing a variable region or a conservative region by a subset of sequences leads to distortion of phylogenetic signals and wrong phylogenetic trees [22, 23]. Dramatic rate heterogeneity among the three codon positions [24, 25], or among genes located in different DNA strands [27, 28] have long been noted. As among-site rate variation is not only common in molecular sequence data but also a known source of phylogenetic bias [19], one should be cautious to compile such data with missing data configuration similar to that in Fig. 1c. As a precaution against such bias, some computer programs, e.g., DAMBE [29], deletes sites containing missing data before likelihood analysis.

One may not consider this as a serious problem because, among all those compiled supermatrices in recent publications [e.g., 1, 2, 3], closely related species tend to share genes (or lack of genes). If we take the data in Fig. 1b as a simple caricature of the supermatrices, S1 and S2 are more likely to be closely related to each other, so are S3 and S4, in real data compilations. This means that the bias above caused by missing data will tend to help recover the true topology or increase the bootstrap support of some true subtrees. This may well have contributed to the increased bootstrap values documented before by Cho et al. [11] who then have argued for the supermatrix approach based on increased bootstrap values for certain taxa. Such an argument is flawed. We may recall an analogous case in the maximum parsimony method, with the inconsistency caused by long-branch attraction. Closely related species generally are more likely to share long branches than remote species, so long-branch attraction could seem a good thing because the bias it causes may lead to more efficient recovering of the true tree or increase bootstrap support for some true subtrees. However, such increased efficiency in recovering the true tree or increased bootstrap support for some true subtrees is purchased with illegal phylogenetic currency and should always be discouraged. In statistical estimation, a bias is a bias and is always undesirable because it often renders results unpredictable. It is fortunate that there has been only one case in which a phylogenetic approach is justified by its bias/inconsistency [30].

In summary, many supermatrices laden with missing sequences have been compiled in recent years while few studies have been carried out on the potential statistical bias that such data may cause. While the likelihood method handles missing data in a sensible way, its implementation may not achieve sufficient precision and may cause phylogenetic bias induced by missing data. In particular, lumping genes with different evolutionary rates runs a high risk of distorting phylogenetic signals and likelihood values and should be strongly discouraged.

Acknowledgements. This study is supported by Discovery Grant from Natural Science and Engineering Research Council (NSERC) of Canada. I thank B. Foley and X. Sun for motivating me to write the paper, and D. Baurain, S. Aris-Brosou, B. Golding, and A. RoyChoudhury for discussion and comments.

References

1. Hackett, S.J., Kimball, R.T., Reddy, S., Bowie, R.C., Braun, E.L., Braun, M.J., Chojnows-ki, J.L., Cox, W.A., Han, K.L., Harshman, J., Huddleston, C.J., Marks, B.D., Miglia, K.J., Moore, W.S., Sheldon, F.H., Steadman, D.W., Witt, C.C., Yuri, T.: A phylogenomic study of birds reveals their evolutionary history. Science 320, 1763–1768 (2008)
2. Perelman, P., Johnson, W.E., Roos, C., Seuanez, H.N., Horvath, J.E., Moreira, M.A., Kessing, B., Pontius, J., Roelke, M., Rumpler, Y., Schneider, M.P., Silva, A., O'Brien, S.J., Pecon-Slattery, J.: A molecular phylogeny of living primates. PLoS Genet. 7, e1001342 (2011)
3. Regier, J.C., Shultz, J.W., Zwick, A., Hussey, A., Ball, B., Wetzer, R., Martin, J.W., Cunningham, C.W.: Arthropod relationships revealed by phylogenomic analysis of nuclear protein-coding sequences. Nature 463, 1079–1083 (2010)
4. Regier, J.C., Shultz, J.W., Ganley, A.R., Hussey, A., Shi, D., Ball, B., Zwick, A., Stajich, J.E., Cummings, M.P., Martin, J.W., Cunningham, C.W.: Resolving arthropod phylogeny: exploring phylogenetic signal within 41 kb of protein-coding nuclear gene sequence. Syst. Biol. 57, 920–938 (2008)
5. Sanderson, M.J., Ane, C., Eulenstein, O., Fernandez-Baca, D., Kim, J., McMahon, M.M., Piaggio-Talice, R.: Fragmentation of large data sets in phylogenetic analysis. In: Gascuel, O., Steel, M. (eds.) Reconstructing Evolution: New Mathematical and Computational Advances, pp. 199–216. Oxford University Press, Oxford (2007)
6. Wiens, J.J., Tiu, J.: Highly incomplete taxa can rescue phylogenetic analyses from the negative impacts of limited taxon sampling. PLoS One 7, e42925 (2012)
7. Lemmon, A.R., Brown, J.M., Stanger-Hall, K., Lemmon, E.M.: The effect of ambiguous data on phylogenetic estimates obtained by maximum likelihood and Bayesian inference. Syst. Biol. 58, 130–145 (2009)
8. Wiens, J.J.: Missing data, incomplete taxa, and phylogenetic accuracy. Syst. Biol. 52, 528–538 (2003)
9. Roure, B., Baurain, D., Philippe, H.: Impact of Missing Data on Phylogenies Inferred from Empirical Phylogenomic Data Sets. Mol. Biol. Evol. 30, 197–214 (2013)
10. Rubin, B.E., Ree, R.H., Moreau, C.S.: Inferring phylogenies from RAD sequence data. PLoS One 7, e33394 (2012)
11. Cho, S., Zwick, A., Regier, J.C., Mitter, C., Cummings, M.P., Yao, J., Du, Z., Zhao, H., Kawahara, A.Y., Weller, S., Davis, D.R., Baixeras, J., Brown, J.W., Parr, C.: Can deliberately incomplete gene sample augmentation improve a phylogeny estimate for the advanced moths and butterflies (Hexapoda: Lepidoptera)? Syst. Biol. 60, 782–796 (2011)
12. Felsenstein, J.: Maximum-likelihood and minimum-steps methods for estimating evolutionary trees from data on discrete characters. Syst. Zool. 22, 240–249 (1973)
13. Felsenstein, J.: Evolutionary trees from DNA sequences: a maximum likelihood approach. J. Mol. Evol. 17, 368–376 (1981)
14. Felsenstein, J.: Inferring phylogenies. Sinauer, Sunderland (2004)
15. Jukes, T.H., Cantor, C.R.: Evolution of protein molecules. In: Munro, H.N. (ed.) Mammalian Protein Metabolism, pp. 21–123. Academic Press, New York (1969)
16. Kishino, H., Hasegawa, M.: Evaluation of the maximum likelihood estimate of the evolutionary tree topologies from DNA sequence data, and the branching order in Hominoidea. J. Mol. Evol. 29, 170–179 (1989)
17. Yang, Z.: Computational molecular evolution. Oxford University Press, Oxford (2006)
18. Shimodaira, H., Hasegawa, M.: Multiple Comparisons of Log-Likelihoods with Applications to Phylogenetic Inference. Mol. Biol. Evol. 16, 1114–1116 (1999)

19. Kuhner, M.K., Felsenstein, J.: A simulation comparison of phylogeny algorithms under equal and unequal evolutionary rates. Mol. Biol. Evol. 11, 459–468 (1994)
20. Xia, X., Lemey, P.: Assessing substitution saturation with DAMBE. In: Lemey, P., Salemi, M., Vandamme, A.M. (eds.) The Phylogenetic Handbook, pp. 615–630. Cambridge University Press, Cambridge (2009)
21. Xia, X.H., Xie, Z., Salemi, M., Chen, L., Wang, Y.: An index of substitution saturation and its application. Mol. Phylogenet. Evol. 26, 1–7 (2003)
22. Van de Peer, Y., Neefs, J.M., De Rijk, P., De Wachter, R.: Reconstructing evolution from eukaryotic small-ribosomal-subunit RNA sequences: calibration of the molecular clock. J. Mol. Evol. 37, 221–232 (1993)
23. Xia, X.H., Xie, Z., Kjer, K.M.: 18S ribosomal RNA and tetrapod phylogeny. Syst. Biol. 52, 283–295 (2003)
24. Xia, X., Hafner, M.S., Sudman, P.D.: On transition bias in mitochondrial genes of pocket gophers. J. Mol. Evol. 43, 32–40 (1996)
25. Xia, X.: The rate heterogeneity of nonsynonymous substitutions in mammalian mitochondrial genes. Mol. Biol. Evol. 15, 336–344 (1998)
26. Marin, A., Xia, X.: GC skew in protein-coding genes between the leading and lagging strands in bacterial genomes: new substitution models incorporating strand bias. J. Theor. Biol. 253, 508–513 (2008)
27. Xia, X.: DNA replication and strand asymmetry in prokaryotic and mitochondrial genomes. Current Genomics 13, 16–27 (2012)
28. Xia, X.: DAMBE5: A comprehensive software package for data analysis in molecular biology and evolution. Mol. Biol. Evol. 30, 1720–1728 (2013)
29. Siddall, M.E.: Success of Parsimony in the Four-Taxon Case: Long-Branch Repulsion by Likelihood in the Farris Zone. Cladistics 14, 209–220 (1998)

An Eigendecomposition Method
for Protein Structure Alignment

Satish Chandra Panigrahi and Asish Mukhopadhyay*

School of Computer Science
University of Windsor
401 Sunset Avenue
Windsor, ON N9B 3P4, Canada
panigra@uwindsor.ca, asishm@cs.uwindsor.ca

Abstract. The alignment of two protein structures is a fundamental problem in structural bioinformatics. Their structural similarity carries with it the connotation of similar functional behavior that could be exploited in various applications. In this paper, we model a protein as a polygonal chain of α carbon residues in three dimension and investigate the application of an eigendecomposition method due to Umeyama to the protein structure alignment problem. This method allows us to reduce the structural alignment problem to an approximate weighted graph matching problem.

The paper introduces two new algorithms, $EDAlign_{res}$ and $EDAlign_{sse}$, for pairwise protein structure alignment. $EDAlign_{res}$ identifies the best structural alignment of two equal length proteins by refining the correspondence obtained from eigendecomposition and to maximize similarity measure, TM-score, for the refined correspondence. $EDAlign_{sse}$, on the other hand, does not require the input proteins to be of equal length. It works in three stages: (1) identifies a correspondence between secondary structure elements (i.e SSE-pairs); (2) identifies a correspondence between residues within SSE-pairs; (3) applies a rigid transformation to report structural alignment in space. The latter two steps are repeated until there is no further improvement in the alignment. We report the TM-score and cRMSD as measures of structural similarity. These new methods are able to report sequence and topology independent alignments, with similarity scores that are comparable to those of the state-of-the-art algorithms such as, TM align and SuperPose.

1 Introduction

Along with DNA and RNA, protein molecules are the main drivers of all life processes at the molecular level. A protein molecule is a linear polypeptide chain, with adjacent pairs of amino acids, joined together by a peptide bond, giving rise to the nomenclature "polypeptide". In order to perform its particular biological function, the linear polypeptide chain folds into a stable, low-energy

* This research is supported by an NSERC Discovery Grant.

M. Basu, Y. Pan, and J. Wang (Eds.): ISBRA 2014, LNBI 8492, pp. 24–37, 2014.
© Springer International Publishing Switzerland 2014

3-dimensional tertiary structure. The latter structure is formed by the joining together of two types of secondary structures, known as α-helices and β-sheets.

The two important aspects of this process are: (1) how the folding takes place; (2) how does the particular structure it assumes allows it to perform its designated function. The first is well-known as the protein folding problem, predicting how a protein will fold, given the amino acid sequence that makes up its polypeptide chain structure. This problem still awaits a comprehensive solution. The second problem is that of predicting function from structure. Here a reductionist approach is a popular one: structural comparisons with proteins of known functions. Thus the problem of structural alignments of proteins, which is the subject of this paper.

As Taylor et al. [1] observed, "The most important things we know about proteins have come therefore not from theory but from observation and comparison of sequences and structures". In view of the importance of the problem, numerous heuristics have been proposed, consequently giving rise to an extensive literature and several large structural databases of proteins [2–4]. These databases help in the classification of the large space of protein sequences into structurally equivalent classes by means of alignment or structure comparison algorithms.

In order to design an alignment algorithm, it is important to enunciate clearly the protein model that will be used. Some of the earliest alignment algorithms assumed a model in which the central α carbon atom of each residue are joined successively to form a polygonal chain in three dimensions. A more primitive model is to view a protein as a collection of points (again the α carbon atoms) in three space, which allows one to view the alignment problem as that of matching two point sets. We must point out that in order to draw biologically meaningful conclusions from an alignment, it is important to supplant these models with features of the proteins like hydrophobicity, exposure to solvents, mutual affinities of amino acids etc.

The alignment of two protein structures is the 3-dimensional analogue of linear sequence alignment of peptide or nucleotide sequences. An initial equivalence set can be obtained by various methods such as comparison of distance matrix [5], maximal common subgraph detection [6], geometric hashing [7], local geometry matching [8], spectral matching [9], contact map overlap [10, 11] and dynamic programming [12,13]. This equivalence set is optimized by different methods such as a Monte Carlo algorithm or simulated annealing [5], dynamic programming [12–15], incremental combinatorial extension of the optimal path [16] and genetic algorithm [17]. Indeed the goal is to determine an alignment of protein residues to measure the extent of structural similarity. To quantify this similarity, various measures have been defined and can be broadly classified into four categories: (1) distance map similarity [5, 18–20] (2) root mean square deviation ($RMSD$) [9, 12, 16, 21] (3) contact map overlap [22] (4) universal similarity matrix [23, 24]. A comprehensive list of different similarity measures are discussed by Hasegawa and Holm [25]. Surprisingly, even after so many years of research there is no universally acknowledged definition of similarity score to measure the extent of structural similarity [25, 26].

In [11], alignment of eigenvectors is used for fast overlapping of contact maps. The paper uses Needleman-Wunch's algorithm to compute a global alignment of two protein sequences, where the cost function is derived from an approximation of the contact map M (of the two protein structures), obtained from the spectral decomposition of M. Using a graph theoretic approach, Taylor et al. [27] obtained a structural similarity measure by matching pairs of secondary structural elements(SSEs) of the input proteins. The set of matching pairs of SSEs is obtained by a bipartite graph-matching algorithm.

In this paper we introduce two new algorithms, $EDAlign_{res}$ and $EDAlign_{sse}$, for the protein structure alignment problem. These algorithms rely on a matrix eigendecomposition approach due to Umeyama [28] for an approximate solution to the weighted graph matching problem. $EDAlign_{res}$ identifies best structural alignment of two equal length proteins by refining the correspondence obtained from the eigendecomposition technique and to maximize similarity measure, TM-score, for the refined correspondence. $EDAlign_{sse}$, on the other hand, does not require the input proteins to be of equal length. It works in three stages: (1) identifies correspondence between secondary structure elements (i.e SSE-pairs); (2) identifies a correspondence between residues within SSE-pairs; (3) applies a rigid transformation to report structural alignment in space. The latter two steps are repeated until there is no improvement in the alignment. These methods are able to provide sequence and topology independent similarities. The reason for this is that the primary equivalence set (residues-pairs for equal length proteins and SSE-pairs for unequal length proteins) depends on the intrinsic geometry of α-the carbon atoms within the tertiary structure that is revealed by eigendecomposition. We report the TM-score and $cRMSD$ as measures of the structural similarity. The similarity scores of both the algorithms are comparable to those of the state-of-the-art algorithms such as, TM align and SuperPose.

2   Preliminaries

2.1   Notations and Definitions

The following definitions help us formulate the problem precisely.

Definition 1. *A protein P is a sequence of points, $P = \{p_i | p_i \in R^3, i = 1, 2, 3, ..., m\}$, in a 3-dimensional Euclidean space, where $m(= |P|)$ is the number residues and p_i represents the coordinates of the central α-carbon atom of the i-th residue.*

Definition 2. *Given two proteins P and Q of length m and n respectively. An alignment of P and Q is:*

- *a sequence of corresponding pairs of points of P and Q, $S(P, Q) = \{(p_{i_1}, q_{j_1}), (p_{i_2}, q_{j_2}), ..., (p_{i_k}, q_{j_k})\}$, where $1 \leq i_1 < i_2 < ... < i_k \leq m$ and $1 \leq j_1 \neq j_2 \neq ... \neq j_k \leq n$, together with*

- *a rigid transformation t, $t(Q) = \{t(q_j) = q'_j | q'_j \in R^3, j = 1, 2, 3, ..., n\}$, that optimizes some similarity measure for the above correspondence.*

Definition 3. *A residue p_i of a protein P is known as a k-neighbor of another residue p_j if $|i - j| = k$, where $1 \leq i, j \leq |P|$.*

2.2 Similarity Measures

To measure the extent of structural similarity of two proteins, the root mean square deviation $(RMSD)$ is widely used [26, 29]. Two different $RMSD$ measures have been proposed in the literature: (1) coordinate root mean square deviation $(cRMSD)$ and (2) distance root mean square deviation $(dRMSD)$. In the proposed algorithms, $EDAlign_{res}$ and $EDAlign_{sse}$, we obtain a correspondence (i.e. residue-pairs for equal length proteins and SSE-pairs for unequal length proteins) that minimizes the $dRMSD$ measure (see equation 5) and finally reports an alignment that minimizes the $cRMSD$ measure and maximizes the TM-score [12, 30]. For completeness, the $cRMSD$ and $dRMSD$ measures are defined below.

Definition 4. *The similarity measure between two aligned substructures of proteins P and Q of length k can be defined as follows*

$$dRMSD = \sqrt{\frac{2}{k^2 - k} \sum_{u=1}^{k-1} \sum_{v=u+1}^{k} (\|p_{i_u} - p_{i_v}\| - (\|q_{j_u} - q_{j_v}\|)^2}; and \quad (1)$$

$$cRMSD = \sqrt{\frac{1}{k} \sum_{u=1}^{k} \|p_{i_u} - t(q_{j_u})\|^2}. \quad (2)$$

Since the similarity measures, $cRMSD$ and $dRMSD$, are in terms of absolute distances, a small presence of outliers may result in a poor $RMSD$ even if the two structures are globally similar. A similar observation has been made by other researchers [12, 26, 31, 32]. To circumvent this problem, Zhang and Skolnick [30] introduced a sequence independent structural alignment measure (TM-score) that is a variation of a measure originally defined by Levitt and Gerstein [33]. A critical assessment of this TM-score has been given by Xu and Zhang [34].

Definition 5. *Given two proteins, a template protein P and a target protein Q, $|P| \geq |Q|$, the structural similarity is obtained by a spatial superposition of P and Q that maximizes the following score*

$$TM\text{-}score = \frac{1}{|Q|} \sum_{i=1}^{k} \frac{1}{1 + (\frac{d_i}{d_0})^2}, \quad (3)$$

where k is the number of aligned residues of P and Q; d_i is the distance between i-th pair of aligned residues and $d_0(= 1.24 \sqrt[3]{|Q| - 15} - 1.8)$ is a normalization factor.

When the value of d_0 in equation (3) is set to $5A^\circ$, the resulting TM-score is known as a raw TM-score (rTM-score). In $EDAlign_{sse}$, to report the TM-score the protein lengths are set to the number of residues in the aligned SSEs, ignoring the residues in the fragments that connects these SSEs. Despite this, the modified score successfully reveals the extent of similarity between the aligned SSEs. Xu and Zhang [34] observed that two proteins are structurally similar and belong to same fold when the TM-score > 0.5.

2.3   Umeyama's Matrix Eigendecomposition Method

The algorithms proposed in this paper rely on Umeyama's matrix eigendecomposition method for weighted graph matching [28] to generate sequence independent alignments, i.e. residue-pairs for equal length proteins and SSE-pairs for unequal length proteins. To make the paper self-contained, we briefly describe Umeyama's technique.

Let P and Q be two proteins of length N each. Let P_G (Q_G) be the adjacency matrix corresponding to a weighted graph $G(H)$ whose vertices are the central α-carbon atoms of P (Q) and $w(p_i, p_j)$ $(w(q_i, q_j))$ is the Euclidean distance between i-th and j-th residues of P (Q). This reduces the protein structure alignment problem to a weighted undirected graph matching problem. This problem is NP-Complete as this is a special case of largest common subgraph problem [35].

In particular, Umeyama's method seeks to obtain a node correspondence

$$S = \{(p_i, \phi(p_i)) \,|\, p_i \in P \text{ and } \phi(p_i) \in Q\} \tag{4}$$

that minimizes the following distance measure

$$J(\phi) = \sum_{i=1}^{N} \sum_{j=1}^{N} ((w(p_i, p_j) - w(\phi(p_i), \phi(p_j)))^2. \tag{5}$$

Umeyama showed that the mapping $\phi()$ can be approximated by a permutation matrix Π, and instead minimizes the following measure:

$$J(\Pi) = \left\| \Pi P_G \Pi^T - Q_G \right\|^2 \tag{6}$$

where $\|.\|$ represents the Euclidean norm.

The proposed approximation algorithm is based on Theorem 3 below, which is proved [28] using the next two theorems.

Theorem 1. *The eigendecompositions of the real symmetric matrices P_G and Q_G are given by*

$$\begin{aligned} P_G &= U_P \Lambda_P U_P^T \\ Q_G &= U_Q \Lambda_Q U_Q^T \end{aligned} \tag{7}$$

where U_P (U_Q) is an orthogonal matrix and Λ_P (Λ_Q) is diagonal. The entries of the diagonal matrix Λ_P (Λ_Q) are the (real) eigenvalues of P_G (Q_G) and the columns of the orthogonal matrix U_P (U_Q) are the eigenvectors of P_G (Q_G).

Theorem 2. *If P_G and Q_G are symmetric matrices then*

$$\|P_G - Q_G\|^2 \geq \sum_{i=1}^{n} (\lambda_i - \mu_i)^2 \tag{8}$$

where λ_i (μ_i), $i = 1, 2, ..n$ are the eigenvalues of P_G (Q_G) with $\lambda_i \geq \lambda_{i+1}$ ($\mu_i \geq \mu_{i+1}$).

Theorem 3. *Let P_G and Q_G two real symmetric matrices with distinct eigenvalues. If O is an orthogonal matrix, ranging over the set of all orthogonal matrices, then $\|OP_GO^T - Q_G\|^2$ attains its minimum when*

$$O = U_Q S U_P^T, \tag{9}$$

where $S = \{s_i | s_i = 1 \ or \ -1, i = 1, 2, ..., n\}$.

If there exists a protein homology, without any conformational changes, between P and Q then the two weighted graphs G and H are isomorphic. Thus from equation (6) we have:

$$\Pi P_G \Pi^T = Q_G. \tag{10}$$

Since the eigenvalues of two isomorphic graphs G and H are the same, from theorem 3 we have

$$OP_GO^T = Q_G. \tag{11}$$

Thus

$$OP_GO^T = \Pi P_G \Pi^T$$
$$U_Q S U_P^T U_P \Lambda_P U_P^T U_P S U_Q^T = \Pi U_P \Lambda_P U_P^T \Pi^T$$
$$\Pi U_P = U_Q S$$
$$\Pi = U_Q S U_P^T.$$

Though the matrix Π in the last line above is orthogonal, it is not necessarily a permutation matrix. Umeyama [28] showed that the desired permutation matrix Π can be obtained using the Hungarian method on a suitably defined matrix as below:

$$\Pi = \text{Hungarian}(|U_Q| |U_P^T|), \tag{12}$$

where $|U_P^T|$ and $|U_Q|$ are matrices whose entries are the absolute values of the corresponding entries of U_P^T and U_Q. This enables us to obtain a residue correspondence $S(P, Q)$.

3   Methods

To design algorithm $EDAlign_{res}$, we first reformulate the pairwise structural alignment problem as a weighted graph matching problem, and apply the matrix eigendecomposition method due to Umeyama [28] to obtain an equivalence set of residues. Next, we use the primary sequences of the proteins to refine the equivalence set by a two-stage strategy: (1) pruning outliers; (2) replacing outliers (patching). During the *pruning* step, we identify $\phi(p_i), 1 < i < N(= |P|)$, as an outlier if it is neither a 1-neighbor of $\phi(p_{i-1})$ nor of $\phi(p_{i+1})$. Similarly $\phi(p_1)$ $(\phi(p_N))$, is an outlier if it is neither a 1-neighbor of $\phi(p_2)$ $(\phi(p_{N-1}))$ nor a 2-neighbor of $\phi(p_3)$ $(\phi(p_{N-2}))$. Once we have identified all outliers, we substitute each suitably, whenever possible. We call this *patching*.

Thus if $\phi(p_i), 1 < i < N$, is an outlier then we identify two non outliers, $q_h \in \{\phi(p_{i-k}) \mid k = 1, 2\}$ and $q_j \in \{\phi(p_{i+l}) \mid l = 1, 2\}$ such that q_h and q_l are $(k+l)$-neighbor along Q. We replace $\phi(p_i)$ with residue q_{h+k} $(= q_{j-l})$. For $i = N$, we have $q_h \in \{\phi(p_{i-k_1}) \mid k_1 = 1, 2, 3, 4\}$ and $q_j \in \{\phi(p_{i-k_2}) \mid k_2 = 1, 2, 3, 4\}$ such that $k_1 \neq k_2$ and q_h and q_j are $|k_1 - k_2|$-neighbor along Q. Thus $\phi(p_N)$ can be replace by q_{h-k_1} $(= q_{j-k_2})$. We replace $\phi(p_1)$ similarly when it is an outlier.

Finally, the aligned residue order with non outliers are used to obtained an alignment that maximizes the TM-score. This involves an application of Kabsch's method to get an initial alignment of two proteins in space. The alignment is refined by repetitive application of dynamic programming followed by Kabsch's rotation that only considers the corresponding pairs, separated by a distance $d_i < d_0$ (see equation 3), $1 \leq i \leq N$.

We note once again that the applicability $EDAlign_{res}$ is limited to equal length proteins. In $EDAlign_{sse}$, we overcome this limitation by using matrix eigendecomposition to obtain SSE-pairs and subsequently residue-pairs from these. The details are as follows.

Identifying SEEs: In this step, we map the residues to secondary structure elements(SSEs) which are limited to α-helices and β-sheets. Based on the hydrogen bond patterns of secondary structure elements (SSEs), Kabsch and Sander [36] came up with following inequalities for assigning a residue to α-helix (β-sheet)

$$\left| d_{j,j+k} - \lambda_k^{\alpha(\beta)} \right| < \delta^{\alpha(\beta)}, \qquad (j = i - 2, i - 1, i; k = 2, 3, 4) \qquad (13)$$

The optimized parameters for the above inequalities [12, 36] are $\lambda_2^\alpha = 5.45A^o$, $\lambda_3^\alpha = 5.18A^o, \lambda_4^\alpha = 6.37A^o, \delta^\alpha = 2.1A^o, \lambda_2^\beta = 6.1A^o, \lambda_3^\beta = 10.4A^o, \lambda_4^\beta = 13A^o$, $\delta^\beta = 1.42A^o$. To identify such structures we have used the DSSP program that implements these inequalities [36, 37].

Representations of SSEs: Let SSE_i^α denote the i-th α-helix of a residue chain with n α-carbon atoms. We use following set of α-carbon atoms to represent the α-helix

$$\left\{ C_k | k = 1, 2, 3, m, n-2, n-1, n \quad \text{and} \quad m = \frac{n+1}{2}, n \geq 7 \right\}. \quad (14)$$

If n is even C_m represents a virtual α-carbon atom whose coordinates are obtained by averaging the coordinates of α-carbon atoms $C_{\frac{n}{2}}$ and $C_{\frac{n}{2}+1}$.

Similarly, to represent a β-sheet, SSE_i^β, with n α-carbon atoms, we use the following representative set of α-carbon atoms

$$\left\{ C_k | k = 1, m_1, m_2, n \quad \text{and} \quad m_1 = \left\lfloor \frac{n}{2} \right\rfloor, m_2 = \left\lceil \frac{n+1}{2} \right\rceil, n \geq 4 \right\}. \quad (15)$$

Since SSEs such as α-helices and β-sheets show regular patterns of hydrogen bonds, the above representation does not affect the overall topology. For this reason such structures have even been represented as vectors in some earlier protein structure alignment algorithms [19, 38].

Identifying SSEs for Alignment: Let protein P (Q) have n_1 (m_1) α-helices and n_2 (m_2) β-sheets. Assume that $n_1 > m_1$ and $n_2 > m_2$. This gives us $\prod_{i=1}^{2} {}^{n_i}C_{m_i}$, possible combinations of SEEs from P that can be aligned with those from Q. This value becomes impractically large when the differences $m_i - n_i$ are large. Fortunately, proteins pairs do not differ much with respect to the number of SSEs. Nevertheless, in cases where the differences exceed a prescribed threshold value, noting that the SSEs are ordered along a protein chain, we allow only the following combinations of SSEs from P as candidates for alignment with those from Q.

$$S_{i,j}^P = \left\{ SSE_{i+1}^\alpha, SSE_{i+2}^\alpha, ..., SSE_{i+m_1}^\alpha, SSE_{j+1}^\beta, SSE_{j+2}^\beta, ..., SSE_{j+m_2}^\beta \right\}, \quad (16)$$

where the values of i, j are in the range of $[0, n_1 - m_1]$ and $[0, n_2 - m_2]$. Such a selection is consistent with other alignment techniques such as dynamic programming and combinatorial extension that also consider residues along the chain with reasonable gaps, and also reduces the number of possible combinations of SSEs to a quadratic order: $\prod_{i=1}^{2}(n_i - m_i)$. The cases where $n_1 < m_1$ and $n_2 > m_2$ or $n_1 > m_1$ and $n_2 < m_2$, with $(n_1 + n_2) > (m_1 + m_2)$ in each case, can be handled in a similar way.

Identifying SSE-Pairs: It is reasonable to assume that regions of P and Q that are perfectly aligned have an equal number of α-helices as well as β-sheets in both. Suppose we know the candidate sets $S_{i,j}^P$ and $S_{i',j'}^Q$ that are to be aligned. We can represent these sets of SSEs, $S_{i,j}^P$ and $S_{i',j'}^Q$, as complete weighted graphs on their constituent α-carbon atoms which are input to Umeyama's method. The output of Umeyama's method is a symmetric matrix M (see equation 12), each entry being the cost of the correspondence between a pair of alpha carbon atoms one in $S_{i,j}^P$ and the other in $S_{i',j'}^Q$. We coalesce the α-carbon atoms that belong to an SSE into a single entity and create a modified cost matrix, M', each entry being the cost of the correspondence of a pair of SSEs, one in $S_{i,j}^P$ and the other in $S_{i',j'}^Q$. When both the SSEs are α-helices the cost is:

$$M'[SSE_P^\alpha, SSE_Q^\alpha] = \frac{\sum_{C_a \in SSE_P^\alpha,\ C_b \in SSE_Q^\alpha} M(C_a, C_b)}{49}, \tag{17}$$

while if both are β-sheets the cost is:

$$M'[SSE_P^\beta, SSE_Q^\beta] = \frac{\sum_{C_a \in SSE_P^\beta,\ C_b \in SSE_Q^\beta} M(C_a, C_b)}{16}. \tag{18}$$

To avoid a correspondence between an α-helix and a β-sheet, we set the cost of such a correspondence to zero. Finally, we apply the Hungarian method to this modified cost matrix M' to get a correspondence that optimizes the total cost. The aligned SSE pairs are used to obtain an initial spatial structural alignment.

Reordering of Residues: In a structural alignment the order of the SSEs may not be same as the primary sequence order. Therefore we reorder the residues, according to the SSE-pairs obtained from the previous step. To make this precise we set the smaller length protein Q as the template and arrange its SSEs as these appear along the chain. We order the SSEs of the P according to their correspondence with the SSEs of Q, ignoring those SSEs that do not have a corresponding SSE in Q. Now a corresponding SSE-pair may align with each other either in forward or reverse direction. To determine the correct direction we have exhaustively checked all possible 2^m combinations where m is the number of SSEs. Of these combinations, we choose the one with minimum $J(\phi)$. Here we also consider the topological ordering of alignment set as one candidate as a majority of protein structural alignment algorithms use the conventional sequence order, primarily for biological reasons [39]. In the above rearrangement, we ignore residues in the loops that connect the SSEs. Finally, we reorder the residues according to their appearance in the ordering of SSEs.

Apply Dynamic Programming: To refine the alignment, the residue order obtained from the previous step is input to a dynamic programming [12, 30, 33] algorithm. The entries of the scoring matrix are defined by

$$S(i,j) = \frac{1}{1 + \left(\frac{d_{ij}}{d_0}\right)^2} \tag{19}$$

where d_{ij} is the distance between the i-th residue in P and the j-th residue in Q and d_0 is scale factor that normalizes the distances between residue pairs of P and Q (see equation 3). Setting an opening gap penalty of -0.6, and considering pair correspondences that are at distances less than d_0, we apply Kabsch's method to superimpose P and Q. The process is repeated until the alignment becomes stable with maximum TM-score. Based on our experiments, it takes typically 2-3 steps to get the best alignment - a fact also observed by Zhang and Skolnick [12].

Table 1. Pairwise structural alignment of equal length proteins

Structure	N	Seq. id	$EDAlign_{res}$		$EDAlign_{sse}$		TM-align		SuperPose	
			cRMSD	TM-score	cRMSD	TM-score*	cRMSD	TM-score	cRMSD	TM-score
Same sequence and similar structure (pair)										
Thioredoxin (2TRXA 2TRXB)	108	100%	0.45 - 105	0.94	0.26 - 53	0.98	0.66 - 108	0.98	0.77 - 108	0.97
Hemoglobin (4HHBA 1DKEA)	141	100%	0.31 - 141	0.99	0.26 - 92	0.99	0.37 - 141	0.99	0.37 - 141	0.99
P21 Oncogene (6Q21A 6Q21B)	171	100%	0.39 - 108	0.61	0.31 - 91	0.99	1.22 - 171	0.96	1.27 - 171	0.96
~ Same sequence and different structure (pair)										
Calmodulin (1A29 1CLL)	144	98.6%	22.81 - 144	0.59	10.19 - 59	0.62	1.91 - 71	0.51	23.83 - 142	0.0002
Maltose Bind Prot. (1OMP 1ANF)	370	100%	3.04 - 327	0.65	2.83 - 176	0.76	3.42 - 364	0.82	8.87 - 369	0.79

cRMSD values are reported as backbone cRMSD in A^o - number of aligned α-carbon atoms
N: Number of residues in protein structure
*: Ignores the residues in the fragment that connects SSEs

Table 2. NMR models 1m2f_A_1-1mef_A_25, compared to an average model 1m2e_A

	Umeyama Method		$EDAlign_{res}$				TM-align		Superpose	
			Pruning Outliers		Replacing Outliers (Patching)					
Model	cRMSD	TM-score	cRMSD	TM-score	cRMSD	TM-score	cRMSD	TM-score	cRMSD	TM-score
1	1.294 - 135	0.93	0.607 - 133	0.91	0.641 - 135	0.93	1.06 - 135	0.95	1.06 - 135	0.95
2	1.032 - 135	0.92	0.542 - 128	0.89	0.585 - 131	0.9	0.91 - 135	0.96	0.91 - 135	0.96
3	3.454 - 135	0.84	0.387 - 111	0.77	0.424 - 124	0.88	1.54 - 134	0.94	1.83 - 135	0.92
4	5.059 - 135	0.78	0.632 - 95	0.65	0.698 - 110	0.76	1.42 - 135	0.94	1.42 - 135	0.93
5	0.891 - 135	0.91	0.755 - 134	0.92	0.755 - 134	0.92	0.89 - 134	0.95	1.11 - 135	0.95
6	6.359 - 135	0.75	0.657 - 90	0.61	0.766 - 113	0.76	1.51 - 135	0.94	1.51 - 135	0.93
7	0.757 - 135	0.92	0.685 - 135	0.92	0.685 - 135	0.92	1.12 - 135	0.95	1.12 - 135	0.95
8	0.814 - 135	0.93	0.604 - 133	0.92	0.612 - 135	0.93	0.79 - 135	0.97	0.79 - 135	0.97
9	4.896 - 135	0.78	0.496 - 98	0.68	0.548 - 115	0.79	1.33 - 135	0.94	1.34 - 135	0.93
10	0.848 - 135	0.91	0.646 - 133	0.91	0.705 - 135	0.92	0.91 - 134	0.95	1.09 - 135	0.95
11	0.638 - 135	0.93	0.629 - 135	0.93	0.629 - 135	0.93	0.89 - 135	0.96	0.89 - 135	0.96
12	1.411 - 135	0.9	0.588 - 128	0.88	0.662 - 135	0.92	1.09 - 135	0.95	1.09 - 135	0.95
13	1.022 - 135	0.89	0.745 - 133	0.91	0.786 - 135	0.92	0.88 - 133	0.95	1.38 - 135	0.93
14	1.318 - 135	0.92	0.532 - 130	0.91	0.629 - 135	0.92	1.08 - 135	0.96	1.08 - 135	0.95
15	1.517 - 135	0.9	0.592 - 129	0.89	0.731 - 135	0.92	1.32 - 135	0.95	1.32 - 135	0.94
16	1.278 - 135	0.91	0.579 - 129	0.88	0.594 - 133	0.92	0.92 - 135	0.96	0.92 - 135	0.96
17	0.964 - 135	0.92	0.594 - 132	0.91	0.64 - 135	0.93	1.03 - 135	0.96	1.03 - 135	0.96
18	0.94 - 135	0.92	0.61 - 132	0.91	0.617 - 135	0.93	0.97 - 135	0.96	0.97 - 135	0.96
19	1.147 - 135	0.91	0.486 - 127	0.89	0.541 - 131	0.91	1.05 - 135	0.95	1.05 - 135	0.95
20	0.723 - 135	0.92	0.636 - 135	0.93	0.636 - 135	0.93	1.05 - 135	0.95	1.05 - 135	0.95
21	2.704 - 135	0.85	0.635 - 120	0.82	0.723 - 135	0.93	1.28 - 134	0.95	1.4 - 135	0.94
22	0.827 - 135	0.91	0.615 - 132	0.91	0.706 - 135	0.92	1.16 - 135	0.95	1.16 - 135	0.95
23	2.138 - 135	0.85	0.598 - 117	0.8	0.718 - 133	0.9	1.19 - 135	0.94	1.19 - 135	0.94
24	4.26 - 135	0.81	0.66 - 104	0.72	0.79 - 122	0.84	1.04 - 135	0.95	1.05 - 135	0.95
25	1.198 - 135	0.9	0.563 - 127	0.88	0.598 - 131	0.9	1.07 - 135	0.95	1.07 - 135	0.95

cRMSD values are reported as backbone cRMSD in A^o - number of aligned α-carbon atoms

4   Results and Discussions

To illustrate the proposed methods, we apply both to pairs of proteins in the following two categories [40]: (1) same primary sequence with slightly different tertiary structures; (2) same primary sequence with vastly different tertiary structures.

As can be see from Table 1, the scores computed by both $EDAlign_{res}$ and $EDAlign_{sse}$ are as good or better than the scores computed by SuperPose [40] and TM-align [12]. The number of residues aligned by $EDAlign_{sse}$ is smaller than that of TM-align as it ignores the residues in the loop fragments that join pairs of SSEs and this fact is reflected in the computation of the TM-score. Nevertheless, the TM-score of $EDAlign_{sse}$ compares remarkably well with that of TM-align. The results also show that $EDAlign_{res}$ is more successful in detecting the structural similarity of homologous proteins of equal length and therefore might be potentially useful during NMR spectroscopy. To further illustrate the effectiveness of $EDAlign_{res}$ and its improvement over the basic Umeyama method, we ran this algorithm on NMR models $1m2f\_A\_1 - 1m2f\_A\_25$ to an average model $1m2e\_A$. Table 2 also includes results from the refinement stages (i.e. pruning outliers and replacing outliers) of $EDAlign_{res}$.

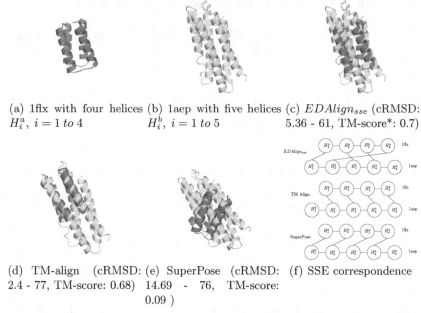

(a) 1flx with four helices
H_i^a, $i = 1\,to\,4$

(b) 1aep with five helices
H_i^b, $i = 1\,to\,5$

(c) $EDAlign_{sse}$ (cRMSD:
5.36 - 61, TM-score*: 0.7)

(d) TM-align (cRMSD: 2.4 - 77, TM-score: 0.68)

(e) SuperPose (cRMSD: 14.69 - 76, TM-score: 0.09)

(f) SSE correspondence

Fig. 1. Structural Alignment of 1flx with 1aep

Unlike $EDAlign_{res}$, $EDAlign_{sse}$ can detect the structural similarity of any pair of proteins. To substantiate this, we have run $EDAlign_{sse}$ on pairs of proteins (see Table 3) that are in the following categories [40]: (1) have modestly dissimilar sequences, lengths and structures (2) have vastly different lengths but similar structures or sequences. Table 3 shows that the TMscore of $EDAlign_{sse}$ is as good as that of TM align and on top of that is able to locate conserved regions between protein pairs.

Table 3. Pairwise structural alignment of unequal length proteins

Structure	N1	N2	Seq. id	$EDAlign_{sse}$		TM-align		SuperPose	
				cRMSD	TM-score*	cRMSD	TM-score	cRMSD	TM-score
modestly dissimilar sequence, length and structure									
Hemoglobin (4HHBA 4HHBB)	141	146	43%	1.11 - 89	0.88	1.41 - 139	0.9	1.61 - 139	0.89
Thioredoxin (3TRX 2TRXA)	105	108	29%	5.97 - 38	0.71	1.4 - 101	0.86	4.72 - 99	0.62
Lysozyme/Lactalbumin (1DPX 1A4V)	129	123	36%	0.62 - 31	0.94	1.48 - 123	0.87	1.63 - 121	0.86
Calmodulin/TnC (1CLL 5TNC)	144	161	47%	4.64 - 79	0.69	3.49 - 107	0.55	6.83 - 144	0.48
different length but similar structure or sequence									
Ubiquitin/Elongin (1UBI 1VCBA)	76	98	26%	1.83 - 28	0.81	1.57 - 75	0.86	3.22 - 72	0.54
Thio/Glutaredoxin (3TRX 3GRXA)	105	82	7%	7.54 - 31	0.64	2.27 - 74	0.67	4.64 - 76	0.33
Hemoglobins (1ASH 2LHB)	147	149	17%	1.75 - 88	0.83	2.15 - 135	0.77	$-^a$	$-^a$
Thioredoxins (1NHOA 1DE2A)	85	87	22%	1.86 - 18	0.81	3.69 - 72	0.47	7.77 - 65	0.19

cRMSD values are reported as backbone cRMSD in A^o - number of aligned α-carbon atoms
N1: Number of residues in 1st protein structure
N2: Number of residues in 2nd protein structure
*: Ignores the residues in the fragment that connects SSEs
$-^a$: Fail to align

   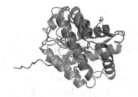

(a) 1e50A with 1pkvA   (b) 2hbg with 2ecd

Fig. 2. $EDAlign_{sse}$ on difficult alignment (a) $EDAlign_{sse}$: cRMSD 1.94 - 20 and TM-score* 0.81 TM-align: cRMSD 4.2 - 68 and TM-score 0.42 SuperPose: cRMSD 13.21 - 72 and TM-score 0.07 (b) $EDAlign_{sse}$: cRMSD 2.42 - 16 and TM-score* 0.77 TM-align: cRMSD 4.23 - 63 and TM-score 0.31 SuperPose: cRMSD 13.77 - 111 and TM-score 0.2

$EDAlign_{sse}$ is also able to detect structural similarity independent of topological order (see Figure 1). To support this claim, we have run $EDAlign_{sse}$ to compare the protein Apolipophorin III (PDB ID:1aep) to a theoretical model of four-helix bundle protein (PDB ID:1flx). As a further demonstration of the versatility of $EDAlign_{sse}$ we consider a difficult case for alignment: (1) Core-binding factor alpha subunit (PDB ID:1e50, Chain A) with Riboflavin synthase alpha chain (PDB ID: 1pkv, Chain A) (2) Hemoglobin (Deoxy) (PDB ID:2hbg) with Tyrosine-protein kinase ABL2 (PDB ID:2ecd). The alignment of SSEs obtained by $EDAlign_{sse}$ is shown in Figure 2. Notably, the two proteins 1e50A and 1pkvA have three aligned β-sheets where as the proteins 2hbg and 2ecd have two aligned α-helices. On top of that, these pairs do not share any structural similarity. This has also been observed by TM align, as reflected in the TM-scores of 0.42 and 0.31 respectively.

5   Conclusions

In this paper we have exploited matrix eigendecomposition to design two new algorithms, $EDAlign_{res}$ and $EDAlign_{see}$, for the structural alignment of two proteins. The former outputs an alignment of residue-pairs, while the latter reports aligned SSE-pairs. $EDAlign_{res}$ can measure the structural similarity of two equal length proteins only; the more general algorithm, $EDAlign_{sse}$, combines eigendecomposition with dynamic programming and TM-score rotation matrix, and is able to handle proteins of unequal lengths. Experimental results show that $EDAlign_{sse}$ is able to align successfully common SSEs of any pair of proteins and also reveal potential conserved regions. Unlike other dynamic programming approaches, $EDAlign_{sse}$ is able to detect alignments that are independent of the order of the SSEs.

References

1. Taylor, W., May, A., Brown, N., Aszódi, A.: Protein structure: geometry, topology and classification. Reports on Progress in Physics 64(4), 517–590 (2001)
2. Murzin, A.G., Brenner, S.E., Hubbard, T., Chothia, C.: SCOP: a structural classification of proteins database for the investigation of sequences and structures. Journal of Molecular Biology 247(4), 536–540 (1995)
3. Orengo, C., Michie, A., Jones, S., Jones, D., Swindells, M., Thornton, J.: CATH - a hierarchic classification of protein domain structures. Structure 5(8), 1093–1108 (1997)
4. Holm, L., Sander, C.: Dali/FSSP classification of three-dimensional protein folds. Nucleic Acids Research 25(1), 231–234 (1997)
5. Holm, L., Sander, C.: Protein Structure Comparison by Alignment of Distance Matrices. Journal of Molecular Biology 233(1), 123–138 (1993)
6. Artymiuk, P.J., Spriggs, R.V., Willett, P.: Graph theoretic methods for the analysis of structural relationships in biological macromolecules: Research articles. J. Am. Soc. Inf. Sci. Technol. 56(5), 518–528 (2005)
7. Nussinov, R., Wolfson, H.J.: Efficient detection of three-dimensional structural motifs in biological macromolecules by computer vision techniques. Proceedings of the National Academy of Sciences 88(23), 10495–10499 (1991)
8. Wu, T.D., Hastie, T., Schmidler, S.C., Brutlag, D.L.: Regression analysis of multiple protein structures. In: RECOMB, pp. 276–284 (1998)
9. Shibberu, Y., Holder, A.: A spectral approach to protein structure alignment. IEEE/ACM Transactions on Computational Biology and Bioinformatics 8, 867–875 (2011)
10. Andonov, R., Malod-Dognin, N., Yanev, N.: Maximum contact map overlap revisited. J. Comput. Biol. 18(1), 27–41 (2011)
11. Di Lena, P., Fariselli, P., Margara, L., Vassura, M., Casadio, R.: Fast overlapping of protein contact maps by alignment of eigenvectors. Bioinformatics 26(18), 2250–2258 (2010)
12. Zhang, Y., Skolnick, J.: TM-align: A protein structure alignment algorithm based on TM-score. Nucleic Acids Research 33, 2302–2309 (2005)
13. Taylor, W.R.: Protein structure comparison using iterated double dynamic programming. Protein Sci. 8(3), 654–665 (1999)
14. Gerstein, M., Levitt, M.: Using iterative dynamic programming to obtain accurate pairwise and multiple alignments of protein structures. In: Proc. Fourth Int. Conf. on Intell. Sys. Mol. Biol., pp. 59–67. AAAI Press (1996)
15. Gerstein, M., Levitt, M.: Comprehensive assessment of automatic structural alignment against a manual standard, the scop classification of proteins. Protein Science 7(2), 445–456 (1998)
16. Shindyalov, I.N., Bourne, P.E.: Protein structure alignment by incremental combinatorial extension (CE) of the optimal path. Protein Engineering 11(9), 739–747 (1998)
17. Szustakowski, J.D., Weng, Z.: Protein structure alignment using a genetic algorithm. Proteins: Structure, Function, and Bioinformatics 38(4), 428–440 (2000)
18. Taylor, W.R., Orengo, C.A.: Protein structure alignment. Journal of Molecular Biology 208(1), 1–22 (1989)
19. Alexandrov, N.N., Takahashi, K., Go, N.: Common spatial arrangements of backbone fragments in homologous and non-homologous proteins. Journal of Molecular Biology 225(1), 5–9 (1992)

20. Orengo, C.A., Taylor, W.R.: SSAP: Sequential structure alignment program for protein structure comparison. In: Doolittle, R.F. (ed.) Computer Methods for Macromolecular Sequence Analysis. Methods in Enzymology, vol. 266, pp. 617–635. Academic Press (1996)
21. Zemla, A.: LGA: a method for finding 3D similarities in protein structures. Nucleic Acids Research 31(13), 3370–3374 (2003)
22. Godzik, A., Kolinski, A., Skolnick, J.: Topology fingerprint approach to the inverse protein folding problem. Journal of Molecular Biology 227(1), 227–238 (1992)
23. Krasnogor, N., Pelta, D.A.: Measuring the similarity of protein structures by means of the universal similarity metric. Bioinformatics 20(7), 1015–1021 (2004)
24. Rahmati, S., Glasgow, J.I.: Comparing protein contact maps via universal similarity metric: an improvement in the noise-tolerance. I. J. Computational Biology and Drug Design 2(2), 149–167 (2009)
25. Hasegawa, H., Holm, L.: Advances and pitfalls of protein structural alignment. Current Opinion in Structural Biology 19(3), 341–348 (2009)
26. Koehl, P.: Protein structure similarities. Current Opinion in Structural Biology 11(3), 348–353 (2001)
27. Taylor, W.R.: Protein structure comparison using bipartite graph matching and its application to protein structure classification. Mol. Cell Proteomics, 334–339 (2002)
28. Umeyama, S.: An eigendecomposition approach to weighted graph matching problems. IEEE Trans. Pattern Anal. Mach. Intell. 10(5), 695–703 (1988)
29. Kolodny, R., Linial, N.: Approximate protein structural alignment in polynomial time. Proc. Natl. Acad. Sci. U. S. A. 101(33), 12201–12206 (2004)
30. Zhang, Y., Skolnick, J.: Scoring function for automated assessment of protein structure template quality. Proteins: Structure, Function, and Bioinformatics 57(4), 702–710 (2004)
31. Wang, S., Ma, J., Peng, J., Xu, J.: Protein structure alignment beyond spatial proximity. Scientific Reports 3 (March 2013)
32. Li, S.C.: The difficulty of protein structure alignment under the RMSD. Algorithms for Molecular Biology 8, 1 (2013)
33. Levitt, M., Gerstein, M.: A unified statistical framework for sequence comparison and structure comparison. Proceedings of the National Academy of Sciences 95(11), 5913–5920 (1998)
34. Xu, J., Zhang, Y.: How significant is a protein structure similarity with TM-score = 0.5? Bioinformatics 26(7), 889–895 (2010)
35. Garey, M.R., Johnson, D.S.: Computers and Intractability; A Guide to the Theory of NP-Completeness. W. H. Freeman & Co., New York (1990)
36. Kabsch, W., Sander, C.: Dictionary of protein secondary structure: pattern recognition of hydrogen-bonded and geometrical features. Biopolymers 22(12), 2577–2637 (1983)
37. Joosten, R.P., te Beek, T.A.H., Krieger, E., Hekkelman, M.L., Hooft, R.W.W., Schneider, R., Sander, C., Vriend, G.: A series of PDB related databases for everyday needs (January 2011)
38. Mizuguchi, K., Go, N.: Comparison of spatial arrangements of secondary structural elements in proteins. Protein Engineering 8(4), 353–362 (1995)
39. Abyzov, A., Ilyin, V.: A comprehensive analysis of non-sequential alignments between all protein structures. BMC Structural Biology 7(1), 1–20 (2007)
40. Maiti, R., Domselaar, G.H.V., Zhang, H., Wishart, D.S.: SuperPose: a simple server for sophisticated structural superposition. Nucleic Acids Research 32(Web-Server-Issue), 590–594 (2004)

Functional Interplay between Hemagglutinin and Neuraminidase of Pandemic 2009 H1N1 from the Perspective of Virus Evolution

Wei Hu

Department of Computer Science, Houghton College, Houghton, NY 14744, USA
wei.hu@houghton.edu

Abstract. Influenza type A viruses are classified into subtypes based on their two surface proteins, hemagglutinin (HA) and neuraminidase (NA). Our time series analysis on the strains of pandemic 2009 H1N1 collected from 2009 to 2013 demonstrated that the HA receptor binding preference of this virus in USA, Europe, and Asia has been the characteristic of swine H1N1 virus since 2009. However, its binding characteristics of seasonal human H1N1 and avian H1N1 both have been on steady rise with American strains having the sharpest surge in 2013. The first increase could enhance the viral transmission and replication in humans and the second boost its ability to cause infection deep in lungs, which might explain the recent human deaths caused by this virus in Texas in December 2013. We further explored the corresponding NA activity of this virus to reveal the functional interdependence between HA and NA during the evolution and adaptation of this virus from 2009 to 2013. To understand the real causality, the amino acid substitutions in HA and NA that actually produced the mutations were also identified.

Keywords: Pandemic 2009 H1N1, Influenza, Hemagglutinin, Neuraminidase, Mutation, Receptor Binding Specificity.

1   Introduction

Influenza A virus is an enveloped virus, subtyped according to its two surface proteins, haemagglutinin (HA) and neuraminidase (NA). Both proteins recognize the glycan receptors on host cells, and they help the viral entry into and release of virions from host cells respectively. The HAs of human viruses preferentially bind to oligo-saccharides that terminate with sialic acid linked to galactose by $\alpha2,6$-linkages (human type receptors), while the HAs of avian influenza favor those by $\alpha2,3$-linkages (avian type receptors). Swine and some avian species can have both $\alpha2,6$ and $\alpha2,3$ receptors. The $\alpha2,6$ receptor cells are usually present in the upper airway of humans, but the $\alpha2,3$ receptor cells are most in the lungs.

The pandemic 2009 H1N1 virus resulted from reassortment of several viruses of swine origin. In particular, its NA gene was derived from Eurasian avian-like swine H1N1, and the HA gene from a triple-reassortant virus circulating in North American swine [1]. Unlike seasonal human H1N1 viruses, which bind mainly to $\alpha2,6$ receptors,

M. Basu, Y. Pan, and J. Wang (Eds.): ISBRA 2014, LNBI 8492, pp. 38–49, 2014.

the pandemic 2009 H1N1 virus could bind to both α2,3 and α2,6 receptors although the primary binding is α2,6 receptors [2].

In comparison, HA could bind to α2,6 or α2,3 receptors, whereas NA has a marked preference for α2,3 receptors although it can cleave both types [3]. However, the NA of the pandemic 2009 H1N1 shows a distinctive enzymatic profile, which hydrolyzes α2,3 receptors as efficiently as avian viruses and hydrolyzes α2,6 receptor as efficiently as classical swine viruses [4].

In the flu season of 2013, the predominant strain of influenza was pandemic 2009 H1N1 in the south central United States, where several human deaths were reported in Texas in December 2013 (www.usatoday.com). In search of any altered molecular features from this virus that might lead to its recent upsurge of its pathogenicity as observed in Texas, we sought to discover, with a computational approach developed in [5,6], any variations in the HA binding patterns of the strains collected from 2009 to 2013. We also quantified the corresponding NA activity of this virus, in association with the HA receptor binding, for demonstrating their functional interdependence during virus evolution and adaptation.

2   Matertials and Methdos

2.1   Sequence Data

The HA and NA protein sequences of influenza viruses used in this study were retrieved from the EpiFlu Database (http://platform.gisaid.org) of GISAID and the Influenza Virus Database of NCBI (http://www.ncbi.nlm.nih.gov/genomes/FLU).

2.2   Informational Spectrum Method

The informational spectrum method (ISM) is a computational technique that can be employed to analyze protein sequences [7,8]. The idea is to transform the protein sequences into numerical sequences based on electron-ion interaction potential (EIIP) of each amino acid. Then the Discrete Fourier Transform (DFT) can be applied to these numerical sequences, and the resulting DFT coefficients are used to produce the energy density spectrum. The informational spectrum (IS) comprises the frequencies and the amplitudes of this energy density spectrum. According to the ISM theory, the peak frequencies of IS of a protein sequence reflect its biological or biochemical functions. The ISM was successfully applied to quantify the effects of HA mutations on the receptor binding preference in [5-8].

3   Results

The theme of this study was using ISM to elucidate any genetic variations in HA and NA of pandemic 2009 H1N1 from 2009 to 2013. For this end, we also included the HA and NA of H1N1 from different species and regions, laying a foundation for our analysis of pandemic 2009 H1N1.

3.1   IS of HA and NA of H1N1 from different Species and Regions

To provide the background for the comparison based on ISM, we calculated the IS of each HA and NA of H1N1 from different species and regions, and took an average of their IS in each. The top two average IS frequencies, primary and secondary, of H1N1 HA and NA respectively from different species and regions were reported (Table 1). The HA frequencies F(0.295) and F(0.055) were first analyzed in study of H1N1 [7], and F(0.236) and F(0.258) of HA were calculated in [8]. Although only the top two IS frequencies were selected in Table 1, the actual top three HA frequencies of pandemic 2009 human H1N1 were F(0.295), F(0.055), and F(0.258) in USA, and F(0.295), F(0.258), and F(0.055) in Europe and Asia, while the swine H1N1 in USA had F(0.295) and F(0.055) as its top two frequencies, showing the HA binding of pandemic 2009 H1N1 was similar to that of North American swine H1N1 [1]. Eurasian swine H1N1 HA binding patterns (F(0.295) and F(0.281)) were all avian like because F(0.281) (an avian H1N1 feature frequency) was their secondary binding frequency. The avian like swine viruses emerged in Europe in the late 1970s after an avian virus was introduced to swine [4]. Seasonal human H1N1 in USA, Europe, and Asia all shared the same top two HA IS frequencies F(0.055) and F(0.236) and the same NA frequencies F(0.373) and F(0.268).

The NA of pandemic 2009 H1N1 had its top three IS frequencies as F(0.074), F(0.347), and F(0.484). F(0.074) was the primary NA frequency of swine and avian H1N1 in Europe and avian H1N1 in Asia. F(0.484) was the primary of swine H1N1 in Asia and secondary of swine H1N1 in Europe. Our analysis suggested that the NA sialidase activity of pandemic 2009 H1N1 was most similar to that of avian like Eurasian swine H1N1, which was in line with the NA origin of this virus [1]. However, a study in [4] implied that the NA activity of this virus was closer to that of classical swine viruses than to that of avian, avian like swine and seasonal human viruses.

Table 1. Primary and secondary average IS frequencies of HA and NA of H1N1 from different species and regions

Virus	Protein	Primary	Secondary	Number of sequences	Years of collection
Pandemic 2009 Human	HA	F(0.295)	F(0.055)	2670	2009-2013
H1N1 in USA	NA	F(0.074)	F(0.346)	2670	2009-2013
Pandemic 2009 Human	HA	F(0.295)	F(0.055)	1689	2009-2013
H1N1 in Europe	NA	F(0.074)	F(0.346)	1689	2009-2013
Pandemic 2009 Human	HA	F(0.295)	F(0.258)	1864	2009-2013
H1N1 in Asia	NA	F(0.074)	F(0.346)	1864	2009-2013
Pandemic 2009 Swine	HA	F(0.295)	F(0.258)	116	2009-2013
H1N1 in USA	NA	F(0.074)	F(0.346)	116	2009-2013
Pandemic 2009 Swine	HA	F(0.295)	F(0.258)	20	2009-2013
H1N1 in Europe	NA	F(0.074)	F(0.346)	17	2009-2011
Pandemic 2009 Swine H1N1 in Asia	HA	F(0.295)	F(0.258)	149	2009-2012
	NA	F(0.074)	F(0.346)	116	2009-2012

Table 1. (*continued*)

Seasonal Human H1N1 in	HA	F(0.055)	F(0.236)	876	2000-2009
USA	NA	F(0.373)	F(0.268)	876	2000-2009
Swine H1N1 in USA	HA	F(0.295)	F(0.055)	668	1930-2013
	NA	F(0.373)	F(0.074)	668	1930-2013
Avian H1N1 in USA	HA	F(0.281)	F(0.295)	190	1980-2012
	NA	F(0.168)	F(0.074)	190	1980-2012
Seasonal Human H1N1 in	HA	F(0.055)	F(0.236)	230	2000-2009
Europe	NA	F(0.373)	F(0.268)	250	2000-2009
Swine H1N1 in Europe	HA	F(0.295)	F(0.281)	116	1939-2012
	NA	F(0.074)	F(0.484)	169	1979-2013
Avian H1N1 in Europe	HA	F(0.295)	F(0.281)	28	1977-2009
	NA	F(0.074)	F(0.490)	35	1983-2009
Seasonal Human H1N1 in	HA	F(0.055)	F(0.236)	1366	2000-2010
Asia	NA	F(0.373)	F(0.268)	756	2001-2010
Swine H1N1 in Asia	HA	F(0.295)	F(0.281)	687	1974-2012
	NA	F(0.484)	F(0.074)	667	1977-2012
Avian H1N1 in Asia	HA	F(0.281)	F(0.295)	24	1976-2011
	NA	F(0.074)	F(0.240)	33	1977-2011

3.2   IS of HA and NA of H1N1 from different Species in USA

Location is critical for the spread and transmission of influenza. Here we conducted ISM on H1N1 HA and NA from various species in USA since the pandemic 2009 H1N1 emerged in North America first (Figure 1). In addition to the IS information in section 3.1, the IS analysis of the H1N1 viruses in USA in this section prepared a groundwork for understanding the HA and NA IS of pandemic 2009 H1N1 presented in the next section. The advantage of our work was able to highlight the changing patterns of HA and NA and their correlation, if any, over a period of various years. We could see that the primary and secondary HA ISs at F(0.055) and F(0.236) of seasonal H1N1 in USA remained at the same level (the top two ISs of NA kept a constant gap), but after 2007 HA IS at F(0.055) played a clear leading role (the leading role of NA IS at F(0.373) started to diminish). Also there was a fluctuation of the top two HA ISs from 2008 to 2009, but their gap remained a constant during these two years (the top two ISs of NA started to reduce their gap). In contrast, the top two ISs of HA and NA of swine H1N1 in USA kept mixing with no clear leader during the whole period of 1930-2013. A random pattern of HA and NA from avian H1N1 in USA could also be seen. One thing worth noting was the first and third NA ISs of avian H1N1 in USA were low in the NA activity of seasonal human and swine H1N1 in USA (Figure 1), which were in contrast to the case of pandemic 2009 H1N1 studied in the next section.

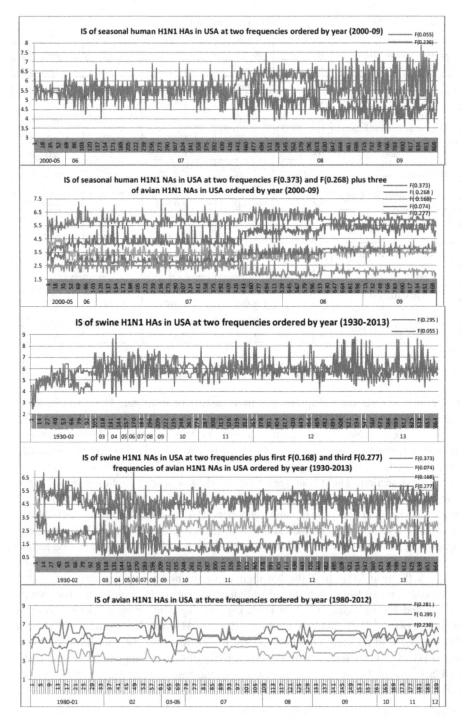

Fig. 1. IS of HA and NA of H1N1 from different origins in USA, where the y-axis represents the amplitude of IS and the x-axis represents the sequence count and year

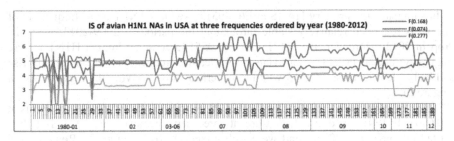

Fig. 1. (*continued*)

3.3   IS of HA and NA of Pandemic 2009 Human H1N1 in different Regions

Our original aim was to discover any molecular changes in HA and NA of pandemic 2009 H1N1 in USA from 2009 to 2013 that might contribute to the recent human deaths by this virus in Texas. We thought it would offer richer understanding if we could include the strains of this virus from Europe and Asia as well to render a bigger picture. To uncover any changing patterns of HA binding specificity, some leading HA IS frequencies of pandemic 2009 human H1N1 in different regions were presented over the period of 2009-2013 (Figure 2). We did not plot the HA IS at F(0.258) in this section because it remained relatively stable from 2009 to 2013 compared to the ISs at other top frequencies, even though it was one of the top frequencies of this virus. There were noticeable number of HA or NA sequences in 2009 that did not have the month and day information, so they were placed in the beginning of 2009 in our plot. Nonetheless, in any case, the HA and NA from the same virus isolate were placed, in order of year, month, and day, at the same position in the plot to visualize any possible correlation between these two proteins over time.

The primary HA IS of this virus in USA and Asia arose in 2009, but this occurred in 2010 in Europe. As in the case of American seasonal H1N1, there seemed a trend, i.e., the primary HA IS went higher wherever the primary NA IS was lower. There was an obvious drop of the HA ISs in 2012, with the drop in USA being the largest. These HA ISs then started to increase, which occurred for American HA ISs in July of 2012 (Figure 2). Minus this drop, the primary HA IS kept stable. However, the HA ISs at F(0.055) and F(0.281) were on steady rise from 2009 to 2013 with a sharp increase in 2013 for American strains. Accompanying this increase of HA IS was the rise of the NA ISs at F(0.346) and F(0.277) in USA, Europe, and Asia. F(0.281) was the primary HA IS frequency of American avian H1N1 (Figure 1). F(0.277) was the third NA IS frequency of American avian H1N1 (Figure 1) and F(0.346) was the second frequency as well as a characteristic frequency of pandemic 2009 H1N1 (Table 1). It was interesting to see that the primary NA IS at F(0.166) of American avian H1N1 and that at F(0.373) of American swine and seasonal human H1N1 remained low in the NA activity of pandemic 2009 H1N1. Both NA ISs at F(0.346) and F(0.277) of pandemic 2009 H1N1 were in a quick decrease in 2013 while its NA ISs at F(0.346) and F(0.277) were on a rise.

In summary, the primary HA IS at F(0.295) (a swine H1N1 feature frequency) of this virus remained relatively stable throughout the period of 2009-2013, with one drop in 2012. Moreover, the HA ISs at F(0.055) (a seasonal human H1N1 feature frequency) and at F(0.281) (an avian H1N1 feature frequency) were on steady rise and had a clear increase in 2013 with the American strains having the biggest growth (Figure 2).

To go together with the variations in HA IS over time, the NA ISs of this virus at F(0.074), F(0.346), and F(0.277) were at three separate levels in 2009. But they started to converge in 2010, and were well mixed in 2013 (Figure 2). Remembered that F(0.277) was a top NA IS frequency of American avian H1N1 (Figure 1). Another evident trend was that the NA ISs started a drop in 2012 at the primary frequency of avian H1N1 F(0.168) and at the primary frequency of seasonal human H1N1 F(0.373). The association between the changing patterns of HA and NA observed here called for an experimental approach to elucidate further their functional interdependence.

The primary NA IS of seasonal human H1N1 in USA had an average of 6.0, whereas that of pandemic 2009 H1N1 in USA, Europe, and Asia all had 4.5, which was close to 4.7 of the classical swine H1N1 in USA but still lower than 5.3 of avian H1N1 in USA (Figures 1 and 2). This finding was matched with the experimental results in [4]. Further, the NA ISs of avian H1N1 at F(0.168) and F(0.277) were both low in swine H1N1 (Figure 1), which demonstrated that swine H1N1 hydrolyzed 2-3-linked sialoside less efficiently than did pandemic 2009 H1N1 [4].

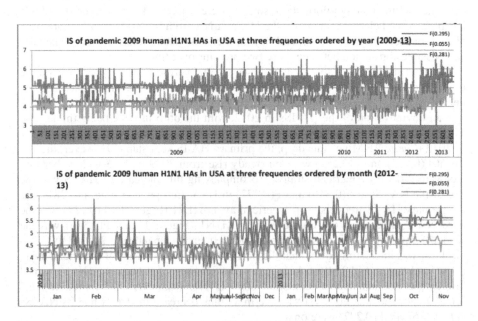

Fig. 2. IS of HA and NA of pandemic 2009 H1N1 in different regions, where the y-axis represents the amplitude of IS and the x-axis represents the sequence count and year

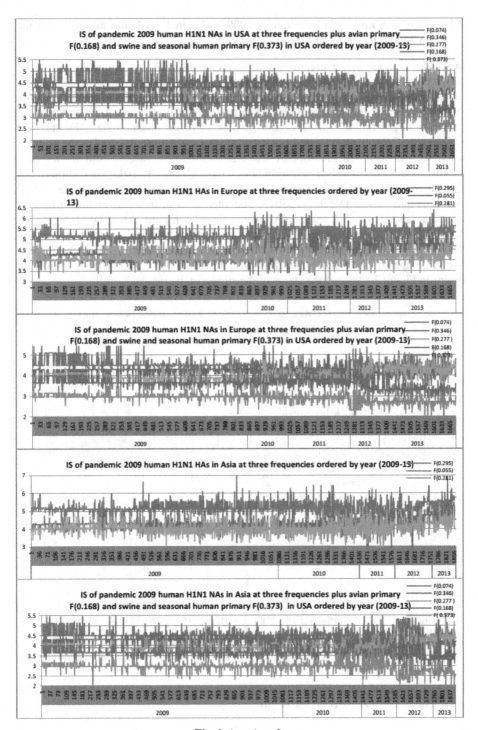

Fig. 2. (*continued*)

3.4   IS of HA and NA of Pandemic 2009 H1N1 in USA From an Early Strain in 2009 and a Recent Strain in 2013

In sections 3.2 and 3.3, a stream of activity patterns of HA and NA from each virus was presented over a period of various years. In this section, we rendered a quick snapshot of the HA and NA IS from pandemic 2009 H1N1 in USA at the two ends of this stream, one at the start of 2009 (A/California/7/2009) and one at the end of 2013 (A/Texas/36/2013), to display the whole IS in each case (Figure 3). A/California/7/2009 was selected because it was one of the strains used in the composition for the Northern Hemisphere 2013-2014 influenza vaccine (http://www.cdc. gov/flu/about/season/vaccine-selection.htm), and A/Texas/36/2013 was chosen because it was collected on November 18, 2013, just before the human deaths caused by this virus in Texas in December, 3013. The alterations of their IS resulting from the slow variations in HA and NA were shown (Figure 3). The HA IS of A/Texas/36/2013 at F(0.281) was increased to the third, and at the same time the NA ISs of A/Texas/36/2013 at F(0.277) and F(0.346) became the first and second respectively, compared to A/California/07/2009. The IS of HA and NA from these two strains in Figure 3 captured the status of the stream at the two ends well.

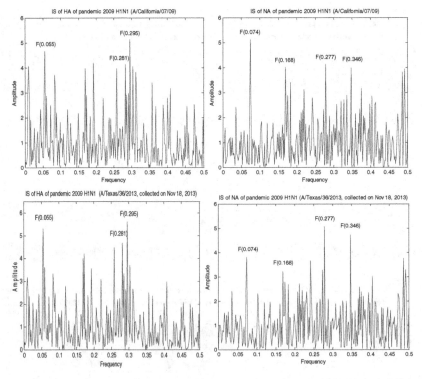

Fig. 3. IS of HA and NA from two representative strains of pandemic 2009 H1N1 in USA, one at the start of 2009 and one at the end of 2013

3.5   Mutations in HA and NA of Pandemic 2009 H1N1 in USA That Caused the Change in Their Activity from 2009 to 2013

Our task in this section was to identify the amino acid substitutions in HA and NA that could cause the gradual change in their activity patterns observed from 2009 to 2013. We reported the top 5 positions in HA and NA calculated by the variable importance rankings of Random Forest [9] to distinshigsh the sequnces in 2009 from those in 2013. The actual amino acids at each of these positions in 2009 and 2013 were displayed (Table 2) and the impact on the IS of HA and NA at leading frequecies by these mutations was also presented (Table 3), which assessed the validity of our finding and affirmed that they did produce the change detected. It was noted that A/California/7/2009 had 106V in its NA already and there were three substitutions of N in NA (Table 2).

Table 2. Important amino acid substitutions in HA and NA that led to the changes in HA and NA activity of pandemic 2009 H1N1 from 2009 to 2013

Position in HA	97	163	185	256	283	Position in NA	44	106	200	241	369
2009	D	K	S	A	K	2009	N	I	N	V	N
2013	N	Q	T	T	E	2013	S	V	S	I	K

Table 3. Impact on IS at leading frequencies of HA and NA of pandemic 2009 H1N1 in USA by the critical amino acid substitutions in HA and NA found by Random Forest (The baseline was A/California/7/2009 and the increase of IS resulted from the mutations was compared to the baseline in percent)

Mutation	HA frequency				Mutation	NA frequency			
	F(0.295)	F(0.055)	F(0.258)	F(0.281)		F(0.074)	F(0.277)	F(0.168)	F(0.346)
Baseline of IS	5.1361	4.6632	3.9709	4.1307	Baseline of IS	5.1212	4.1260	4.0124	4.0083
D97N, K163Q, S185T, A256T, K283E	5.6422	5.7017	4.1145	4.7916	N44S, N200S, V241I, N369K	4.9836	4.5973	3.6310	3.8265
Increase of IS	9%	18%	0.04%	14%	Increase of IS	-2.7%	10.3%	-9.5%	-4.5%

The HA of pandemic 2009 H1N1 was found to harbor mutations D222G or D222E in the receptor binding site sporadically. D222G was correlated with cases of severe or fatal disease and it altered the HA binding preference [10]. When applied to an early strain of pandemic 2009 H1N1, A/California/04/09, it showed a modest reduction in the binding avidity to α2,6 receptors and an increase in the binding to α2,3 receptors in comparison with the wild type virus [10]. Therefore, the HA containing D222G acquired dual binding specificity for α2,3 and α2,6 receptors, which made the virus more efficiency in human transmission and enhanced its ability to infect human lungs. This finding might help explain why some patients that contracted it also developed more serious lung infections.

We applied D222G to a collection of HA sequences from pandemic 2009 H1N1 in USA, of which 2903 HAs contained 222D and 41 had 222G (Table 4). Our finding on this mutation D222G demonstrated its effect to decrease human binding and increase avian binding, matching the experimental results in [10] perfectly.

Table 4. Impact on the average HA IS at leading frequencies of pandemic 2009 H1N1 sequences in USA by mutation D22G

Receptor binding type associated with the leading frequencies	Human Type	Human Type	Human Type	Avian Type
Four leading frequencies	F(0.295)	F(0.055)	F(0.258)	F(0.281)
HAs containing 222D (n=2903)	5.1800	4.3514	4.3432	4.0843
HAs containing 222G (n=41)	4.9417	3.9848	3.807	4.7867
After applying D222G to HAs containing 222D (n=2903)	5.0809	3.9399	3.845	4.7092

4   Conclusion

The recent human deaths in Texas caused by pandemic 2009 H1N1 in December 2013 alerted us to take a closer look at this virus again. We wondered what mutations had taken place in this virus from 2009 to 2013 that might contribute to the recent escalation of its virulence. We first analyzed the HA and NA sequences from seasonal human, swine, avian H1N1 in USA, Europe, and Asia collected from a period of various years. Then we evaluated the HA and NA sequences of pandemic 2009 H1N1 collected from 2009 to 2013 in USA, Europe, and Asia. Our time series analysis concluded that the HA binding preference of this virus has been the characteristic of swine H1N1 virus since 2009. However its binding characteristics of seasonal human H1N1 and avian H1N1 both have been on stable rise and had an increase in 2013 with American strains having the largest surge. The first increase could enhance the viral transmission and replication in humans and the second improve its ability to cause infection deep in lungs, which might account for the escalated pathogenicity as reported in Texas. In light of the closely interacting roles of HA and NA, we further studied the corresponding NA activity of this virus to expose the interdependence between HA and NA during the evolution and adaptation of this virus from 2009 to 2013. To understand the real causality, we also identified amino acid substitutions in HA and NA of the virus that actually caused the observed change. Some of our findings on HA and NA from this virus were matched perfectly with experimental results.

References

1. Smith, G.J., Vijaykrishna, D., Bahl, J., et al.: Origins and evolutionary genomics of the 2009 swine-origin H1N1 influenza A epidemic. Nature 459(7250), 1122–1125 (2009)
2. Childs, R.A., Palma, A.S., Wharton, S., et al.: Receptor-binding specificity of pandemic influenza A (H1N1) 2009 virus determined by carbohydrate microarray. Nat Biotechnol. 27, 797–799 (2009)

3. Gulati, U., Wu, W., Gulati, S., Kumari, K., Waner, J.L., Air, G.M.: Mismatched hemagglutinin and neuraminidase specificities in recent human H3N2 influenza viruses. Virology 339(1), 12–20 (2005)
4. Gerlach, T., Kühling, L., Uhlendorff, J., et al.: Characterization of the neuraminidase of the H1N1/09 pandemic influenza virus. Vaccine 30(51), 7348–7352 (2012)
5. Hu, W.: Receptor binding specificity and sequence comparison of a novel avian-origin H7N9 virus in China. Journal of Biomedical Science and Engineering 6(5), 533–542 (2013)
6. Hu, W.: Mutations in Hemagglutinin of a Novel Avian-Origin H7N9 Virus That Are Critical for Receptor Binding Specificity. Tsinghua Science and Technology 18(5), 522–529 (2013)
7. Veljkovic, V., Niman, H.L., Glisic, S., et al.: Identification of hemagglutinin structural domain and polymorphisms which may modulate swine H1N1 interactions with human receptor. BMC Structural Biology 9, 62 (2009)
8. Veljkovic, V., Veljkovic, N., Muller, C.P., et al.: Characterization of conserved properties of hemagglutinin of H5N1 and human influenza viruses: possible consequences for therapy and infection control. BMC Struct. Biol. 7, 9–21 (2009)
9. Breiman, L.: Random forests. Machine Learning 45, 5–32 (2001)
10. Belser, J.A., Jayaraman, A., Raman, R., et al.: Effect of D222G Mutation in the Hemagglutinin Protein on Receptor Binding, Pathogenesis and Transmissibility of the 2009 Pandemic H1N1 Influenza Virus. PLoS ONE 6(9), e25091 (2011)

Predicting Protein Submitochondrial Locations Using a K-Nearest Neighbors Method Based on the Bit-Score Weighted Euclidean Distance

Jing Hu[1] and Xianghe Yan[2]

[1] Department of Mathematics and Computer Science, Franklin & Marshall College,
415 Harrisburg Ave., Lancaster, PA 17603, USA
jing.hu@fandm.edu
[2] Eastern Regional Research Center, Agricultural Research Service, USDA,
600 E. Mermaid Lane, Wyndmoor, PA 19038, USA
xianghe.yan@ars.usda.gov

Abstract. Mitochondria are essential subcellular organelles found in eukaryotic cells. Knowing information on a protein's subcellular or sub-subcellular location provides in-depth insights about the microenvironment where it interacts with other molecules and is crucial for inferring the protein's function. Therefore, it is important to predict the submitochondrial localization of mitochondrial proteins. In this study, we introduced MitoBSKnn, a K-nearest neighbor method based on a bit-score weighted Euclidean distance, which is calculated from an extended version of pseudo-amino acid composition. We then improved the method by applying a heuristic feature selection process. Using the selected features, the final method achieved a 93% overall accuracy on the benchmarking dataset, which is higher than or comparable to other state-of-art methods. On a larger recently curated dataset, the method also achieved a consistent performance of 90% overall accuracy. MitoBSKnn is available at http://edisk.fandm.edu/jing.hu/mitobsknn/mitobsknn.html.

Keywords: Bit-score weighted Euclidean distance, K-nearest neighbors method, Submitochondrial location, Feature selection.

1   Introduction

Mitochondria are subcellular organelles found in most eukaryotic cells [1]. They play important roles in many biological processes and tasks, such as supplying cellular energy, signaling, ionic homeostasis, cell differentiation, controlling the cell cycle, cell growth and cell death [2, 3]. Recent studies have indicated that mitochondria are involved in several human diseases such as mitochondrial disorders and cardiac dysfunction, and may play key roles in the aging process [4, 5]. Just like a cell that can be divided into several subcellular locations, mitochondria can also be divided into three submitochondrial locations, which are inner membrane, matrix and outer membrane. Knowing the submitochondrial location of a mitochondrial protein provides insights about the protein's roles in these biological processes and tasks. Therefore, it is

M. Basu, Y. Pan, and J. Wang (Eds.): ISBRA 2014, LNBI 8492, pp. 50–58, 2014.
© Springer International Publishing Switzerland 2014

important to identify mitochondrial protein's submitochondrial locations to study their unknown functions for various applications.

Recent advancements in high-throughput sequencing technologies have resulted in a huge number of proteins accumulated in public databases. Because of their speed and cost, experimental approaches can hardly keep up with the accumulation of new biological data. Therefore, it is necessary to develop computational methods that can predict submitochondrial locations of mitochondrial proteins.

Several computational methods have been developed to predict proteins' submitochondrial locations. SubMito [6] is a Support Vector Machine (SVM) method based on an extended version of pseudo-amino acid composition. In their study they have curated a benchmarking dataset for mitochondrial proteins. Nanni and Lumini developed GP-Loc [7]. The method uses a genetic programming to choose pseudo-amino acid based features, which are obtained by combining pseudo-amino acid compositions with hundreds of amino-acid indices and amino acid substitution matrices. Zeng et al. proposed a sequence-based algorithm combining the augmented Chou's pseudo amino acid composition based on auto covariance (AC) to predict protein submitochondrial locations [8]. SubIdent [9] utilizes a novel pseudo amino acid approach based on discrete wavelet transform feature extraction. MitoLoc [10] is a SVM method taking a hybrid approach to incorporate ten models of protein samples such as pseudo amino acid composition, dipeptide composition, functional domain composition, secondary structure information, etc. Fan and Li combined six types of features such as amino acid composition, dipeptide composition, reduced physicochemical properties, gene ontology, evolutionary information, and pseudo-average chemical shift, and developed an algorithm of the increment of diversity combined with the SVM [11]. TetraMito [12] is a SVM method using over-represented tetra-peptides selected by binomial distribution. MK-TLM [13] is a multi-kernel transfer-learning model for protein submitochondrial localization that uses proteins' gene ontology (GO) information. Recently Du and Yu introduced a new method, SubMito-PSPCP, to predict protein submitochondrial locations by introducing a new concept: positional specific physicochemical properties [14].

In this study, we present MitoBSKnn, a K-Nearest Neighbor method to predict protein's submitochondrial locations. The method is based on the bit-score weighted Euclidean distance, which is calculated from the composition of an extended version of pseudo-amino acids based on the study of Du and Li [6]. By applying a heuristic feature selection process, the final method achieved 93.1% overall accuracy on the benchmarking dataset of SubMito [6] and 90.1% overall accuracy on a larger dataset of SubMito-PSPCP [14] by leave-one-out cross-validation.

2   Materials and Methods

2.1   Datasets

Two datasets published in previous studies were adopted in this study. The first dataset was obtained from SubMito-PSPCP [14], which was curated recently by Du and Li to reflect the most recent advances in the collection of mitochondrial proteins. It contains 938 proteins, including 661 proteins from inner membrane, 177 proteins from matrix, and 145 proteins from outer membrane. This dataset was denoted as Mito-938. The second dataset was from the study of SubMito [6], which was used as the benchmarking dataset by most studies. The SubMito dataset has 317 proteins, among which 131 proteins are from inner membrane, 145 proteins are from matrix, and 41 proteins are from outer membrane. We named this dataset as Mito-317 in this study. Both datasets have been removed redundancy by CD-Hit [17] using a sequence similarity threshold of 40%.

2.2   Features

The model of pseudo-amino acid (PseAA) composition was first developed by Chou [15] and was used by many previous studies to predict protein subcellular localizations. PseAA consists of the composition of 20 amino acids in a protein and λ different ranks of sequence-order correlation factors. Chou and Cai later extended the model by including two sets of sequence-order correlation factors: delta-function set (λ discrete numbers) and hydrophobicity set (μ discrete numbers) by [16]. In Du and Li's study, they further extended the definition of pseudo-amino acid composition by developing 9 sets of various physicochemical properties [6].

In this study, the features we investigated include the composition of 20 amino acids, a set of delta-function factors, and 9 sets of physicochemical factors.

Delta function set was calculated as in [16]. Suppose a protein X with a sequence of L amino acid residues: $R_1 R_2 R_3 R_4 . . . R_L$, where R_1 represents the amino acid at sequence position 1, R_2 the amino acid at position 2, and so on. The first set, delta-function set, consists of λ sequence-order-correlated factors, which are given by

$$\delta_i = \frac{1}{L-i} \sum_{j=1}^{L-i} \Delta_{j,j+i} \tag{1}$$

where $i = 1,2,3...\lambda$, $\lambda < L$, and $\Delta_{j,j+i} = \Delta(R_j, R_{j+i}) = 1$ if $R_j = R_{j+i}$, 0 otherwise. These features were named as $\{\delta_1, \delta_2...\delta_\lambda\}$.

The remaining 9 sets of physicochemical properties were based on the study of [6]. Same as [6], the following AAIndex [18] indices were used: BULH740101 (transfer free energy to surface), EISD840101 (consensus normalized hydrophobicity), HOPT810101 (hydrophilicity value), RADA880108 (mean polarity), ZIMJ680104 (isoelectric point), MCMT640101 (refractivity), BHAR880101 (average flexibility indices), CHOC750101 (average volume of buried residue), COSI940101 (electron-ion interaction potential values). For each of 9 AAindex indices, we obtained μ sequence-order-correlated factors given by

$$h_i = \frac{1}{L-i} \sum_{j=1}^{L-i} H_{j,j+i} \tag{2}$$

where $i = 1,2,3...\mu$, $\mu < L$, and $H_{i,j} = H(R_i) \cdot H(R_j)$. In this study, $H(R_i)$ and $H(R_j)$ are the normalized AAindex values of residues R_i and R_j respectively. The normalized AAindex values of each amino acid is calculated by

$$H(AA_i) = (H^0(AA_i) - \overline{H^0}) \Big/ \sqrt{\{\sum_{j=1}^{20}(H^0(AA_j) - \overline{H^0})^2\}/20} \tag{3}$$

where $i=1,2,3,...,20$. $H^0(AA_i)$ is the original AAindex value of amino acid i, and $\overline{H^0}$ is the average AAindex value of 20 amino acids. For each of 9 AAIndex types (i.e, BULH740101, EISD840101, etc.), we obtained μ features using (2) and (3). In total there are 9μ features. We named these features as {BULH740101_1, BULH740101_2, ... BULH740101_μ, EISD840101_1, EISD840101_2, ... EISD840101_μ, ... COSI940101_1, COSI940101_2, ... COSI940101_μ}.

Therefore, the features investigated in this study consist of 20 (classic amino acid compositions) + λ (delta-function factors) + 9μ (9 sets of physicochemical factors) numbers.

2.3   Distance between Proteins

The distance between two proteins (i.e., protein t and T) is calculated as

$$D_{t,T} = \sqrt{\sum_{i=1}^{20}(t_i - T_i)^2} \Big/ BS(t,T) \tag{4}$$

where t_i and T_i are the ith feature of the protein t and T respectively. $BS(t, T)$ is the bit score computed by the blastp program of Blast package [19] when comparing the pairwise local sequence similarity between protein t and T. To ensure that blastp can still output bit scores for protein pairs which share very low sequence similarity, we set the E value threshold as 1,000. A higher bit score indicates two protein sequences are more similar, and vice versa. Notice that $\sqrt{\sum_{i=1}^{20}(t_i - T_i)^2}$ gives the Euclidean distance (ED) between two proteins. Here, the distance is weighted by a factor. Therefore, the distance is named as the bit-score weighted Euclidean distance (BS-WED).

2.4   K-Nearest Neighbors Method

A typical K-nearest neighbors method (K-NN) is an instance-learning algorithm making classification by the majority voting strategy. Our K-NN method differs in that for each query protein, it finds K nearest neighbors (measured by BS-WED) from each class of the training samples. For each location (i.e., inner membrane, matrix, or outer membrane), the average of these K BS-WEDs is used as the distance between the query protein to that location. The distances between the query protein to all locations are compared and the query protein was assigned to a location to which the distance is the shortest.

2.5   Performance Measurement

Leave-one-out cross-validation was used to evaluate the performance of the algorithm. Performances were measured using accuracy (ACC) and Matthew's Correlation Coefficient (MCC) for each submitochondrial location i:

$$ACC_i = TP_i / (TP_i + FN_i) \tag{5}$$

$$MCC_i = \frac{TP_i \times TN_i - FP_i \times FN_i}{\sqrt{(TP_i + FN_i)(TP_i + FP_i)(TN_i + FP_i)(TN_i + FN_i)}} \tag{6}$$

where TP_i, TN_i, FP_i, and FN_i were numbers of true positives, true negatives, false positives, and false negatives of proteins for location i (i.e., inner membrane, matrix, outer membrane). In addition, the overall accuracy (OA) and normalized overall accuracy (NOA) [20] were also used to evaluate the overall performance.

$$OA = \sum_{i=1}^{K} TP_i / \sum_{i=1}^{K} N_i \tag{7}$$

$$NOA = \sum_{i=1}^{K} ACC_i / K \tag{8}$$

where N_i was the number of proteins of location i and K was the total number of locations.

2.6   Heuristic Feature Selection

In total, 20 (amino acid compositions) + λ (delta-function factors) + 9μ (9 sets of physicochemical factors) features were extracted for each protein. In this study, λ and μ were both set to 10. Therefore, there were 120 features investigated in total. A previously published feature selection process [21] was used to identify the most informative subset of features. The search initially started with an empty feature set. Then, one feature was added and leave-one-out cross-validation was used to evaluate the performance (i.e., overall accuracy). This step was repeated 120 times so that every feature was attempted individually. After the 120 accuracies were calculated, the feature with the highest prediction performance was added to the feature set. In the next iteration, every feature from the remaining available features was tested and the feature that showed the largest improvement in the performance when combined with the current selected feature set was added to the feature set. The size of feature set was then increased by 1. This feature selection process continued until adding any of the remaining features to the feature set would decrease the performance.

3   Results and Discussion

3.1   Bit-Score Weighted Euclidean Distance vs. Standard Euclidean Distance

We first investigated the performance of the proposed K-NN method using only the composition of 20 amino acids. Various K values ranging from 1 to 20 were tested. Leave-one-out cross-validation was used to evaluate the performance. Bit-score weighted Euclidean distance was used as distance measurement between proteins. As can be seen from Figure 1, the best performance was achieved when $K = 2$ for both datasets (i.e., Mito-938 and Mito-317). The proposed method achieved 88.3% overall accuracy on Mito-938 dataset and 84.5% overall accuracy on Mito-317 dataset when assigning proteins to 3 submitochondrial locations (i.e., inner membrane, matrix, and outer membrane). For comparison, we also investigated the performance of K-NN method using standard Euclidean distance (ED). As can been seen from Figure 1, BS-WED is better than ED in measuring the distances between proteins.

3.2   Prediction Performance Using Selected Features

We then tried to improve the prediction performance of the proposed method by including the composition of pseudo-amino acids of the extended version. In total, there are 120 features investigated in this study, which include the composition of 20 amino acids, 10 delta-function factors, and 90 physicochemical factors. We applied the greedy feature selection process as described in Method and Material to search for a combination of features that was most useful for the prediction. In the end, there were 14 features (i.e., D, C, Q, I, Y, BULH740101_1, EISD840101_5, EISD840101_7, HOPT810101_2, HOPT810101_8,   RADA880108_4,   RADA880108_8,   MCMT640101_1, BHAR880101_9) selected for the Mito-938 dataset and 15 features (i.e., R, C, Q, K, S, W, V, seqOrder_7, seqOrder_8, BULH740101_2, EISD840101_2, EISD840101_5, RADA880108_4, MCMT640101_3, CHOC750101_1) chosen for the Mito-317 dataset. Trained on the selected features, MitoBSKnn achieved 90.1% overall accuracy and 84.3% normalized overall accuracy on Mito-938 dataset, and 93.1% overall accuracy and 90.5% normalized overall accuracy on Mito-317 dataset (see Table 1).

Table 1. The performance of the proposed method (MitoBSKnn) using selected features

Submitochondria location	Mito-938 dataset		Mito-317 dataset	
	ACC	MCC	ACC	MCC
Inner membrane	95.9%	0.796	89.3%	0.890
Matrix	75.7%	0.774	99.3%	0.885
Outer membrane	81.4%	0.818	82.9%	0.884
Overall accuracy	90.1%		93.1%	
Normalized overall accuracy	84.3%		90.5%	

3.3   Comparison with Other State-of-Art Methods

The proposed MitoBSKnn method was compared with other published methods (i.e., SubMito [6], GPLoc [7], Predict_SubMito [8], SubIdent [9], MitoLoc [10], Fan and Li's method [11], TetraMito [12], SubMito-PSPCP [14]) on the benchmark dataset (Mito-317) and the newer dataset (Mito-938). Prediction results of these methods were obtained directly from their studies. We did not compare with MK-TLM [13] because it needs gene ontology (GO) information of the query protein or its homologous proteins, which are not directly available for all proteins. As can been seen from Table 2, on Mito-317 dataset, our method outperformed most previously published methods and was comparable to some of the highest performed methods. Also, no method achieved better performance than our method on all three locations, which indicates that a possible ensemble method with higher performance could be developed by integrating our method with some of other methods. On the Mito-938 dataset, only SubmitoPSPCP [14] has prediction results and our method has achieved better performance (Table 3).

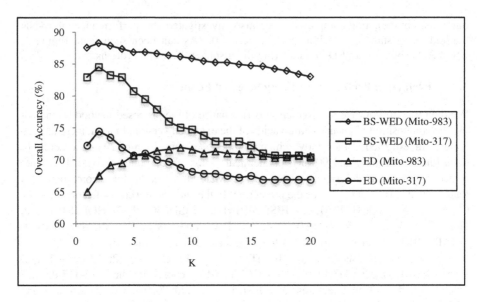

Fig. 1. Prediction performances of K-NN method of various K values (1-20) based on Euclidean distance (ED) vs. Bit-Score Weighted Euclidean distance (BS-WED) on Mito-983 and Mito-317 datasets

Table 2. Comparison of performances on the Mito-317 dataset

Methods	Inner membrane		Matrix		Outer membrane		OA	NOA
	ACC	MCC	ACC	MCC	ACC	MCC		
SubMito [6]	85.5%	0.79	94.5%	0.77	51.2%	0.64	85.2%	77.1%
GPLoc [7]	83.2%	0.80	97.2%	0.85	78.1%	0.77	89.0%	86.2%
Predict_SubMito [8]	91.8%	0.79	96.4%	0.79	66.1%	0.63	89.7%	84.8%
SubIdent [9]	91.6%	0.86	97.3%	0.79	82.9%	0.88	93.1%	90.6%
MitoLoc [10]	97.7%	0.94	99.0%	0.93	68.3%	0.81	94.7%	83.3%
Fan and Li [11]	94.7%	0.91	99.3%	0.96	80.5%	0.84	94.9%	91.5%
TetraMito [12]	100%	0.90	96.6%	0.95	65.9%	0.79	94.0%	87.5%
SubmitoPSPCP [14]	98.6%	0.92	93.9%	0.89	70.7%	0.79	93.1%	87.7%
MitoBSKnn	**89.3%**	**0.89**	**99.3%**	**0.89**	**82.9%**	**0.88**	**93.1%**	**90.5%**

Table 3. Comparison of performances on the Mito-938 dataset

Methods	Inner membrane		Matrix		Outer membrane		OA	NOA
	ACC	MCC	ACC	MCC	ACC	MCC		
SubmitoPSPCP [14]	95.5%	0.77	74.0%	0.73	77.9%	0.83	89.0%	82.5%
MitoBSKnn	**95.9%**	**0.80**	**75.7%**	**0.77**	**81.4%**	**0.82**	**90.1%**	**84.3%**

4   Conclusion

In this study, we presented a K-nearest neighbors method, MitoBSKnn, to predict protein's submitochondrial localizations. The method is based on a novel bit-score weighted Euclidean distance (BS-WED), which is calculated from selected compositions of an extended version of pseudo-amino acids. In this study, pseudo-amino acids include 20 classic amino acids, a set of delta-function factors, and 9 sets of physicochemical factors. Analyses indicated that BS-WED is a better way in measuring the similarity between proteins than standard Euclidean distance. Comparison with previously published methods showed that MitoBSKnn is among the best methods in predicting protein's submitochondrial localizations. In conclusion, BitBSKnn is an efficient method for predicting proteins' submitochondrial locations.

Acknowledgments. This research was partially supported by funding from Franklin & Marshall College to JH.

References

1. Henze, K., Martin, W.: Evolutionary Biology: Essence of Mitochondria. Nature 426, 127–128 (2003)
2. McBride, H.M., Neuspiel, M., Wasiak, S.: Mitochondria: More Than Just a Powerhouse. Curr. Biol. 16, R551–R560 (2006)
3. Gottlieb, R.A.: Programmed cell death. Drug News Perspect 13, 471–476 (2000)
4. Gardner, A., Boles, R.G.: Is a "Mitochondrial Psychiatry" in the Future? A Review. Curr. Psychiatry Review 1, 255–271 (2005)
5. Lesnefsky, E.J., Moghaddas, S., Tandler, B., Kerner, J., Hoppel, C.L.: Mitochondrial Dysfunction in Cardiac Disease: Ischemia—Reperfusion, Aging, and Heart failure. J. Mol. Cell Cardiol. 33, 1065–1089 (2001)
6. Du, P., Li, Y.: Prediction of protein submitochondria locations by hybridizing pseudo-amino acid composition with various physicochemical features of segmented sequence. BMC Bioinformatics 7, 518 (2006)
7. Nanni, L., Lumini, A.: Genetic Programming for Creating Chou's Pseudo Amino Acid Based Features for Submitochondria Localization. Amino Acids 34, 653–660 (2008)
8. Zeng, Y.H., Guo, Y.Z., Xiao, R.Q., Yang, L., Yu, L.Z., Li, M.L.: Using The Augmented Chou's Pseudo Amino Acid Composition for Predicting Protein Submitochondria Locations Based on Auto Covariance Approach. J. Theor. Biol. 259, 366–372 (2009)
9. Shi, S.P., Qiu, J.D., Sun, X.Y., Huang, J.H., Huang, S.Y., Suo, S.B., Liang, R.P., Zhang, L.: Identify Submitochondria and Subchloroplast Locations with Pseudo Amino Acid Composition: Approach from the Strategy of Discrete Wavelet Transform Feature Extraction. Biochim. Biophys. Acta 1813, 424–430 (2011)
10. Zakeri, P., Moshiri, B., Sadeghi, M.: Prediction of Protein Submitochondria Locations Based on Data Fusion of Various Features of Sequences. J. Theor. Biol. 269, 208–216 (2011)
11. Fan, G.L., Li, Q.Z.: Predicting Protein Submitochondria Locations by Combining Different Descriptors Into the General Form of Chou's Pseudo Amino Acid Composition. Amino Acids 43, 545–555 (2012)

12. Lin, H., Chen, W., Yuan, L.F., Li, Z.Q., Ding, H.: Using Over-Represented Tetrapeptides to Predict Protein Submitochondria Locations. Acta Biotheor. 61, 259–268 (2013)
13. Mei, S.: Multi-kernel Transfer Learning Based on Chou's PseAAC Formulation for Protein Submitochondria Localization. J. Theor. Biol. 293, 121–130 (2012)
14. Du, P., Yu, Y.: SubMito-PSPCP: Predicting Protein Submitochondrial Locations by Hybrid-izing Positional Specific Physicochemical Properties with Pseudoamino Acid Compositions. Biomed Res. Int. 2013, 263829 (2013)
15. Chou, K.C.: Prediction of Protein Cellular Attributes Using Pseudo-Amino Acid Composition. Proteins 43, 246–255 (2001)
16. Chou, K.C., Cai, Y.D.: Prediction and Classification of Protein Subcellular Location-Sequence-Order Effect and Pseudo Amino Acid Composition. J. Cell Biochem. 90, 1250–1260 (2003)
17. Li, W., Jaroszewski, L., Godzik, A.: Clustering of Highly Homologous Sequences to Reduce the Size of Large Protein Databases. Bioinformatics 17, 282–283 (2001)
18. Kawashima, S., Kanehisa, M.: AAindex: Amino Acid Index Database. Nucleic Acids Res. 28, 374 (2000)
19. Altschul, S.F., Madden, T.L., Schäffer, A.A., Zhang, J., Zhang, Z., Miller, W., Lipman, D.J.: Gapped BLAST and PSI-BLAST: A New Generation of Protein Database Search Programs. Nucleic Acids Res. 25, 3389–3402 (1997)
20. Pierleoni, A., Martelli, P.L., Fariselli, P., Casadio, R.: BaCelLo: A Balanced Subcellular Localization Predictor. Bioinformatics 22, e408–e416 (2006)
21. Hu, J., Ng, P.C.: Predicting the Effects of Frameshifting Indels. Genome Biology 13, R9 (2008)

Algorithms Implemented for Cancer Gene Searching and Classifications

Murad M. Al-Rajab and Joan Lu

School of Computing and Engineering, University of Huddersfield
Huddersfield, UK
{U1174101,j.lu}@hud.ac.uk

Abstract. Understanding the gene expression is an important factor to cancer diagnosis. One target of this understanding is implementing cancer gene search and classification methods. However, cancer gene search and classification is a challenge in that there is no an obvious exact algorithm that can be implemented individually for various cancer cells. In this paper a research is conducted through the most common top ranked algorithms implemented for cancer gene search and classification, and how they are implemented to reach a better performance. The paper will distinguish algorithms implemented for Bio image analysis for cancer cells and algorithms implemented based on DNA array data. The main purpose of this paper is to explore a road map towards presenting the most current algorithms implemented for cancer gene search and classification.

Keywords: cancer, genes, searching algorithms, classification algorithms.

1 Introduction

Cancer is one of the world's most serious diseases in modern society and a major cause of death worldwide. Traditional diagnostics methods are based mainly on the morphological and clinical appearance of cancer, but have limited contributions as cancer usually results from other environmental factors. There are several causes of cancer (carcinogens) such as smoke, radiation, synthetic chemicals, polluted water, and others that may accelerate the mutations and many undiscovered causes. On the other hand, a need to select the most informative genes from wide data sets, removal of uninformative genes and decreases noise, confusion and complexity and increase the chances for identification of diseases and prediction of various outcomes like cancer types is mandatory [1]. One of the challenging tasks in cancer diagnosis is how to identify salient expression genes from thousands of genes in microarray data that can directly contribute to the phenotype or symptom of disease [3]. The development of array technologies indicates the possibility of early detection and accurate prediction of cancer. Through these technologies, it is possible to get thousands of gene expression levels simultaneously through arrays, and also the ability to make use to know and find out whether it is cancer or not, and classify cancer [5]. Thus, there is a need to identify the informative genes that contribute to a cancerous state. An informative

M. Basu, Y. Pan, and J. Wang (Eds.): ISBRA 2014, LNBI 8492, pp. 59–70, 2014.
© Springer International Publishing Switzerland 2014

gene is a gene that is useful and relevant for cancer classification [6]. Cancer classification, which can help to improve health care of patients and the quality of life of individuals, is essential for cancer diagnosis and drug discovery [3]. Cancer classification or prediction refers to the process of constructing a model on the microarray dataset and then distinguishing one type of samples from other types within the induced model [7]. Microarray is a device or a technology used to measure expression levels of thousands of genes simultaneously in a cell mixture, and finally produces a microarray data, which is also known as gene expression data. The task of cancer classification using microarray data is to classify tissue samples into related classes of phenotypes such as cancer versus normal [8]. A major problem in these microarray data is the high redundancy and the noisy nature of many genes or irrelevant information for accurate classification of cancer. Only a small number of genes may be important [9]. Early and accurate detection and classification of cancer is critical to the wellbeing of patients. The need for a method or algorithms for cancer identification is important and has a great value in providing better treatment and this can be done through analysis of genetic data. For practical use an algorithm has to be fast and accurate as well as easy to implement, test, and maintain. The optimal algorithm for a given task would have adequate performance with minimal implementation complexity [10]. To study the algorithms implemented for cancer gene search and classification, a long path of solid literature review must be constructed from Bioinformatics understanding passing through Bio-image processing and algorithms analysis toward cancer gene searching and selection algorithms implemented in the field and how these algorithms can be applied to classify cancer cells and how efficient they are. Due to the emergence of new technologies such as the micro array data, these new technologies produce large datasets characterized by a large number of features (genes); this is why feature selection (gene selection) has become very important in several fields such as Bioinformatics. Authors in [6, 11] introduced a new hybrid feature selection method that combines the advantages of filter strategy based on the Laplacian Score joint with a simple wrapper strategy. The suggested algorithm resulted in a fast hybrid feature selectors that can solve feature selection problems in high dimensional datasets and select a small subset full of informative genes that is most relative to cancer classification. Another research developed an automated system for robust and reliable cancer diagnoses based on gene microarray data as stated by the authors in [9]. They investigated that support vector machine classifier algorithms outperforms other algorithms such as K nearest neighbors, naive Bayes, neural networks and decision tree; and thus they could adopt the important genes for cancer tumor classifications. On the other hand the authors in [12], found the smallest set of genes that can ensure highly accurate cancer classifications from microarray data by using supervised machine learning algorithms. Moreover, the authors in [13], survived different feature selection techniques and their application for gene array data, they found two optimal search methods for cancer classification which are Genetic Algorithms (GA) and Tabu search (TS) to generate candidate genes for classifications. They argued that GA is an optimal search method that behaves like evolution processes in nature, while TS is a heuristic method that guides the search for optimal solution making use of flexible memory.

The main purpose of this paper is to explore a road map towards presenting the most current algorithms implemented for cancer gene search and classification.

The remainder of this paper will be structured as follows; Section 2 will discuss the common algorithms implemented in the research topic, on the other hand, section 3will give an overview of the algorithms, while, results and discussion will be presented in section 4. Finally, section 5 will conclude the paper.

2   Common Algorithms for Cancer Gene Search and Classification

The study of the algorithms is classified into two categories; first the algorithms that focus on gene expression analysis for cancer gene selection, and second, the algorithms that focus on Bio-Image analysis and performs cancer classification. These categories are discussed below:

2.1   Analysis of Cancer Gene Selection and Classification Algorithms

Microarray data is being an influence to cancer diagnostics. Its accurate prediction to the type or size of tumors based on reliable and efficient classification algorithms, so that patient can be provided with better treatment or therapy response. The main issue behind microarray data is its high dimensionality which may lead to low efficiency in cancer gene classification and also makes it difficult to classify the related genes. Among thousands of genes whose expression levels are measured, not all are needed for classification [5]. Thus, one challenging task in cancer diagnosis is how to identify silent expression genes from thousands of genes in microarray data and how to select informative genes for classification that can assist to the symptom of disease [7]. Below is a summary of the most well implemented classification algorithms applied in the field and argued to be efficient for diverse cancer type's diagnosis and treatment.

Integrated Gene-Search Algorithm
The integrated algorithm is based on Genetic Algorithm (GA) and Correlation-based heuristics [1]. (Correlation-based feature selection) (CFS) for data preprocessing and data mining (decision tree and support vector machine algorithms) for making predictions. Thereafter, bagging and stacking algorithms were applied for further enhancement classification accuracy and the analysis of data was performed by WEKA data mining software. This work was proposed and successfully applied to the training and testing genetic expression data sets of ovarian, prostate, and lung cancers but also can be successfully applied to any other cancer like colon, breasted, bladder, leukemia, and so on. The Algorithm consists of two phases as shown in Figure 1, the iterative phase I, where data partitioning, execution of Decision Tree (DT) algorithm or any other data mining algorithms applied to the data set, then GA and CFS for gene reduction take place. After that, in phase II, data-mining algorithms are applied to the training and testing data sets generated from phase I and their results will be evaluated to determine the most significant gene set.

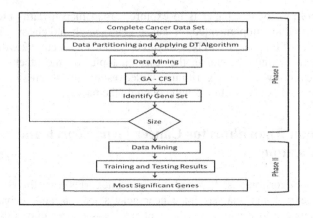

Fig. 1. Integrated Gene Search Algorithm

An Integrated Algorithm for Gene Selection and Classification Applied to Microarray Data for Ovarian Cancer

By applying a hybrid of algorithms (Genetic Algorithm "GA", Particle Swarm Optimization "PSO", Support Vector Machine "SVM", and Analysis of Variance "ANOVA") to select gene markers from target genes, finally fuzzy model is applied to classify cancer tissues [2]. Due to the huge amount of data types generated from gene expression and lack of systematic procedure to analyze the information instantaneously, in addition to avoid higher computational complexity, the need to select the most likely differential gene markers to explain the effects on ovarian cancer. It is concluded that the proposed algorithm has superior performance over ovarian cancer and can be applied and performed on other cancer diagnosis studies, and that is noticed from table 1.

Table 1. The Proposed Algorithm Accuracy of classification for various approaches

	The hybrid process of SVM and GA (%)	The hybrid process of SVM and PSO (%)	The proposed algorithm (%)
Colon	95.65%	97.13%	99.13
Breast	96.23%	97.95%	98.55

Source: Zne-Jung Lee, An integrated algorithm for gene selection and classification applied to microarray data of ovarian cancer, International Journal Artificial Intelligence in Medicine 42 (2008) 91.

A Bootstrapped Genetic Algorithm and Support Vector Machine to Select Genes for Cancer Classification

The algorithm states that gene expression data obtained from microarrays have shown to be useful in cancer classification. A novel system is suggested for selecting a set of genes for cancer classification. The system is based on linear support vector machine and a genetic algorithm The proposed system considers two databases for the solution, one for the colon cancer and the other for the leukemia. It is argued that this proposed system of hybridization of genetic algorithm, support vector machine and bootstrapped methods is very efficient for classification problems. [4].

A Novel Embedded Approach Composed of Two Main Phases to the Problem of Cancer Classification Using Gene Expression Data

Phase one includes the use of gene selection to select the important predictive genes which make it later easier to be correctly classified. The second phase is to build powerful classifier models. For gene selection, a proposed of three filter approaches are analyzed, Information Gain (IG), Relief Algorithm (RA), and t-statistics (TA) to obtain a predictive reduced feature (gene) space containing the most informative genes. Later five well known classifier algorithms are utilized (Support Vector Machine (SVM), K Nearest Neighbor (KNN), Naïve Bayes (NB), Neural Network (NN), and Decision Tree (DT)) to classify nine famous available gene expression datasets. After the experiments, it was resulted that in 8 out of 9 datasets, SVMs classifier outperforms KNN, NB, NN and DT obviously in all cases [9].

Genetic Algorithm (GA) with an Initial Solution Provided by t-statistics (t-GA) for Selecting a Group of Informative Genes from Cancer Microarray Data

The Decision Tree classifier (DT) is then built on the top of these selected genes. The performance of the proposed approach among other selection methods and indicated that t-GA has the highest accurate rate among different methods [14].

2.2   Cancer Classification through Bio Image Analysis Algorithms

CAIMAN system (CAncer IMage ANalysis) [15] is an online algorithm repository that analyze the image produced by experiments relevant to cancer research (www.caiman.org.uk), three algorithms have been implemented to this project, an algorithm for measuring cellular migration, other one for vasculature analysis and an algorithm for image shading correction. The following table was a result of the estimation performance of the CAIMAN system (CAncer IMage ANalysis) , the three proposed algorithms were tested with two groups of five images each one of approximately 10kb in size and the other more than 1 Mb.The times are recorded from the moment the user opens the web page to the time the email with the results are received, as in Table 2 below:

Table 2. Proposed Algorithm Performance Estimation

Algorithm	Dimension (pixels)	Size (kb)	Time ± (s)
Migration	285 x 203	1001	62.6 ± 9.6
	127 x 900	1700	81.4 ± 16.7
Tracing	220 x 164	108	66.2 ± 20.3
	768 x 576	1300	207.4 ± 14.6
Shading	285 x 203	100	59.5 ± 14.1
	1270 x 900	1700	65.0 ± 15.6

Source: Constantino Carlos Reyes-Aldasoro, Michael K. Griffiths, Deniz Savas, Gillian M. Tozer, CAIMAN: An online algorithm repository for Cancer Image Analysis, Computer Methods and Programs in Biomedicine, Volume 103, Issue 2, August 2011, Page 103, ISSN 0169-2607, 10.1016/j.cmpb.2010.07.007.

Fig. 2. Integrated Cancer Selection and Classification criteria

3   Algorithms Overview

It is noticed that to classify cancer cells into normal cells or cancerous cells, Selection and Searching Algorithms must be implemented first as shown in figure 2.

3.1   Searching and Selection Algorithms

Genetic Algorithm (GA) is a search algorithm. A GA is initiated with a set of solutions (chromosomes) called the population [1, 16]. Solutions from one population are taken and used to form a new population. This is motivated by a hope that the new population will be better than the old one. Solutions which are selected to form new solution are selected according to their fitness – the more suitable they are, the more chances they have to reproduce [14, 16], the chart of GA is presented in Figure 3.*Correlation-based feature selection (CFS) i*t is a process of choosing or selecting a subset of original features so that the feature space is optimally reduced according to a certain evaluation criterion [17]. It reduces the number of features, removes irrelevant, redundant, or noisy data, and brings the immediate effects for applications [18]. *Particle Swarm Optimization (PSO)* is a population based search algorithm based on the simulation of the social behavior [19]. PSO is similar to GA in that the system is initialized with a population of random solutions. It is unlike GA, however, in that each potential solution is also assigned a randomized velocity, and the potential solutions "particles", are then "flown" through the problem space [20]. *Analysis of variance (ANOVA)* is an extremely important method in exploratory and confirmatory data analysis [21]. *Information Gain (IG)* is a method that attempts to quantify the best possible class predictability that can be obtained by dividing the full range of a given gene expression into two disjoint intervals corresponding to the down-regulation of the gene. It predicts samples in one interval to normal and samples in another interval to cancer [14].

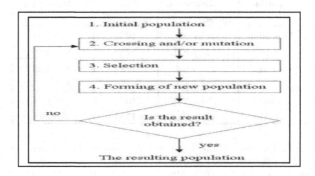

Fig. 3. Block Diagram of Genetic Algorithm

3.2   Classification Algorithms

Support Vector Machine (SVM) is considered popular classifier for microarray data [22]. It has an advantage applied in cancer diagnostic in that its performance appears not to be affected by using the set of full genes [9]. k- Nearest Neighbor (KNN) is one of the simplest learning algorithms, and applied to a variety of problem. It is used as a classifier among a given set of data and uses class labels of the most similar neighbor to predict the new class [9]. Naïve Bayes (NB) is a classifier that can achieve relatively good performance on classification tasks, based on the elementary Bayes' theory [9]. Decision Tree (DT) different methods exit to build a DT, in which a given data in a tree structure, with each branch representing an association between attribute values and a class label [9]. The most famous DT methods is the C4.5 algorithm, which partition the training data set according to tests on the potential of attribute values in separating the classes.

Table 3. Feature Selection Algorithms Specifications

Methods/ Technology involved	Importance	Area/s	Advantages	Disadvantages	Problems
Filter Selection Techniques	Compute the importance of each feature (gene) and then select the top ranked	Gene Selection	Simple Fast Easy scales to very high dimensional data	Univarate that means each feature is considered and treated separately, ignoring any correlation between features	Low classification performance
Wrapper Selection Technique	Selects subset of features that is useful to build a good classifier or predictor	Gene Selection	The ability to take into account the correlation between features and the interaction with the classifier	Prone to high risk of over fitting It require very intensive computation	Unfeasible for feature selection in high-dimensional data More complex

4   Results and Discussion

In this paper, various algorithms were analyzed that perform the task of cancer gene search and classification by first selecting the informative genes and reducing the size and then distinguish the type of the cell tumor or not. Cancer gene selection is a pre-processing step used to find a reduced-sample size of microarray data. This can be achieved by two feature (gene) selection approaches as stated in Table 3. From the table it is found that both filter and wrapper models play a role in feature (gene) selection, but each has its pros and cons. Filter model is noticed to be fast but may give a low classification performance result, while the wrapper model takes time and more complex, but may give somehow a high performance result. Furthermore,it is noticed from Table 4 (see appendix 1), that multiple algorithms implemented in integration and hybridization to analyze multiple kinds of cancer type. In addition, the efficiency of the algorithms was based on the cancer type and the algorithm implemented. The need for a scientific methodology to determine the efficient algorithm or integration of algorithms for cancer types was missed. We mean by algorithm efficiency how fast the algorithm to be implemented in terms of time and speed in order to analyze the cancer cells. Furthermore, Table 5 (see appendix 1) gives a summary for each individual algorithm and to which cancer type it was implemented. It is concluded from table 5 (see appendix 1), that Genetic Algorithm as a selection algorithm was implemented to almost all cancer types for a high performance, except the brain cancer, while Decision Tree and Support Vector Machine Algorithms were implemented to almost all types of cancer for high performance results. In addition figure 4 shows that the Integrated Algorithm for gene selection and classification has the highest accuracy 99% for colon and breast cancers, while the Bootstrapped Genetic Algorithm and Support Vector Machine give good performance accuracy without indicating the percentage. Also the Integrated Gene Search Algorithm has the second high performance up to 98% in accuracy results.

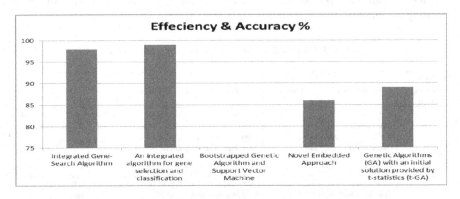

Fig. 4. Algorithm Efficiency and Accuracy

On the other hand, from the detailed review to many researchers' contributions, Table 6 (see appendix 1) summarizes out the most common Algorithms used for cancer gene search and classifications, most of these algorithms where implemented in an integrated model or hybridization methods as discussed, in order to give out an optimum desired result.

The main issue with the previous algorithms is the efficiency in performance, due that most of the suggested algorithms and technologies followed the hybridization methodology in order to achieve better in terms of efficiency and accuracy. When we talk about efficiency we mean less time and less memory, but the main concern will be saving time.

5   Conclusion and Future Work

It is concluded that there are multiple computational algorithms applied for cancer gene selection that are either filter or wrapper methods, each has its own advantages or disadvantages and trying to reach a well performance result. On the other hand and in order to classify cancer cells, selection algorithms must be implemented first to reduce the microarray sample size and reach informative genes, then it would be easier to implement classifier algorithms to distinguish out tumor from normal cells.

Moreover, the paper showed that most algorithms are implemented in an integration methodology and in a harmony in order to achieve a better performance result. Nevertheless, it was clear that the dominant algorithm applied in integration with other algorithms for gene selection was the Genetic Algorithm, while for classification was the Support Vector Machine; as both reached better results. The future work will be to analyze the processing time of each of the algorithms implemented in order to decide the best performance algorithm.

References

1. Shah, S., Kusiak, A.: Cancer gene search with data mining and genetic algorithms. Computers in Biology and Medicine 37(2), 251–261 (2007)
2. Lee, Z.-J.: An integrated algorithm for gene selection and classification applied to microarray data of ovarian cancer. International Journal Artificial Intelligence in Medicine 42, 81–93 (2008)
3. Liu, H., Liu, L., Zhang, H.: Ensemble gene selection for cancer classi-cation. Pattern Recognition 43(8), 2763–2772 (2010) ISSN 0031-3203, 10.1016/j.patcog.2010.02.008
4. Chen, X.-W.: Gene selection for cancer classification using bootstrapped genetic algorithms and support vector machines. In: Proceedings of the 2003 IEEE Bioinformatics Conference, CSB 2003, August 11-14, pp. 504–505 (2003)
5. Park, C., Cho, S.-B.: Evolutionary ensemble classifier for lymphoma and colon cancer classification. In: The 2003 Congress on Evolutionary Computation, CEC 2003, December 8-12, vol. 4, pp. 2378–2385 (2003)
6. Mohamad, M.S., Omatu, S., Yoshioka, M., Deris, S.: An Approach Using Hybrid Methods to Select Informative Genes from Microarray Data for Cancer Classification. In: Second Asia International Conference on Modeling & Simulation, AICMS 2008, May 13-15, pp. 603–608 (2008)
7. Liu, H., Liu, L., Zhang, H.: Ensemble gene selection for cancer classification. Pattern Recognition 43(8), 2763–2772 (2010) ISSN 0031-3203
8. Mohamad, M.S., Omatu, S., Deris, S., Hashim, S.Z.M.: A Model for Gene Selection and Classification of Gene Expression Data. International Journal of Artificial Life & Robotics 11(2), 219–222 (2007)
9. Osareh, A., Shadgar, B.: Microarray data analysis for cancer classification. In: 2010 5th International Symposium on Health Informatics and Bioinformatics (HIBIT), April 20-22, pp. 125–132 (2010)
10. Nurminen, J.K.: Using software complexity measures to analyze algorithms—an experiment with the shortest-paths algorithms. Computers & Operations Research 30(8), 1121–1134 (2003) ISSN 0305-0548, 10.1016/S0305-0548(02)00060-6
11. Solorio-Fernandez, S., Martinez-Trinidad, J.F., Carrasco-Ochoa, J.A., Zhang, Y.-Q.: Hybrid feature selection method for biomedical datasets. In: 2012 IEEE Symposium on Computational Intelligence in Bioinformatics and Computational Biology (CIBCB), May 9-12, pp. 150–155 (2012)
12. Wang, L., Chu, F., Xie, W.: Accurate Cancer Classification Using Expres-sions of Very Few Genes. IEEE/ACM Transactions on Computational Biology and Bioinformatics 4(1), 40–53 (2007)
13. Li, J., Su, H., Chen, H., Futscher, B.W.: Optimal Search-Based Gene Subset Selection for Gene Array Cancer Classification. IEEE Transactions on Information Technology in Biomedicine 11(4), 398–405 (2007)

14. Yeh, J.-Y., Wu, T.-S., Wu, M.-C., Chang, D.-M.: Applying Data Mining Techniques for Cancer Classification from Gene Expression Data. In: International Conference on Convergence Information Technology, November 21-23, pp. 703–708 (2007)
15. Reyes-Aldasoro, C.C., Griffiths, M.K., Savas, D., Tozer, G.M.: CAIMAN: An online algorithm repository for Cancer Image Analysis. Computer Methods and Programs in Biomedicine 103(2), 97–103 (2011) ISSN 0169-2607, 10.1016/j.cmpb.2010.07.007
16. Goldberg, D.E.: Genetic algorithms in search, optimization and machine learning. Addison Wesley, MA (1989)
17. Yu, L., Liu, H.: Feature selection for high-dimensional data: A fast correlation-based filter solution. In: ICML, pp. 856–863 (2003)
18. Tiwari, R., Singh, M.P.: Correlation-based Attribute Selection using Genetic Algorithm. International Journal of Computer Applications (0975 – 8887) 4(8), 28–34 (2010)
19. Khanesar, M.A., Teshnehlab, M., Shoorehdeli, M.A.: A novel binary particle swarm optimization. In: Mediterranean Conference on Control & Automation, MED 2007, June 27-29, pp. 1–6 (2007)
20. Eberhart, R.C., Shi, Y.: Particle swarm optimization: developments, applications and resources. In: Proceedings of the 2001 Congress on Evolutionary Computation, vol. 1, pp. 81–86 (2001)
21. Gelman, A.: Analysis of Variance - Why it is More Important Than Ever. The Annals of Statistics 33(1), 1–53 (2005)
22. Vapnik, V.: Statistical learning theory. Wiley (1998)

Appendix

Table 4. Efficient Algorithms for various cancer types

Algorithm	Embedded Algorithms	Cancer Type	Comments
Integrated Gene-Search Algorithm[1]	Genetic Algorithm Correlation-based heuristics Decision tree Support vector machine	Ovarian Prostate Lung Can be successfully applied to any other cancer like colon, breasted, bladder, leukemia, and so on.	High classification accuracy (94 – 98%)
An integrated algorithm for gene selection and classification [2]	Genetic Algorithm Particle Swarm Optimization Support Vector Machine Analysis of Variance Fuzzy Model	Ovarian Colon Breast	Superior performance for gene selection and classification (colon and breast 99% accuracy)
Bootstrapped Genetic Algorithm and Support Vector Machine [4]	Genetic Algorithm Support vector machine	Colon Leukemia	Well suited for feature (gene) selection problems
Novel Embedded Approach [9]	Information Gain Relief Algorithm t-statistics Support Vector Machine K Nearest Neighbour Naïve Bayes Neural Network Decision Tree	Lung Prostate Breast Leukemia Brain Colon Ovarian	Suport Vector Machines peroforms accuracies > 85% with the combination of Information Gain Decision Tree are the worst model in accuracy
Genetic Algorithms (GA) with an initial solution provided by t-statistics (t-GA) [14]	Genetic Algorithm T-statistics Decision Tree	Colon Leukemia Lymphoma Lung Central Nervous System (CNS)	Colon accuracy 89% Leukemia accuracy 94% Lymphona accuracy 92% Lung accuracy 98% CNS accuracy 77%
CAIMAN system (CAncer IMage ANalysis) [15]	Migration measurement Vasculature tracing Shading correction	Cancer related images	More algorithms can be implemented

Table 5. Cancer Types Algorithms

Cancer / Algorithm	Ovarian	Prostate	Lung	Colon	Breast	Bladder	Leukemia	Brain	Lymphoma	CNS
Genetic Algorithm	✓	✓	✓	✓	✓	✓	✓		✓	✓
Correlation based heuristics	✓	✓	✓	✓	✓	✓	✓			
Decision tree	✓	✓	✓	✓	✓	✓	✓	✓	✓	✓
Support Vector Machine	✓	✓	✓	✓	✓	✓	✓	✓		
Particle Swarm Optimization	✓			✓	✓					
Analysis of variance	✓			✓	✓					
Fuzzy Model	✓			✓	✓					
Information Gain	✓	✓	✓	✓	✓		✓	✓		
Relief Algorithm	✓	✓	✓	✓	✓		✓	✓		
t-statistics	✓	✓	✓	✓	✓		✓	✓	✓	✓
K nearest Neighbor	✓	✓	✓	✓	✓		✓	✓		
Naïve Bayes	✓	✓	✓	✓	✓		✓	✓		
Neural Network	✓	✓	✓	✓	✓		✓	✓		

Table 6. Common Feature Selection and Classifications Algorithms

Selection Algorithms	Classification Algorithms
Genetic Algorithm (GA)	Support Vector Machine (SVM)
Correlation-based heuristics (Correlation-based feature selection) (CFS)	Bootstrapped SVM
Particle Swarm Optimization (PSO)	K-Nearest Neighbors (KNN)
Analysis of Variance (ANOVA)	Naïve Bayes
Information Gain (IG)	Neural Networks (NN)
Relief Algorithm (RA)	Decision Tree (DT)
t-statistics (TA)	Bagging and Stacking Algorithms
	Fuzzy Model

Dysregulated microRNA Profile in HeLa Cell Lines Induced by Lupeol

Xiyuan Lu, Cuihong Dai, Aiju Hou, Jie Cui, Dayou Cheng, and Dechang Xu[*]

School of Food Science and Engineering,
Harbin Institute of Technology, Harbin, China
dcxu@hit.edu.cn

Abstract. Lupeol attracted lots of research attention because of its anticancer activity. This work presents the complete microRNA profile between lupeol treated HeLa cell lines and control group, and investigates the complete small RNA sequencing data analysis process, including microRNA annotation, novel microRNA prediction, dysregulated microRNA identification and microRNA target prediction. Based on single replicate data, we applied generalized fold change (GFOLD) algorithm to detect significant regulated microRNAs. Furthermore, we adopted GOmir to predict targets of some microRNAs which have received fully attention and perform ontology analysis. The experimental results indicate that the predicted microRNAs are highly correlated with carcinogenesis.

Keywords: small RNA sequencing, lupeol, HeLa, microRNA.

1   Introduction

As the microRNA (miRNA) field is still in its relative infancy, there is currently a lack of consensus regarding optimal methodologies for microRNA quantification, data analysis and data standardization [1]. Robust methods for investigating the expression levels of microRNAs are required. The current commercially available methodologies include cloning, northern blot, microRNA microarray, and high throughput sequencing. Cloning methods and northern blot methods are not high-throughput and consume far more RNA samples [2]. Microarray relies on designed probes to detect verified reference microRNA sequences residing in miRBase [3], although this application features good reproducibility [4]. Sequencing based approaches are laborious and high expense techniques [5]. Considering the flaws of aforementioned method, next generation sequencing technologies offer a more complete view of the microRNA transcriptome [6].

The output of a next generation small RNA sequencing typically contains millions of short reads. Before selecting out specific microRNAs for further applications, several numerical analysis steps need to be processed from primary raw data [7]. After removal of reads containing ambiguous base calls, unique sequences remained

---

[*] Corresponding author.

M. Basu, Y. Pan, and J. Wang (Eds.): ISBRA 2014, LNBI 8492, pp. 71–80, 2014.

with read counts are mapped to corresponding genome and referential sequences respectively for novel microRNA prediction and annotation. There are several easy-to-use software and packages to facilitate the downstream differential expression analysis, such as miRDeep, miRanalyzer, and microRNAkey[8]. Each software is designed via different principles bearing its advantages and some limitations.

Our previous research using microarrays [9] indicated that several microRNAs could be significantly regulated by lupeol, a triterpene found in various plants [10], in HeLa cell lines. This work addresses the numerical analysis from small RNA sequencing of HeLa cells induced by lupeol.

2   Sample Preparation and Sequencing Procedures

HeLa cell lines (American type culture collection) are grown in monolayer culture in RPMI 1640 (Thermo Scientific HyClone®) supplemented with 10% fetal bovine serum (Zhejiang Tianhang Biological Technology Co. Sijiqing®, mycoplasma free) at 37°C in a humidified atmosphere consisting of 5% CO_2 and 95% air. HeLa cell lines are treated with 100μmol/L of lupeol (Shanghai Ronghe Medical Technology Co.) dissolved by DMSO for 12 hours (EX). For the control treatment (CON), cells are treated with 1% DMSO (Sigma) for 12 hours. Treated cells are prepared in TRIzol (Invitrogen) for RNA and protein isolation followed by high-throughput sequencing.

Total RNA are subjected to microRNA sequencing library preparation, before they are sequenced using Illumina HiSeq^{TM} 2000 platform (Illumina Inc., USA). The small RNA sequencing is undertaken by Beijing Genomics Institution (BGI) with the method of sequencing by synthesis (SBS), followed by primary digitalization analysis and raw reads filtering. Data filtering includes removing adapter sequences, contamination and low-quality reads from raw reads.

3   Analytical Methods

3.1   Small RNA Sequencing Data Analysis

We first utilize miRanalyzer version 0.2 [11] to align reads to Homo sapiens (hg18) and predict novel microRNAs. Unique reads and their corresponding copy number are uploaded into miRanalyzer using following settings: 3 mismatches (known microRNA), 2 mismatches (libraries), 0 mismatches (genome), 0.99 threshold of the posterior probability, and minimum 5 models positive.

Filtered reads are then mapped to fasta format sequences of all Homo sapiens microRNA downloaded from miRBase [3] by National Center for Biotechnology Information (NCBI) Windows BLAST-2.2.28+ software [12], to annotate reads from CON and EX samples. For producing biologically meaningful rankings of differentially expressed microRNAs from single replicate data, generalized fold change (GFOLD) algorithm is adopted [13]. The criteria of significant cutoff for fold change is 0.01. To strength the soundness of the results, we apply DESeq [14] package to calculate the p value, and cut off the selected microRNAs with p value lower than 0.1. Fig. 1 shows the whole microRNA sequencing data analysis process.

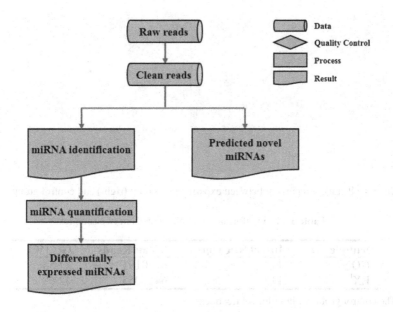

Fig. 1. Data processing flow chart from small RNA sequencing

3.2   MicroRNA Target Prediction and Gene Ontology Analysis

There are many GO related software including Semantic Similarity analysis tool recently published on ISBRA 2013[15].GOmir [16] is utilized for human microRNAs target prediction and ontology clustering. GOmir run on Windows. By running JTarget, we find targets from miRanda database. TAGGO[17] provides the platform for performing the gene ontology clustering, with default Information Content (IC) threshold, and does not exclude non-desired GO categories. The gene ontology file is downloaded from Gene Ontology Consortium.

4   Experimental Results

4.1   Samples

After 12 hours of lupeol treatment, shrinkage and detachment could be observed in HeLa cells (Fig. 2).

4.2   MicroRNA Profile

After sequencing, the raw reads are filtered, and we obtain the statistics of data production. Table 1 shows the basic statistical results after data treatment. Insert size represents the length of sequencing fragment.

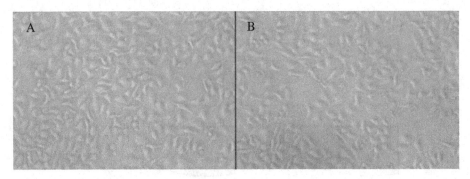

Fig. 2. Cell state comparison between experimental group (right) and control group (left)

Table 1. Basic information of clean reads in two samples

Sample ID	Insert Size (bp)	Clean Reads	GC (%)
CON[a]	112	6290191	42.55
EX[b]	113	6450055	42.68

[a] The control group without lupeol treatment.

[b] The experiment group with lupeol treatment.

Clean reads are aligned to the known microRNAs sequences for microRNA annotation and determination of expression levels. Within 2,578 known microRNAs, we identify 884 microRNAs by combining the data from CON and EX, including 642 microRNAs found in both samples. The overlap of microRNAs identified by CON and EX is approximately three-quarters (73%) (Fig. 3).

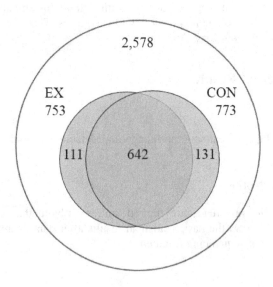

Fig. 3. Venn diagram of identified microRNA in two samples

Besides the known microRNAs, 489 novel microRNA sequence candidates are detected in control group, while 478 novel microRNA sequence candidates are in experimental group in our study. Table 2 shows a part of the predicted microRNAs in CON group. Based on the alignment algorithm, some of them were expressed at relatively high levels, and not every candidate exists in both samples. Further investigation is needed for confirmation, expression patterns and possible roles of these microRNA candidates.

Table 2. Novel microRNAs predicted in CON group (Incomplete)

No.	Chrom	ChrStart	ChrEnd	Strand	Unique Reads	Read Count	Read Cluster Sequence
1	chr20	3846116	3846242	+	128	33679	AAGCAGCATTGTACA GGGCTATGAAA
2	chr10	677603	677737	-	4	5	ACAAGGAAGGACAAG AGGTGTGAGC
3	chr12	526521	526629	+	4	12	GGGCGTTGCTGGGCGT TGCTGG
4	chr7	5501952	5502092	-	5	7	TCAGAACAAATGCCG GTTCCCAG

Known microRNA expression files between CON group and EX group are compared to find out the differentially expressed microRNAs. The expression differentiation of microRNAs in paired samples is calculated by GFOLD (0.01) $\log_2$Ratio (CON/EX) and p-value. We determine 15 differentially expressed known Homo sapiens microRNAs (Table 3), from which nine microRNAs are up-regulated and six microRNAs are down-regulated. Hsa-miR-152-5p (GFOLD=7.61743) is the most significant up-regulated microRNA, and hsa-miR-152-3p (GFOLD=-8.48248) is the most significant down-regulated microRNA.

Table 3. A collection of deregulated microRNAs detected by deep sequencing in lupeol treated cervical cancer cells

microRNA symbol	GFOLD (0.01)	$\log_2$Ratio[a]	p value
hsa-miR-152/-3p	-8.48248	-11.2151	5.54E-81**
hsa-miR-550/a-3p	-0.405168	-3.51402	0.067599
hsa-miR-183/-5p	-0.90727	-1.42298	0.00336**
hsa-miR-27a/-3p	-0.585789	-0.836515	0.015453*
hsa-miR-182/-5p	-0.455418	-0.793427	0.044341*
hsa-miR-99b-3p/*	-0.400464	-0.662778	0.063915
hsa-miR-1307/-3p	0.429864	0.582841	0.028779*
hsa-miR-193b/-3p	0.457787	0.637289	0.021246*

Table 3. (*continued*)

hsa-miR-140-3p	0.591885	0.64262	0.009476**
hsa-miR-940	0.260533	0.73896	0.086117
hsa-miR-889/-3p	0.50091	0.775092	0.014771*
hsa-miR-1307-5p	0.643696	0.856144	0.0035**
hsa-miR-1293	0.532685	1.80791	0.026839*
hsa-miR-3124/-5p	0.29599	3.40484	0.069227
hsa-miR-152-5p	7.61743	9.12905	9.45E-77**

[a] $\log_2$Ratio>0 means up-regulation; $\log_2$Ratio<0 means down-regulation

* means $|GFOLD|{\neq}0$, $|\log_2$Ratio$|>0.5$, and $p<0.05$

** means $|GFOLD|{\neq}0$, $|\log_2$Ratio$|>0.5$, and $p<0.01$; No mark means others

Significant regulated microRNAs are operated to target detection. The concordance between microRNAs and mRNA targets is shown in Table 4. As some microRNAs have not been included in miRanda database, we could not find targets from them, but still the quantity of targets for other microRNAs are relatively huge. MicroRNAs share several targets, for example PLCH2 is the target for both has-miR-550 and has-miR-183.

Table 4. Predicted targets of significantly regulated microRNAs

microRNA symbol	Targets
hsa-miR-152/-3p	ATAD3C, PRDM16, ERRFI1, etc. (1570 in total)
hsa-miR-550/a-3p	LOC643837, RP3-395M20.1, PLCH2, etc. (1346 in total)
hsa-miR-183/-5p	KLHL17, RP3-395M20.1, PLCH2, etc. (1321 in total)
hsa-miR-27a/-3p	CR601056, KLHL17, CAMTA1, etc. (1834 in total)
hsa-miR-182/-5p	RP3-395M20.1, PLCH2, C1orf188, etc. (2058 in total)
hsa-miR-99b-3p/*	0 target found
hsa-miR-1307/-3p	UBE2J2, UBE4B, KIAA0684, etc. (272 in total)
hsa-miR-193b/-3p	BC032353, GNB1, CAMTA1, etc. (916 in total)
hsa-miR-140-3p	BC032353, KLHL17, CCNL2, etc. (1439 in total)
hsa-miR-940	BC032353, BC102012, BC042880, etc. (1973 in total)
hsa-miR-889/-3p	LOC643837, PER3, PDPN, etc. (918 in total)
hsa-miR-1307-5p	0 target found
hsa-miR-1293	LOC643837, DKFZp434A1923, ATAD3C, etc. (1383 in total)
hsa-miR-3124/-5p	0 target found
hsa-miR-152-5p	0 target found

After eliminating the duplicate targets, we obtain 7968 proteins in total. Most of the proteins do not exist in GO database, and only 2803 items are left for further classification. We make GO analysis based on all identified proteins that could be sorted. The ontology covers three domains that are the cellular component, molecular function, and biological process. Meanwhile, pie charts for every ontology show the match ratio (Fig. 4, Fig. 5, and Fig. 6).

From the figures, we could conclude that most targets of dysregulated microRNAs belong to the membrane part and extracellular region part, leading us to the conjecture that lupeol treatment mainly alters the status of cell surface.

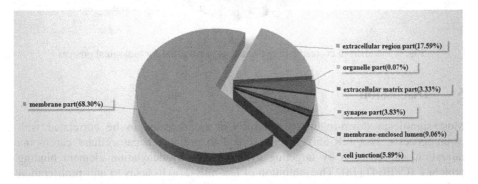

Fig. 4. Pie chart of corresponding GO terms distribution for cellular component

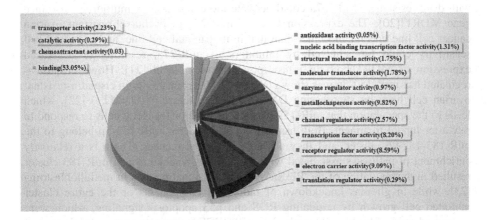

Fig. 5. Pie chart of corresponding GO terms distribution for molecular function

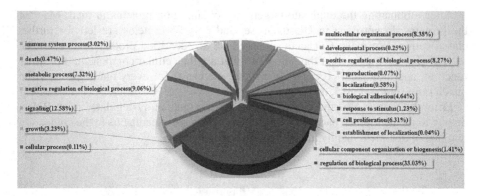

Fig. 6. Pie chart of corresponding GO terms distribution for biological process

5   Discussion

Most significant down-regulated microRNAs are revealed to be correlated with carcinogenesis. Hsa-miR-550a-3p was found to promote hepatocellular carcinoma migration and invasion by targeting cytoplasmic polyadenylation element binding protein 4 (CPEB4) [18]. The inhibition of hsa-miR-27a-3p leads to anti-proliferation and colony restriction in pancreatic cancer cells by binding 3'UTR of Sprouty2 (Spry2) [19]. Promisingly, accumulated evidence shows that miR-27a is also related with drug resistance. MiR-27a could activate the expression of multidrug resistance gene MDR1 [20]. The expression level of miR-27a in paclitaxel-resistant ovarian cancer cell line was much higher than that in its parental cell line, and increased cell sensitivity could be achieved by inhibitors of miR-27a, resulting in decreased expression of MDR1 both in transcript and protein level [21]. However, down-regulated hsa-miR-152-5p was shown to improve drug sensitivity in cisplatin-resistant ovarian cell lines by suppressing DNA methyltransferase 1 (DNMT1) [22], which leads to the speculation that miR-152 was an early factor for HeLa cells to respond to risk after lupeol treatment. Both hsa-miR-182-5p and has-miR-183-5p belong to the miR-183-182-96 cluster, and they all have similar sequences. There is a relatively enrichment of pathways associated with this cluster. Knockdown of miR-183 cluster resulted in dysregulation of the PI3K/AKT/mTOR signaling axis in medulloblastoma [23], while in gliomas knockdown of individual components or the entire cluster inhibited cell growth by regulating ROS and p53 apoptosis signaling, which were associated with fibroblast growth factor 9 (FGF9), cytoplasmic polyadenylation element binding protein 1 (CPEB1) and Forkhead box protein O1 (FOXO1) genes [24]. MiR-183 cluster could also suppress the expression of FOXO1 in breast cancer cells [25]. Separately, several biomarker, including PDCD4 [26], β-catenin [27], and RECK [28], are involved in miR-183 and miR-182 upstream and downstream. Among the nine significant up-regulated microRNAs, only hsa-miR-193b-3p and hsa-miR-140-3p have get cancer related research, and shown to be tumor suppressors. MicroRNA-140 suppressed NF-κB activity by regulating nuclear receptor-interacting protein 1(NRIP1) [25]. The suppression of miR-193b leads to up-regulation of urokinase-type plasminogen activator (uPA) protein expression, and breast cancer cell invasion enhancement [29]. MiR-193b also represses cell proliferation and regulates cyclin D1 in melanoma cells, suggesting miR-193b participates in different signal pathways [30].

Acknowledgements. This work is partially supported by the China Natural Science Foundation (Grant Number: 30771371, 31271781), the National High-tech R&D Program of China (863 Program) (Grant Number: 2001AA231091, 2004AA231071), Heilongjiang Province Science Foundation (Grant Number: 2004C0314), Heilongjiang Province key scientific and technological project (Grant Number: WB07C02), HIT Science Foundation (Grant Number: HIT. 2003. 38), National MOST special fund (Grant Number: KCSTE- 2000-JKZX- 021, NCSTE- 2007-JKZX- 022, 2012EG111228).

References

1. Redshaw, N., Wilkes, T., Whale, A., Cowen, S., Huggett, J., Foy, C.A.: A comparison of microRNA isolation and RT-qPCR technologies and their effects on quantification accuracy and repeatability. BioTechniques 54(3), 155–164 (2013)
2. Mattie, M.D., Benz, C.C., Bowers, J., Sensinger, K., Wong, L., Scott, G.K., Fedele, V., Ginzinger, D., Getts, R., Haqq, C.: Optimized high-throughput microRNA expression profiling provides novel biomarker assessment of clinical prostate and breast cancer biopsies. Mol. Cancer 5, 24 (2006)
3. Griffiths-Jones, S.: miRBase: the microRNA sequence database. Methods Mol. Biol. 342, 129–138 (2006)
4. Davison, T.S., Johnson, C.D., Andruss, B.F.: Analyzing micro-RNA expression using microarrays. Methods Enzymol. 411, 14–34 (2006)
5. Cummins, J.M., He, Y., Leary, R.J., Pagliarini, R., Diaz, L.A.: The colorectal microRNAome. Proc. Natl. Acad. Sci. U. S. A. 103(10), 3687–3692 (2006)
6. Morin, R.D., O'Connor, M.D., Griffith, M., Kuchenbauer, F., Delaney, A.: Application of massively parallel sequencing to microRNA profiling and discovery in human embryonic stem cells. Genome Res. 18(4), 610–621 (2008)
7. Hesse, J.E., Liu, L., Innes, C.L., Cui, Y., Palii, S.S., Paules, R.S.: Genome-Wide Small RNA Sequencing and Gene Expression Analysis Reveals a microRNA Profile of Cancer Susceptibility in ATM-Deficient Human Mammary Epithelial Cells. PLoS One 8(5), 1–9 (2013)
8. Li, Y., Zhang, Z., Liu, F., Vongsangnak, W., Jing, Q., Shen, B.R.: Performance comparison and evaluation of software tools for microRNA deep-sequencing data analysis. Nucleic Acids Research 40(10), 4298–4305 (2012)
9. Liu, D.C., Dai, C.H., Xu, D.C.: Cluster analysis of microRNAs microarray data and prediction of lupeol's anti-cancer path. In: 2011 IEEE International Conference on Bioinformatics and Biomedicine Workshops, pp. 440–445. IEEE Press, Atlanta (2011)
10. Saleem, M.: Lupeol, a novel anti-inflammatory and anti-cancer dietary triterpene. Cancer Lett. 285, 109–115 (2009)
11. Hackenberg, M., Sturm, M., Langenberger, D., Falcón-Pérez, J.M., Aransay, A.M.: miRanalyzer: a microRNA detection and analysis tool for next-generation sequencing experiments. Nucleic Acids Res. 37(Web Server issue), W68–W76 (2009)
12. National Center for Biotechnology Information, http://www.ncbi.nlm.nih.gov

13. Feng, J., Meyer, C.A., Wang, Q., Liu, J.S., Shirley Liu, X., Zhang, Y.: GFOLD: a generalized fold change for ranking differentially expressed genes from RNA-seq data. Bioinformatics 28(21), 2782–2788 (2012)

14. Anders, S., Huber, W.: Differential expression analysis for sequence count data. Genome Biol. 11(10), R106 (2010)

15. Cai, Z., Eulenstein, O., Gibas, C.: Guest Editors Introduction to the Special Section on Bioinformatics Research and Applications. IEEE/ACM Transactions on Computational Biology and Bioinformatics 11(2), 1–2 (2014)

16. GOmir, http://www.bioacademy.gr/bioinformatics/projects/GOmir/

17. TAGGO, http://www.bioacademy.gr/bioinformatics/TAGGO/

18. Tian, Q., Liang, L., Ding, J., Zha, R., Shi, H., Wang, Q., Huang, S., Guo, W., Ge, C., Chen, T., Li, J., He, X.: MicroRNA-550a acts as a pro-metastatic gene and directly targets cytoplasmic polyadenylation element-binding protein 4 in hepatocellular carcinoma. PLoS One 7(11), 1–9 (2012)

19. Ma, Y., Yu, S., Zhao, W., Lu, Z., Chen, J.: miR-27a regulates the growth, colony formation and migration of pancreatic cancer cells by targeting Sprouty2. Cancer Lett. 298, 150–158 (2010)

20. Zhu, H., Wu, H., Liu, X., Evans, B.R., Medina, D.J., Liu, C.G., Yang, J.M.: Role of MicroRNA miR-27a and miR-451 in the regulation of MDR1/P-glycoprotein expression in human cancer cells. Biochem. Pharmacol. 76, 582–588 (2008)

21. Li, Z., Hu, S., Wang, J., Cai, J., Xiao, L., Yu, L., Wang, Z.: MiR-27a modulates MDR1/P-glycoprotein expression by targeting HIPK2 in human ovarian cancer cells. Gynecol. Oncol. 119, 125–130 (2010)

22. Xiang, Y., Ma, N., Wang, D., Zhang, Y., Zhou, J., Wu, G., Zhao, R., Huang, H., Wang, X., Qiao, Y., Li, F., Han, D., Wang, L., Zhang, G., Gao, X.: MiR-152 and miR-185 co-contribute to ovarian cancer cells cisplatin sensitivity by targeting DNMT1 directly: a novel epigenetic therapy independent of decitabine. Oncogene, 1–9 (2013)

23. Weeraratne, S.D., Amani, V., Teider, N., Pierre-Francois, J., Winter, D.: Pleiotropic effects of miR-183~96~182 converge to regulate cell survival, proliferation and migration in medulloblastoma. Acta, Neuropathol. 123, 539–552 (2012)

24. Tang, H., Bian, Y., Tu, C., Wang, Z., Yu, Z., Liu, Q., Xu, G., Wu, M., Li, G.: The miR-183/96/182 cluster regulates oxidative apoptosis and sensitizes cells to chemotherapy in gliomas. Curr. Cancer Drug Targets 13(2), 221–231 (2013)

25. Guttilla, I.K., White, B.A.: Coordinate Regulation of FOXO1 by miR-27a, miR-96, and miR-182 in Breast Cancer Cells. J. Biol. Chem. 284, 23204–23216 (2009)

26. Li, J., Fu, H., Xu, C., Tie, Y., Xing, R., Zhu, J., Qin, Y., Sun, Z., Zheng, X.: miR-183 inhibits TGF-β1-induced apoptosis by downregulation of PDCD4 expression in human hepatocellular carcinoma cells. BMC Cancer 10, 354 (2010)

27. Chiang, C.H., Hou, M.F., Hung, W.C.: Up-regulation of miR-182 by β-catenin in breast cancer increases tumorigenicity and invasiveness by targeting the matrix metalloproteinase inhibitor RECK. Biochim. Biophys. Acta. 1830, 3067–3076 (2013)

28. Takata, A., Otsuka, M., Kojima, K., Yoshikawa, T., Kishikawa, T., Yoshida, H., Koike, K.: MicroRNA-22 and microRNA-140 suppress NF-κB activity by regulating the expression of NF-κB coactivators. Biochem. Biophys. Res. Commun. 411, 826–831 (2011)

29. Li, X.F., Yan, P.J., Shao, Z.M.: Downregulation of miR-193b contributes to enhance urokinase-type plasminogen activator (uPA) expression and tumor progression and invasion in human breast cancer. Oncogene 28, 3937–3948 (2009)

30. Chen, J., Zhang, X., Lentz, C., Abi-Daoud, M., Paré, G.C., Yang, X., Feilotter, H.E., Tron, V.A.: miR-193b Regulates Mcl-1 in Melanoma. Am. J. Pathol. 179(5), 2162–2168 (2011)

A Simulation for Proportional Biological Operational Mu-Circuit

Dechang Xu[1], Zhipeng Cai[2], Ke Liu[1], Xiangmiao Zeng[1], Yujing Ouyang[1],
Cuihong Dai[1], Aiju Hou[1], Dayou Cheng[1], and Jianzhong Li[1]

[1] Harbin Institute of Technology, Harbin, China
[2] Georgia State University, Atlanta, USA

Abstract. To quantitatively control the expression of target gene is challenging but highly desired in practice. We design a device-Biological Proportional Operational Mu-circuit (P-BOM) incorporating AND/OR gate and operational amplifier into one circuit and explore its behaviors through simulation. The results imply that we can control input-output proportionly by manipulating the RBS of *hrp*R, *hrp*S, *tet*R and output gene.

Keywords: Simulation, Proportional regulation, Modeling.

1 Introduction

In 1990s, Eric Mjolsness and several researchers developed the concept of genetic circuit from electronic circuits, to facilitate modeling gene expression and regulation (Mjolsness, Sharp et al. 1991[1], McAdams and Shapiro 1995[2], Reinitz and Sharp 1996[3], Sharp and Reinitz 1998[4]). Upon the inspiration, Ron Weiss, a computer engineer of MIT, constructed an AND-gate genetic circuit in 2001 (Weiss, Knight et al. 2001[5]). Based on Weiss' pioneering result, the team of Jeff Hasty took IPTG (isopropyl β-D-1-Thiogalactopyranoside) and aTc (anhydrotetracycline) as the inputs and GFP (green fluorescent protein) as the output and demonstrated the AND-gate circuit in a more lucid and applicable way (Hasty, McMillen et al. 2002[6]). In 2005, to investigate the prospect of genetic circuit, Oliver Rackham proved that other kinds of logic circuits are feasible by orthogonal topological structures (Rackham and Chin 2005[7]). In this paper, we inherit the idea of resembling electronic circuits. By incorporating AND/OR gate and operational amplifier into one circuit, we create our device-Biological Proportional Operational Mu-circuit (P-BOM).

The composition of P-BOM is that *hrp*R's promoter depends on the input, however, *hrp*S' promoter is always P*tet* and *tet* owns P*hrp*L, while the output gene follows *tet* and shares P*hrp*L. Once the input is sensed, the input promoter triggers *hrp*R's transcription. The activity of P*tet* is constitutive, which means HrpS protein is ample. As HrpR accumulates, HrpS binds to HrpR and form HrpRS which then triggers P*hrp*L, and *tet*R and output begin to accumulate. *tet*R can inhibit P*tet*. As a feedback, HrpS and HrpRS will decrease. P*hrp*L will be of lower activity so that the amount of *tet*R and the level of output will decline. The decrease of *tet*R will enhance the *hrp*S' expression. All these construct a feedback cycle. Finally, the output will stabilize and be

M. Basu, Y. Pan, and J. Wang (Eds.): ISBRA 2014, LNBI 8492, pp. 81–91, 2014.

in a certain proportion with the input. By manipulating the RBS (ribosome binding site) of *hrp*R, *hrp*S, *tet*R and output gene, we can control input-output proportion.

2 Premises

At the beginning of the project, a mathematical model would be necessary to obtain a clear depict of the system as well as a reasonable expectation of the subsequent experiments. However, to start with, several premises shall be settled in the first step.

2.1 The Circuit

(1) The transcription follows the Hill equation, while other reactions follow the mass action principle and the saturation kinetics.

(2) Some of the parameter values are obtained according to the published results, while others are assigned in a reasonable way.

(3) The amount of protein binding to promoter is negligible compared with the total amount of protein.

(4) The simulation was conducted using SBToolbox[8] for matlab and R, a free software environment for statistical computing and graphics.

The following picture is the whole design of our circuit.

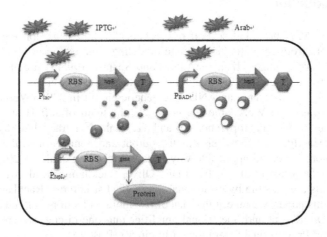

Fig. 1. A Schematic of the Biological And-gate

As showen in Fig. 1, a protein complex HrpRS is formed by a combination of HrpS and HrpR. HrpRS then binds to the promoter of *hrp*L, triggering transcription of the downstream gene. Two inducible promoters, the isopropylthiogalactoside (IPTG)-inducible P*lac* and the arabinose-inducible P$_{BAD}$, are selected as the AND gate inputs. A model based on protein and protein-promoter interactions for the *hrp* gene regulation is described below:

2.2   The Transcription of htpR and hrpS mRNA

In this case, Hill function [9] is selected as the transcription function to explain quantitatively how IPTG, Arabinose and HrpRS influence the promoters. The Hill function and its curve are shown as

$$h^+(x_j,\theta_j,m) = \frac{x_j^m}{x_j^m + \theta_j^m}, h^-(x_j,\theta_j,m) = 1 - h^+(x_j,\theta_j,m).$$

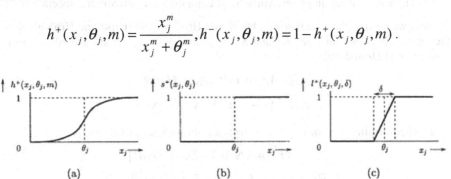

Fig. 2. Example of regulation functions (a)Hill function;(b)Step function;(c)logoid function

Where θ_j appears as the threshold for the regulatory influence of x_j on a target gene and m the steepness parameter. h^+, in the function, refers to the positive regulation on the target gene while h^- shows the negative one.

IPTG binds to promoter Plac and triggers the transcription of *hrp*R_mRNA at the maximum rate constant *vmr*,

$$n_1 \cdot IPTG + P_{lac} \rightleftharpoons n_1 \cdot IPTG \cdot P_{lac}$$
$$n_1 \cdot IPTG \cdot P_{lac} \rightarrow mRNAr + n_1 \cdot IPTG \cdot P_{lac}$$

While the amount of *hrp*R_mRNA degrades at a rate constant *Drnar*. Hence, the change rate for *hrp*R_mRNA is

$$\cdot\frac{d}{dt}[mRNAr] = \frac{vmr \cdot (\frac{[IPTG]}{n})^{n_1}}{(\frac{[IPTG]}{n})^{n_1} + \theta_1^{n_1}} - Drnar \cdot [mRNAr].$$

Similarly, we could obtain the reaction for P_{BAD},

$$n_A \cdot Arab + P_{BAD} \rightleftharpoons n_A \cdot Arab \cdot P_{BAD}$$
$$n_A \cdot Arab \cdot P_{BAD} \rightarrow mRNAs + n_A \cdot Arab \cdot P_{BAD}$$

$$\frac{d}{dt}[mRNAs] = \frac{vms \cdot (\frac{[Arab]}{n})^{n_A}}{(\frac{[Arab]}{n})^{n_A} + \theta_A^{n_A}} - Drnas \cdot [mRNAs].$$

2.3   Translation of hrpR and hrpS Protein

To simplify our model, we have the following assumptions:

(1) The amount of translation product is proportional to the mRNA amount;
(2) There is no delay in either synthesis of components or protein transportation.

Next we discuss the translation of HrpR and HrpS protein. HrpR and HrpS are synthesized at the constant rate Kr and Ks respectively, and degrade according to the constant rate Dr and Ds.

$$mRNAr \rightarrow mRNAr + HrpR$$

$$mRNAs \rightarrow mRNAs + HrpS$$

The net forming rates of HrpR and HrpS are shown respectively as

$$Rr = Kr \cdot [mRNAr] - Dr \cdot [HrpR]$$

$$Rs = Ks \cdot [mRNAs] - Ds \cdot [HrpS]$$

2.4   Interaction between Protein hrpRS and Promoter hrpL

At the beginning, we define the binding rate constant of HrpR and HrpS as Krs, their dissociating rate constant K_rs and the degrading rate constant of HrpRS as we have mentioned previously,

$$HrpR + HrpS \rightleftharpoons hrpRS .$$

The net forming rate of HrpRS is shown as

$$Rrs = Krs \cdot [HrpR] \cdot [HrpS] - K\_rs \cdot [HrpRS] - Drs \cdot [HrpRS]$$

Activation of the promoter $hrpL$ is similar to that of Plac and P$_{BAD}$

$$n_{RS} \cdot HrpRS + P_{hrpL} \rightleftharpoons n_{RS} \cdot HrpRS \cdot P_{hrpL}$$

$$n_{RS} \cdot HrpRS \cdot P_{hrpL} \rightarrow mRNAl + n_1 \cdot HrpRS \cdot P_{hrpL}$$

Promoter $hrpL$ triggers the transcription of our target protein. Ko and Do denote the rate constants of synthesis and degradation of output protein.

All equations for the modeling of hrp AND gate are summarized below

$$\frac{d}{dt}[mRNAr] = \frac{vmr \cdot (\dfrac{[IPTG]}{n})^{n_1}}{(\dfrac{[IPTG]}{n})^{n_1} + \theta_1^{n_1}} - Drnar \cdot [mRNAr]$$

$$\frac{d}{dt}[mRNAs] = \frac{vms \cdot (\dfrac{[Arab]}{n})^{n_A}}{(\dfrac{[Arab]}{n})^{n_A} + \theta_A^{n_A}} - Drnas \cdot [mRNAs]$$

$$\frac{d}{dt}[HrpR] = Kr \cdot [mRNAr] - Dr \cdot [HrpR] - Krs \cdot [HrpR] \cdot [HrpS] + K\_rs \cdot [HrpRS]$$

$$\frac{d}{dt}[HrpS] = Ks \cdot [mRNAs] - Ds \cdot [HrpS] - Krs \cdot [HrpR] \cdot [HrpS] + K\_rs \cdot [HrpRS]$$

$$\frac{d}{dt}[HrpRS] = Krs \cdot [HrpR] \cdot [HrpS] - K\_rs \cdot [HrpRS] - Drs \cdot [HrpRS]$$

$$\frac{d}{dt}[mRNAo] = \frac{vmo \cdot [HrpRS]^{n_{RS}}}{[HrpRS]^{n_{RS}} + \theta_{RS}^{n_{RS}}} - Drnao \cdot [mRNAo]$$

$$\frac{d}{dt}[output] = n \cdot (Ko \cdot [mRNAo] - Do \bullet [output])$$

3   Simulation Results

Firstly, we test the model when n=1 (in one cell). Supposing $Kt=Ko=Kr=Ks=30$ (Kt, Ko, Kr, Ks are the parameters connected to RBS strength. In our model, we assume that $K=3$ denotes weak RBS. $K=30$ denotes middle RBS while $K=60$ denotes strong RBS). By varying the amount of IPTG and Arab, we obtained the output according to the inputs.

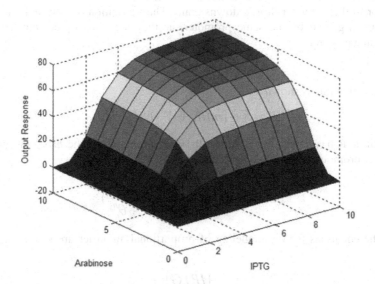

Fig. 3. Simulation Results of hrpL AND gate

As we can work out from Fig. 3, the output response stays at a low level when at least one of the two inputs (IPTG and Arabinose) stays low in quantity. The response reaches to a high level only when both inputs appear high in quantity.

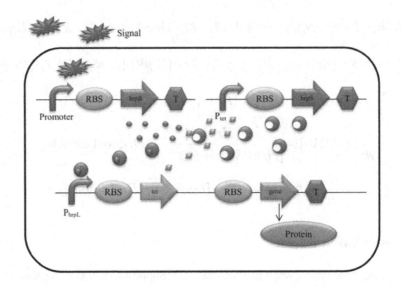

Fig. 4. A Schematic of the Biological Proportional Operational MU-circuit

Based on the hrp AND gate, we substituted the promoter P*tet*R for P*lac* (P*tet*R is constitutively on and could be inhibited by the tetR protein), and add a *tet*R protein generator to the promoter hrpL's downstream. The translation of tetR and our future output are triggered by the same promoter, as they are transcribed into the same mRNA single strand.

$$mRNAo \rightarrow mRNAo + tetR$$

The reaction rate is shown as

$$Rt = Ks \cdot [mRNAo] - Ds[tetR]$$

Protein *tet*R has a negative effect on the P*lac* promoter, so the *hrpS* transcription rate under promoter P*tet*R could be depicted as

$$Rst = vmst \cdot (1 - \frac{[\text{tetR}]^{n_2}}{[\text{tetR}]^{n_2} + \theta_t^{n_2}})$$

All the equations for the modeling of proportional amplifier are summarized below:

$$\frac{d}{dt}[mRNAr] = \frac{vmr \cdot (\frac{[IPTG]}{n})^{n_1}}{(\frac{[IPTG]}{n})^{n_1} + \theta_1^{n_1}} - Drnar \cdot [mRNAr]$$

$$\frac{d}{dt}[mRNAs] = vmst \cdot (1 - \frac{[tetR]^{n_2}}{[tetR]^{n_2} + \theta_t^{n_2}}) - Drnas \cdot [mRNAs]$$

$$\frac{d}{dt}[HrpR] = Kr \cdot [mRNAr] - Dr \cdot [HrpR] - Krs \cdot [HrpR] \cdot [HrpS] + K\_rs \cdot [HrpRS]$$

$$\frac{d}{dt}[HrpS] = Ks \cdot [mRNAs] - Ds \cdot [HrpS] - Krs \cdot [HrpR] \cdot [HrpS] + K\_rs \cdot [HrpRS]$$

$$\frac{d}{dt}[HrpRS] = Krs \cdot [HrpR] \cdot [HrpS] - K\_rs \cdot [HrpRS] - Drs \cdot [HrpRS]$$

$$\frac{d}{dt}[mRNAo] = \frac{vmo \cdot [HrpRS]^{n_{RS}}}{[HrpRS]^{n_{RS}} + \theta_{RS}^{n_{RS}}} - Drnao \cdot [mRNAo]$$

$$\frac{d}{dt}[tetR] = Kt \cdot [mRNAo] - Dt \cdot [tetR]$$

$$\frac{d}{dt}[output] = n \cdot (Ko \cdot [mRNAo] - Do \bullet [output])$$

We test the output response by varying IPTG combination of RBS strength. To reduce simulation times, Taguchi Design is adopted in our model. The Taguchi Design table for Ko, Kt, Kr, Ks is shown in Table 1

Table 1. Experiment for Taguchi Design

Order	Ko	Kt	Kr	Ks	m
1	60(Strong)	3(Weak)	60	30(Middle)	0.1333
2	3	3	3	3	0.8
3	30	60	3	30	0.5333
4	30	3	30	60	0.2
5	3	30	30	30	0.1333
6	60	30	3	60	0.8667
7	30	30	60	3	0.1333
8	3	60	60	60	0.0667
9	60	60	30	3	0.0667

Then, we got a series of data of the steady states of input and output and plotted them together. The results are shown in Fig. 5.

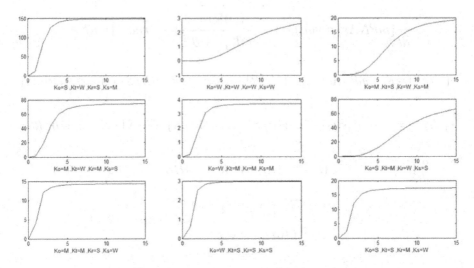

Fig. 5. Simulation Results for Output Response VS IPTG input

Fig. 5 demonstrates that *(Ko=3, kt=3, Kr=3, Ks=3)*, *(Ko=30, Kt=60, Kr=3, Ks=30)* and *(Ko=60, Kt=30, Kr=3, Ks=60)* present a better linear relationship between input and output than that of other experiments. Linear regression is applied to describe the relationship between input IPTG concentration and output response.

Table 2. Linear Regression of Simulation Result for Experiments 2, 3, and 6

Order	Relationship	R^2
2	$[Output] = 0.2060 \cdot [IPTG] - 0.3617 = 0.2060 \cdot ([IPTG] - 1.7558)$	0.9713
3	$[Output] = 1.5725 \cdot [IPTG] - 1.2230 = 1.5725 \cdot ([IPTG] - 0.7777)$	0.9397
6	$[Output] = 5.2721 \cdot [IPTG] - 8.3323 = 5.2721 \cdot ([IPTG] - 1.5805)$	0.9697

Here we design a coefficient m $(m=b/a)$ where b is the concentration range of linear area between output response induced by IPTG and a is the whole range of IPTG concentration.

In Fig. 6, we found that the width of linear section reaches its maximum when *Ko=30, Kt=3* (or *Kt=30*), *Kr=3*, and *Ks=30*.

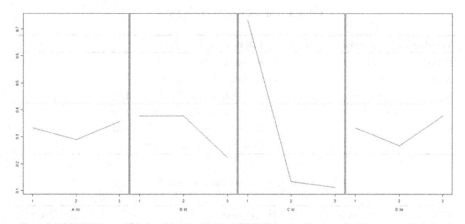

Fig. 6. Effect Plot for m

The parameters and variables of the modeling are listed below.

Table 3. Parameters used in the Model

Parameter	Dimension	Description	Value
n	$1 \times$cell	number of cells	
vmr	$\text{nmol} \cdot \text{cell}^{-1} \cdot \text{time}^{-1}$	maximum rate of mRNAr transcription	5
vms	$\text{nmol} \cdot \text{cell}^{-1} \cdot \text{time}^{-1}$	maximum rate of mRNAs transcription	5
vmo	$\text{nmol} \cdot \text{cell}^{-1} \cdot \text{time}^{-1}$	maximum rate of mRNAo transcription	5
n_I	1	steepness parameter	2
n_A	1	steepness parameter	2
n_{RS}	1	steepness parameter	2
θ_I	$\text{nmol} \cdot \text{cell}^{-1}$	threshold for the IPTG regulatory influence	10
θ_A	$\text{nmol} \cdot \text{cell}^{-1}$	threshold for the Arabinose regulatory influence	10
θ_{RS}	$\text{nmol} \cdot \text{cell}^{-1}$	threshold for the HrpRS regulatory influence	10
Drnar	time^{-1}	degradation rate constant for mRNAr	2
Drnas	time^{-1}	degradation rate constant for mRNAs	2
Drnao	time^{-1}	degradation rate constant for mRNAo	2
Dr	time^{-1}	degradation rate constant for protein HrpR	1
Ds	time^{-1}	degradation rate constant for protein HrpS	1
Dt	time^{-1}	degradation rate constant for tetR protein	1
Do	time^{-1}	degradation rate constant for output protein	1
Krs	$\text{nmol}^{-1} \cdot \text{cell} \cdot \text{time}^{-1}$	association rate constant for HrpR and HrpS	5
K_rs	time^{-1}	disassociation rate constant for complex HrpRS	3
Drs	time^{-1}	degradation rate constant for complex HrpRS	0.5

Table 4. Variables used in the Model

Variable	Dimension	Description
IPTG	nmol	Total amount of IPTG
Arab	nmol	Total amount of Arabinose
Kr	time^{-1}	synthesis rate constant for protein HrpR
Ks	time^{-1}	synthesis rate constant for protein HrpS
Kt	time^{-1}	synthesis rate constant for tetR protein
Ko	time^{-1}	synthesis rate constant for output protein

Table 5. States used in the Model

State	Dimension	Description	Initial Value
mRNAr	nmol•cell^{-1}	Total amount of hrpR_mRNA	0
mRNAs	nmol•cell^{-1}	Total amount of hrpS_mRNA	0
mRNAo	nmol•cell^{-1}	Total amount of output_mRNA	0
HrpR	nmol•cell^{-1}	Total amount of HrpR	0
HrpS	nmol•cell^{-1}	Total amount of HrpS	0
HrpRS	nmol•cell^{-1}	Total amount of HrpRS	0
tetR	nmol•cell^{-1}	Total amount of tetR protein	0

Table 6. Observation used in the Model

Observation	Dimension	Description
output	nmol	Total amount of output protein

PS : The parameters are constants in our model. The variables are what we controlled in one experiment. The states are dependent variable changing in one experiment. The observation is our target output.

4   Discussion

In comparison with Weiss' invention, P-BOM has only one homogeneous or heterogeneous input and output. Besides, multiple magnitudes of input are distinguishable by the new circuit, and a given input corresponds to a certain output. Moreover, whereas general genetic circuits are vulnerable to the noise of cellular background, P-BOM seems more interference-free, implicating the prominent stability of our device.

In some applications, this device could be used to enhance Bio-electric Interface, the device designed by 2012 Edinburg iGEM team. If the input is biochemical signal molecules and the output becomes electrons, P-BOM can be coupled with Bio-electric Interface, that ephemeral processes in cells will be precisely measurable by simple electronic methods and analyzable by computer, which will be potentially used in pathogen surveillance situations[10]. P-BOM is also helpful in yoghurt producing, where the control of pH is inaccurate. When P-BOM is transplanted into yoghurt-producing bacteria, let hydrogen ions be the input and lacR be the output, and select the proper P-BOM parameters, then the pH will be steady around 5.5. In short, our inspired and reliable device is promising in various fields.

Acknowledgements. This work is partially supported by the Natural Science Foundation of China (Grant Number: 30771371, 31271781), the National High-tech R&D Program of China (863 Program) (Grant Number: 2001AA231091, 2004AA231071), Heilongjiang Province Science Foundation (Grant Number: 2004C0314), Heilongjiang Province key scientific and technological project (Grant Number: WB07C02), HIT Science Foundation (Grant Number: HIT. 2003. 38), National MOST special fund (Grant Number: KCSTE- 2000-JKZX- 021, NCSTE- 2007- JKZX- 022, 2012EG111228).

References

[1] Mjolsness, E., Sharp, D.H., Reinitz, J.: A connectionist model of development. J. Theor. Biol. 152, 429–453 (1991)

[2] McAdams, H.H., Shapiro, L.: Circuit simulation of genetic networks. Science 269, 650–656 (1995)

[3] Reinitz, J., Sharp, D.H.: Gene circuits and their uses. In: Collado-Vides, J., Magasanik, B., Smith, T.F. (eds.) Integrative Approaches to Molecular Biology, pp. 253–272. MIT Press, Cambridge (1996)

[4] Sharp, D.H., Reinitz, J.: Prediction of mutant expression patterns using gene circuits. BioSystems 47, 79–90 (1998)

[5] Weiss, R., Knight, T.F.: Engineered Communications for Microbial Robotics. In: Condon, A., Rozenberg, G. (eds.) DNA 2000. LNCS, vol. 2054, pp. 1–16. Springer, Heidelberg (2001)

[6] Hasty, J., McMillen, D., Collins, J.J.: Engineered gene circuits. Nature 420, 224–230 (2002)

[7] Rackham, O., Chin, J.W.: A network of orthogonal ribosome·mRNA pairs. Nature Chemical Biology 1, 159–166 (2005)

[8] Schmidt, H., Jirstrand, M.: Systems Biology Toolbox for MATLAB: a computa-tional platform for research in systems biology. Bioinformatics 22, 514–515 (2006)

[9] de Jong, H.: Modeling and Simulation of Genetic Regulatory Systems: A Li-terature Review. Journal of Computational Biology 9(1), 67–103 (2002)

[10] Kayali, G., Kandeil, A., El-Shesheny, R., Kayed, A., Gomaa, M.M., Maatouq, A., Shehata, M.M., Moatasim, Y., Bagato, O., Cai, Z., Rubrum, A., Kutkat, M., McKen-zie, P., Webster, R., Webby, R., Ali, M.: Active Surveillance for Avian Influenza Virus, Egypt, 2010 - 2012. Emerging Infectious Diseases Journal 20(4) (2014)

Computational Prediction
of Human Saliva-Secreted Proteins

Ying Sun[1], Chunguang Zhou[1], Jiaxin Wang[2], Zhongbo Cao[1],
Wei Du[1,*], and Yan Wang[1,*]

[1] Key Laboratory of Symbolic Computation
and Knowledge Engineering of Ministry of Education,
College of Computer Science and Technology,
Jilin University, Changchun 130012, China
[2] Nari Group Corporation/State Grid Electric Power Research Institute,
Nanjing, Jiangsu 211000, China
weidu@jlu.edu.cn, wy6868@hotmail.com

Abstract. Using proteins in saliva as biomarkers has great advantage
in early diagnosis and prognosis evaluation of health conditions or diseases. In this article, we present a computational method for predicting
secreted proteins in human saliva. Firstly, we collected currently known
saliva-secreted proteins and the representatives that deem to be not extracellular secretion into saliva. Secondly, we pruned the negative data
concerned the imbalance condition, and then extracted the relevant features from the physicochemical and sequence properties of all remained
proteins. After that, a support vector machine classifier was built which
got performance of average sensitivity, specificity, precision, accuracy
and Matthews correlation coefficient value to 80.67%, 90.56%, 90.09%,
85.53% and 0.7168, respectively. These results indicated that the selected
features and the model are effective. Finally, a screening test was implemented to all human proteins in UniProt and acquired 5811 proteins
as predicted saliva-secreted proteins which may be used as biomarker
candidates for further salivary diagnosis.

Keywords: saliva-secreted protein, biomarker, salivary diagnosis, bioinformatics.

1 Introduction

Saliva is one of the most important components in oral environment, which plays
many important biological roles of lubrication, buffering, assists digestion, antibac-
terial and maintaining the integrity of mucosa function. Saliva is mainly secreted
by three pairs of major salivary glands (parotid, submandibular gland, sublin-
gual gland) and numerous minor salivary glands spreading over oral cavity [19,24].
These glands can secrete specific biomolecules in response to different factors such
as hormone, lymphatic factor and cell factors, which may be released by various or-
gans into blood circulation due to their disease states, as previously demonstrated

---

* Corresponding author.

M. Basu, Y. Pan, and J. Wang (Eds.): ISBRA 2014, LNBI 8492, pp. 92–101, 2014.

[17]. Saliva also includes the gingival crevicular fluid and oral mucosa. The composition of saliva is very complicated. Besides water, saliva is composed of inorganic substance, such as sodium, potassium, calcium, plasma, and organic matter, such as salivary amylase, mucopolysaccharide, mucin and lysozyme. The traditional biochemistry method can only detect a few proteins in saliva, along with the extensive application of high throughput, high precision protein detection technology, more and more proteins in saliva were detected. So that the saliva proteome became clear gradually [14]. It is found that the changes of protein content in the saliva were closely related to many oral and systemic diseases, such as oral disease [27,21], rheumatic diseases [16,10], periodontal disease [9], alzheimer's disease [2], head and neck squamous cell carcinomas [3], diabetes [18], prostate adenocarcinoma [20], breast cancer [23], and pancreastic cancer [25].

Compared with the serum extraction, the collection of saliva is simple, adequate, non-invasive to the body, able to sample repeatedly, easy for preserving, available for screening large-scale of samples. Using proteins in saliva as biomarkers has great advantages in early diagnosis and prognosis evaluation of health conditions or diseases, especially suitable for the medical condition of limited area, or infant disease detection [27]. Salivary diagnosis, has gradually become the research target and caused widespread concern [14]. With the activities related to the development of proteomics technology, more and more researchers found saliva proteins by proteomics experiments [28,15]. However, these experiments are expensive and lots of inconsistent proteins were obtained from different experiments in different laboratories. Meanwhile, there is no computational method yet for detecting and validating proteins which can be secreted into saliva by now. Our previous work has been successfully established the prediction model which can be used to distinguish the proteins in saliva coming from blood [26]. And in this paper, we present a computational method for predicting secreted proteins in human saliva based on two sets of human proteins from published literatures and public databases. One set contains known proteins which can be secreted into saliva, and along with another set of proteins that deem to be not extracellular secretion. Then, we extracted the relevant features from the physicochemical and sequence properties using support vector machine based recursive feature elimination algorithm (SVM-RFE) [11]. After that, a support vector machine (SVM) classifier was built on our two sets using the relevant features. The results indicated that the selected features and the model are effective. Finally, among all human proteins, this model was used to identify saliva-secreted proteins which might be used as biomarker candidates for further salivary diagnosis.

2   Materials and Methods

2.1   Collection of Saliva-secreted Proteins and Non-saliva-secreted Proteins

In order to obtain the saliva-secreted proteins, we searched secretory and salivary proteins respectively from both public databases and published literatures.

Secretory proteins could be collected from the secreted protein database (SPD) [4], the mammalian protein subcellular localization database (LOCATE) [22] and the Universal Protein Resource (UniProt) [8]. Salivary proteins were collected from sys-body fluid database [15]. In recent years, there were many reports about salivary proteome and these proteins didn't belong to any database. Therefore, we also selected salivary proteins from the references [13] and [7], in which the number of proteins reported were more. The duplicate proteins were removed from our collection, giving rise to 4312 secretory and 1987 salivary proteins. The set of saliva-secreted proteins as the positive set was the intersection of salivary and secretory proteins, containing a total of 557 proteins. The detail information is shown in Table 1.

Table 1. The information of data sources and the protein number

Secretory Proteins	Number	Salivary Proteins	Number
SPD	2194	Sys-body Fluid	2161
LOCATE	3376	HU et al	331
UniProt KB	1847	Denny et al	1166
Deduplicate	4312	Deduplicate	1987

In order to generate a negative set of proteins for the classification problem, we should select representatives from non-saliva-secreted proteins, including both proteins unrelated to secretory pathway and secreted proteins not involved in the circulatory system. There were 5029 Pfam protein families [1] containing no previously mentioned saliva-secreted proteins. About 7000 proteins in these families, far more than the protein number of the positive set. To overcome influence on the classifier caused by the imbalance of tow sets, we selected ten representatives from each of these families containing more than ten proteins as the final negative set, which contained 2560 proteins.

2.2 Methods

Features Construction and Selection. We employed the proteins features that has been used by references [5,6] for prediction of blood-secreted proteins. There are 34 properties initially with 1523 features for each protein which can be categorized into four groups:

(i) General sequence features, such as sequence length, amino acid composition and di-peptide composition;
(ii) Physicochemical properties, such as hydrophobicity, normalized Van der Waals volume, polarity, polarizability, charges, solubility, unfoldability and disordered regions;
(iii) Domains/Motifs, such as signal peptides, transmembrane domains and twin-arginine signal peptides motif (TAT);

(iv) Structural properties, such as secondary structural content, radius and radius of gyration.

In order to get a better classification effect, we needed to remove irrelevant, redundant features. We used the student T test to evaluate the statistically significant of each features between the positive and negative set. The features with *p-value* ≥ *0.05* were eliminated. Then, we used SVM-RFE method to sort the features according to their contribution values. The minimal set of features with the best classification performance was obtained.

Classifier and Evaluation Criteria. Using the minimal set of features with the best classification performance, we trained a SVM classifier based on RBF kernel. The classification performance is measured by the *sensitivity, specificity, precision, accuracy* and Matthews correlation coefficient value (*MCC*). The formulas are shown as Eq. 1 - 5.

$$sensitivity = \frac{TP}{TP+FN} \qquad (1)$$

$$specificity = \frac{TN}{TN+FP} \qquad (2)$$

$$precision = \frac{TP}{TP+FP} \qquad (3)$$

$$accuracy = \frac{TP+TN}{N} \qquad (4)$$

$$MCC = \frac{TP*TN - FP*FN}{\sqrt{(TP+FN)(TP+FP)(TN+FP)(TN+FN)}} \qquad (5)$$

where TP, TN, FP and FN are the number of true positive, true negative, false positive and false negative, respectively, and $N = TP + FN + TN + FP$ is the total number of proteins in the dataset.

The distance d between the positions of a prediction protein in the feature space and the optimal separating hyper plane can be derived by the parameters of the classifier. If a protein is predicted as saliva-secreted protein, its distance d is a positive number, otherwise it is negative. And if the absolute value of d is larger, classification result is more reliable. Ranking the proteins according to d value and along with the class labels, the operating characteristics (ROC) curve was drawn.

3   Results

3.1   The Features of Saliva-secreted Proteins

There are total 1523 features for each protein. However, there were lots of irrelevant features which only add noises to the training of the SVM classifier.

Therefore, we needed to remove the irrelevant features with an effective feature selection method, to optimize the performance of the SVM-based classifier. With the utilization of the feature selection process outlined in Materials and Methods, a minimal set contained 68 features with the best classification performance was obtained. The following features are the most important ones for our classification, which are listed below: transmembrane domains, signal peptides, dipeptide composition, normalized Moreau-Broto autocorrelation, transition hydrophobicity, amino acid composition, distribution secondary structure, composition secondary structure, moran autocorrelation, geary autocorrelation, distribution hydrophobicity, sequence order, distribution polarity, radius, distribution polarizability, pseudo-AA descriptor, distribution solvent accessibility, isoelectric point, distribution charge, and transition solvent accessibility. Among the selected features, the most discriminatory one is the number of transmembrane domains that have been reported to be important factors for prediction of extracellularly secreted proteins [6]. Another important feature is the presence of signal peptides. Most proteins which are secreted through the Endoplasmic Reticulum (ER) have signal peptides. These proteins are trafficked to their destination by the specific signal peptides [12].

3.2   Performance of the Classifier

Based on the 68 selected features, we evaluated the performance of the SVM classifier by RBF kernel using 10-fold cross validation repeated 100 times. It can be found that the level of performance of these 100 classifiers is generally desirable. The sensitivity values were ranging from 70.18% to 92.98% and the specificity values were ranging from 78.18% to 100%. The average sensitivity, specificity, precision, accuracy and MCC value of our prediction performance were 80.67%, 90.56%, 90.09%, 85.53% and 0.7168, respectively. These results (Table 2) indicated that our selected features are informative.

Table 2. The performance of the classifiers based on 68 features using 10-fold cross validation repeated 100 times

	sensitivity(%)	specificity(%)	precision(%)	accuracy(%)	MCC
Best	92.98	100	100	93.75	0.8821
Worst	70.18	78.18	80.00	80.36	0.6077
Average	80.67	90.56	90.09	85.53	0.7168

We used the whole dataset that we collected to train a SVM classifier by RBF kernel. The distance d between the positions of a prediction protein in the feature space and the optimal separating hyper plane could be derived by the parameters of the classifier. Ranking the proteins according to d value and along with the class labels, the operating characteristics (ROC) curve was drawn, as shown in Fig. 1.

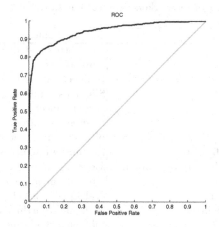

Fig. 1. ROC curve of the SVM model

We did another round of literature search for proteins that were associated with human diseases but not in our training set. A total of 58 proteins were found. These proteins are relevant to different diseases such as breast cancer, prostate cancer, bladder cancer, Burkitts lymphoma, rheumatoid arthritis, sleep bruxism, head and neck squamous cell carcinomas (HNSCC), periodontal disease, oral squamous cell carcinoma, Sjögren's syndrome, malignant tumors, malignant ovarian tumors and so on. Our model predicted 46 (79.31%) proteins as saliva-secreted proteins. The list of prediction saliva-secreted proteins is shown in Table 3. Many of them were already found in saliva, such as salivary cystatin S, Prostate specific antigen. These biomarkers could become powerful tools that can be collected from saliva for disease diagnostic and prognostic.

A screening test was done to predict all human proteins in the UniProt database. There are a total of 20270 human proteins, and 20186 proteins with complete properties information. By our SVM model, 5811 were predicted as saliva-secreted proteins. By the distance d, all proteins were ranked. We compared the ranking result with 58 human biomarkers we collected, and the $p-values$ for having such rankings if assuming that the ranking is random are calculated (as shown in Table 4). This test result indicated that many known cancer biomarkers are probably to be secreted into saliva. Once verified, they will become salivary diagnostic targets, which could be used for disease diagnosis.

4   Concluding Remarks

We attempted to solve a problem in the salivary diagnosis field, which was formulated into identification of proteins that can be secreted into saliva. In this study, we provided a new and good classification model for scientists to assist them to conduct the analysis. We selected 68 features related with saliva-secreted proteins from the global and local characteristics containing the physical properties, chemical properties, amino acid sequence and structural features of proteins.

Table 3. The list of biomarkers predicted as saliva-secreted proteins

UniProt ID	Protein Name	Disease Name
Q02747	Guanylin	pleomorphic adenoma warthin tumors
Q9HC47	cTAGE-1	lymphomas
P06731	CEA	Colorectal cancer
P61626	Lysozyme C	Sjögren's syndrome
P15941	CA15-3	Breast cancer
Q8WWA0	lactoferrin	Sjögren's syndrome
Q16661	Guanylate cyclase activator 2B	pleomorphic adenoma warthin tumors
P80511	Protein S100-A12	rheumatoid arthritis
P54107	Cysteine-rich secretory protein 1	Sjogren's syndrome
P02766	transthyretin	familial amyloidotic polyneuropathy
P01037	cystatin SA-I	oral squamous cell carcinoma
P02647	Apolipoprotein A-I	rheumatoid arthritis
P60568	Interleukin-2	Sjögren's syndrome
P01036	salivary cystatin S	periodontitis
P54108	CRISP-)	Sjogren's syndrome
P12104	Fatty acid-binding protein	rheumatoid arthritis
P01137	TGF	Malignant tumours
P09228	salivary cystatin S	periodontitis
P22894	Matrix metalloproteinase-8	oral disease, rheumatoid arthritis
P07288	Prostate specific antigen	Prostate cancer
Q01469	Fatty acid-binding protein, epidermal	rheumatoid arthritis
P05109	Calgranulin-A	rheumatoid arthritis
P31151	psoriasin	Lung involvement in systemic sclerosis
P16562	Cysteine-rich secretory protein 2	Sjögren's syndrome
P01033	TIMP-1	cancer cardiovascular diseases diabetes
P10645	Chromogranin-A	sleep bruxism
P01266	Thyroglobulin	Papillary and follicular thyroid cancer
Q12794	Hyaluronidase	HNSCC
P05231	Interleukin-6	Sjögren's syndrome
P01034	Cystatin-C	Sjögren's syndrome
O43820	Hyaluronidase-3	Sjogren's syndrome
P02144	Myoglobin	acute myocardial infarction
P02771	AFP	Hepatocellular carcinomas
P29622	Kallikrein	Sjögren's syndrome
P31947	14-3-3 protein	rheumatoid arthritis
Q12891	Hyaluronidase-2	Sjögren's syndrome
P07339	Cathepsin D	breast cancer
Q04917	14-3-3 protein eta	rheumatoid arthritis
P21217	Leb antigen	pancreatic cancer
P43080	GCAP 1	pleomorphic adenoma warthin tumors
P61981	14-3-3 protein gamma	rheumatoid arthritis
P63104	14-3-3 protein zeta/delta	rheumatoid arthritis
P51587	BRCA-2	Breast cancer
P04637	Cellular tumor antigen p53	HNSCC
Q2M3T9	Hyaluronidase-4	HNSCC
P05305	Endothelin-1	oral lichen planus or oral cancer

Table 4. Comparison of the ranking result with biomarkers for many sorts of diseases

Total protein number	Known biomarker number	Top number	biomarkers included in top number	$p-value$
20186	58	500	9	1.03997E-05
20186	58	1000	17	1.45463E-09
20186	58	1500	21	3.42156E-10
20186	58	2000	30	1.01306E-15
20186	58	2500	31	4.94602E-14
20186	58	3000	35	2.00691E-15
20186	58	3500	37	3.78696E-15
20186	58	4000	37	2.86256E-13
20186	58	4500	42	5.27231E-16
20186	58	5000	43	3.25018E-15
20186	58	5500	44	1.52709E-14
20186	58	6000	46	6.80921E-15
20186	58	8000	49	2.13251E-12
20186	58	10000	53	9.62474E-12

Using these features, an SVM classifier was established to predict proteins that were likely to be secreted into the saliva. And we identified 5811 saliva-secreted proteins from all human proteins. Many of these proteins are already used to salivary diagnosis as biomarkers for diseases. Thus indirectly proves the feasibility of salivary diagnosis for systemic diseases because the biomarkers such as breast cancer, pancreatic cancer could be secreted into saliva. This work may provide a useful tool to identify the salivary biomarkers for various human diseases and promote the development of salivary diagnosis.

Acknowledgments. We would like to thank Prof. Juan Cui (UNL) for the assistance in data collection. This work was support by NSFC (61175023, 61202309), the Ph.D. Program Foundation of MOE of China (20120061120106), the China Postdoctoral Science Foundation (2012M520678), and the Science-Technology Development Project of Jilin Province of China (20130522111JH, 20130522114JH, 20140101180JC).

References

1. Bateman, A., Coin, L., Durbin, R., Finn, R.D., Hollich, V., Griffiths-Jones, S., Khanna, A., Marshall, M., Moxon, S., Sonnhammer, E.L., et al.: The pfam protein families database. Nucleic Acids Research 32(suppl. 1), D138–D141 (2004)
2. Bermejo-Pareja, F., Antequera, D., Vargas, T., Molina, J.A., Carro, E.: Saliva levels of abeta1-42 as potential biomarker of Alzheimer's disease: a pilot study. BMC Neurology 10(1), 108 (2010)
3. Boyle, J.O., Mao, L., Brennan, J.A., Koch, W.M., Eisele, D.W., Saunders, J.R., Sidransky, D.: Gene mutations in saliva as molecular markers for head and neck squamous cell carcinomas. The American Journal of Surgery 168(5), 429–432 (1994)

4. Chen, Y., Zhang, Y., Yin, Y., Gao, G., Li, S., Jiang, Y., Gu, X., Luo, J.: Spda web-based secreted protein database. Nucleic Acids Research 33(suppl. 1), D169–D173 (2005)
5. Cui, J., Han, L., Lin, H., Tang, Z., Ji, Z., Cao, Z., Li, Y., Chen, Y.: Advances in exploration of machine learning methods for predicting functional class and interaction profiles of proteins and peptides irrespective of sequence homology. Current Bioinformatics 2(2), 95–112 (2007)
6. Cui, J., Liu, Q., Puett, D., Xu, Y.: Computational prediction of human proteins that can be secreted into the bloodstream. Bioinformatics 24(20), 2370–2375 (2008)
7. Denny, P., Hagen, F.K., Hardt, M., Liao, L., Yan, W., Arellanno, M., Bassilian, S., Bedi, G.S., Boontheung, P., Cociorva, D., et al.: The proteomes of human parotid and submandibular/sublingual gland salivas collected as the ductal secretions. Journal of Proteome Research 7(5), 1994–2006 (2008)
8. Dimmer, E.C., Huntley, R.P., Alam-Faruque, Y., Sawford, T., O'Donovan, C., Martin, M.J., Bely, B., Browne, P., Chan, W.M., Eberhardt, R., et al.: The uniprot-go annotation database in 2011. Nucleic Acids Research 40(D1), D565–D570 (2012)
9. Fine, D.H., Markowitz, K., Furgang, D., Fairlie, K., Ferrandiz, J., Nasri, C., McKiernan, M., Donnelly, R., Gunsolley, J.: Macrophage inflammatory protein-1α: a salivary biomarker of bone loss in a longitudinal cohort study of children at risk for aggressive periodontal disease? Journal of Periodontology 80(1), 106–113 (2009)
10. Giusti, L., Baldini, C., Bazzichi, L., Ciregia, F., Tonazzini, I., Mascia, G., Giannaccini, G., Bombardieri, S., Lucacchini, A.: Proteome analysis of whole saliva: a new tool for rheumatic diseases-the example of sjögren's syndrome. Proteomics 7(10), 1634–1643 (2007)
11. Guyon, I., Weston, J., Barnhill, S., Vapnik, V.: Gene selection for cancer classification using support vector machines. Machine Learning 46(1-3), 389–422 (2002)
12. Hong, C.S., Cui, J., Ni, Z., Su, Y., Puett, D., Li, F., Xu, Y.: A computational method for prediction of excretory proteins and application to identification of gastric cancer markers in urine. PloS One 6(2), e16875 (2011)
13. Hu, S., Loo, J.A., Wong, D.T.: Human saliva proteome analysis and disease biomarker discovery. Expert Review of Proteomics 4(4), 531–538 (2007)
14. Kaufman, E., Lamster, I.B.: The diagnostic applications of salivaa review. Critical Reviews in Oral Biology & Medicine 13(2), 197–212 (2002)
15. Li, S.J., Peng, M., Li, H., Liu, B.S., Wang, C., Wu, J.R., Li, Y.X., Zeng, R.: Sys-bodyfluid: a systematical database for human body fluid proteome research. Nucleic Acids Research 37(suppl. 1), D907–D912 (2009)
16. Mirrielees, J., Crofford, L.J., Lin, Y., Kryscio, R.J., Dawson III, D.R., Ebersole, J.L., Miller, C.S.: Rheumatoid arthritis and salivary biomarkers of periodontal disease. Journal of Clinical Periodontology 37(12), 1068–1074 (2010)
17. Pfaffe, T., Cooper-White, J., Beyerlein, P., Kostner, K., Punyadeera, C.: Diagnostic potential of saliva: current state and future applications. Clinical Chemistry 57(5), 675–687 (2011)
18. Rao, P.V., Reddy, A.P., Lu, X., Dasari, S., Krishnaprasad, A., Biggs, E., Roberts Jr., C.T., Nagalla, S.R.: Proteomic identification of salivary biomarkers of type-2 diabetes. Journal of Proteome Research 8(1), 239–245 (2009)
19. Sas, R., Dawes, C.: The intra-oral distribution of unstimulated and chewing-gum-stimulated parotid saliva. Archives of Oral Biology 42(7), 469–474 (1997)
20. Shiiki, N., Tokuyama, S., Sato, C., Kondo, Y., Saruta, J., Mori, Y., Shiiki, K., Miyoshi, Y., Tsukinoki, K.: Association between saliva psa and serum psa in conditions with prostate adenocarcinoma. Biomarkers 16(6), 498–503 (2011)

21. Shintani, S., Hamakawa, H., Ueyama, Y., Hatori, M., Toyoshima, T.: Identification of a truncated cystatin sa-i as a saliva biomarker for oral squamous cell carcinoma using the seldi proteinchip platform. International Journal of Oral and Maxillofacial Surgery 39(1), 68–74 (2010)
22. Sprenger, J., Fink, J.L., Karunaratne, S., Hanson, K., Hamilton, N.A., Teasdale, R.D.: Locate: a mammalian protein subcellular localization database. Nucleic Acids Research 36(suppl. 1), D230–D233 (2008)
23. Streckfus, C.F., Mayorga-Wark, O., Arreola, D., Edwards, C., Bigler, L., Dubinsky, W.P.: Breast cancer related proteins are present in saliva and are modulated secondary to ductal carcinoma in situ of the breast. Cancer Investigation 26(2), 159–167 (2008)
24. Sun, Q.F., Sun, Q.H., Du, J., Wang, S.: Differential gene expression profiles of normal human parotid and submandibular glands. Oral Diseases 14(6), 500–509 (2008)
25. Tempero, M.A., Uchida, E., Takasaki, H., Burnett, D.A., Steplewski, Z., Pour, P.M.: Relationship of carbohydrate antigen 19-9 and lewis antigens in pancreatic cancer. Cancer Research 47(20), 5501–5503 (1987)
26. Wang, J., Liang, Y., Wang, Y., Cui, J., Liu, M., Du, W., Xu, Y.: Computational prediction of human salivary proteins from blood circulation and application to diagnostic biomarker identification. PloS One 8(11), e80211 (2013)
27. Wong, D.T.: Salivary diagnostics for oral cancer. Journal of the California Dental Association 34(4), 303–308 (2006)
28. Wong, D.T.: Salivary diagnostics powered by nanotechnologies, proteomics and genomics. The Journal of the American Dental Association 137(3), 313–321 (2006)

A Parallel Scheme for Three-Dimensional Reconstruction in Large-Field Electron Tomography

Jingrong Zhang[1,2], Xiaohua Wan[1], Fa Zhang[1], Fei Ren[1],
Xuan Wang[3], and Zhiyong Liu[1]

[1] Key Lab. of Intelligent Information Processing
and Advanced Computing Research Lab., Institute of Computing Technology,
Chinese Academy of Sciences, Beijing, China
[2] University of Chinese Academy of Sciences, Beijing, China
[3] Yanshan University
zhangjingrong1990@gmail.com, {wanxiaohua,zhangfa,renfei,zyliu}@ict.ac.cn,
wangxuan@ysu.edu.cn

Abstract. Large-field high-resolution electron tomography enables visualizing detailed mechanisms under global structure. As field enlarges, the processing time increases and the distortions in reconstruction become more critical. Adopting a nonlinear projection model instead of a linear one can compensate for curvilinear trajectories, nonlinear electron optics and sample warping. But the processing time for the reconstruction with nonlinear projection model is rather considerable. In this work, we propose a new parallel strategy for block iterative reconstruction algorithms. We also adopt a page-based data transfer in this strategy so as to dramatically reduce the processing time for data transfer and communication. We have tested this parallel strategy and it can yield speedups of approximate 40 times according to our experimental results.

Keywords: Electron tomography, Three-dimensional reconstruction, Iterative methods, Nonlinear projection model, TxBR.

1 Introduction

In electron tomography (ET), the specimen is tilted within a limited range $[-60°, 60°]$ or $[-70°, 70°]$ in small increment of $1 - 2°$ or so. During the process of taking projection images, electron beams impinge upon the specimen and penetrate it. ET can reconstruct a specimen's three-dimension (3D) internal structure from these projections. Now ET plays a crucial role in studying macromolecular assemblies. Especially, large-field high-resolution ET allows visualizing and understanding global structure such as organelles, membranes and microfiber networks extending throughout the cell and into intercellular spaces [1]. The development of hardware and techniques has made large-field high-resolution ET possible [2,3].

M. Basu, Y. Pan, and J. Wang (Eds.): ISBRA 2014, LNBI 8492, pp. 102–113, 2014.

Now the size of projection image has reached to 8192*8192 or larger. Then the size of final reconstruction volume will reach several GBytes [4]. In large-field ET, there are still problems to acquire high-quality reconstruction results. Projection images in ET are extremely noisy because of low signal-to-noise ratio (SNR). Furthermore, projection images are not complete at limited angle, which can lead to artifact during reconstruction. Different from traditional back-projection reconstruction algorithms, iterative methods have good performance in handling incomplete and noisy data. Many iterative methods, e.g. SIRT [5], BICAV [6] and ASART [7] have been adopted in three-dimensional (3D) reconstruction of ET. As image sizes increase, distortions in reconstructions become more pronounced because electron trajectories are helical under the influence of magnetic fields. To decrease these distortions, a global nonlinear projection model is proposed and has been already used in TxBR [8] and iterative methods [2].

However, the curvilinear projection model increases the complexity of calculation and extends the processing time. Meanwhile, iterative methods for 3D reconstruction in large-field ET are time-consuming compared with back-projection algorithms. Here to cope with the computational problem, parallel processing has been applied. Parallel strategies on clusters [2, 3, 8] have been widely used in 3D reconstruction of ET with curvilinear projection model to reduce turnaround times. These parallel strategies consider each projection map as a set of nonlinear transforms on z-sections which sum to produce the image, and the sections along Z-axis are reconstructed separately on different processors.

Graphic Processing Units (GPUs) have been widely used to accelerate scientific applications. Unlike clusters, these desktop supercomputers can obtain significant speed-ups on relatively inexpensive hardware with impressive performance-per-watt. The parallelization on GPUs is difficult for large-field data owing to the limited storage of memory in GPUs. For example, the global memory of GTX 480 graphic card is 1.5GB. However, the raw projection data and the final 3D reconstruction are approaching 50GB and 200GB respectively, if each image size is 8K*8K. TxBR adopts a parallel scheme to calculate 3D reconstruction using curvilinear projection models and backprojection algorithms on GPUs [3]. In this parallel scheme, all the volume is divided into several slabs including several Z-sections along Z-axis and each slabs are reconstructed on GPUs sequentially. But for iterative reconstruction methods, we usually need many iterations to achieve good reconstructions. Using the previous parallel schemes, we have to repeatly transfer all the sections into the memory of GPUs for each iterative step, which is rather time-consuming.

In this work, we describe a variant block-iterative version of SIRT method and implement its parallelization on GPUs. Our contribution includes two aspects. First, we propose a new Block-iterative SIRT parallel algorithm (BSIRT) with curvilinear projection model. We analyze the locality of curvilinear trajectory and then divide the data vertically according to the locality. Secondly, we adopt a page-based data transfer scheme in order to reduce the time for data transfer.

The paper is laid out as follows. Section 2 overviews iterative reconstruction algorithms, curvilinear model and previous GPU parallel strategies. Section 3 describes variant block-iterative version of SIRT methods (BSIRT) with curvilinear model. Then the implementation details of page-based data transfer scheme will be introduced in Section 4. Finally, we will show and evaluate our experimental results in Section 5.

2   Related Work

As our work focus on the iterative methods and its parallelization on GPU with curvilinear projection model. In this section, we first overview the iterative algorithms and introduce the curvilinear projection model. Then, we will review the previous parallel strategies.

2.1   Iterative Reconstruction Methods and Curvilinear Projection Model

In ET, the reconstruction problem is to obtain the internal structure of specimen by projection series. Working in real-space, the iterative methods solve the 3D reconstruction problem by formulating it as a large system of linear equations. Assuming the voxel as basis function to represent the volume, we present the result by the value of N ($N = n_{\text{width}} * n_{\text{length}} * n_{\text{height}}$) voxels. We suppose that the total number of projection pixels is M ($M = n_{\text{pro\_width}} * n_{\text{pro\_length}} * n_{\text{ang}}$), the projection procedure can be simply represented as follows:

$$p_i = \sum_{j=1}^{N} A_{ij} s_j \qquad 1 \leq i \leq M \qquad (2.1)$$

where p_i is the value of ith projection pixel, s_j is the value of jth voxel and A_{ij} in matrix A indicates the contribution of the voxel j to the projection i. We can calculate matrix A according to projection model. We use (x, y, g) to indicate the projection point, where g is the index of orientation and $i = g * n_{\text{pro\_width}} * n_{\text{pro\_length}} + y * n_{\text{pro\_width}} + x$. We can use iterative algorithms to calculate the value of voxels $S = \{s_1, s_2 \ldots s_N\}$.

Iterative methods can be generally classified into sequential, block-iterative and simultaneous methods [9]. In essence, the sequential and simultaneous algorithms are special cases of block iterative reconstruction [10]. Suppose all the equations of linear system may be subdivided into B blocks each of size T, we use a generalized version of iterative methods to describe the iterative step:

$$s_j^{k+1} = s_j^k + \lambda_k \sum_{i \in BLOCK_b} \frac{A_{ij}}{\sum_{v=1}^{N} w_v^b A_{iv}^2} (p_i - \sum_{w=1}^{N} A_{iw} s_w^k) \qquad 1 \leq j \leq N \quad (2.2)$$

where $b = k \bmod B$ is the index of block, i is the index of equation of system and w_v^b is the weighting factor [11]. The relaxation parameter λ_k is critical for convergence speed, usually it is found by training or experimenting [6, 12].

For the sequential iterative algorithms ($B = M, T = 1$), the equations are considered one-by-one in a circular manner. If the block number $B = 1$, the algorithm turns into simultaneous iterative algorithm. Simultaneous Iterative Reconstruction Technique (SIRT) [5] is typical simultaneous iterative algorithm. Block iterative algorithms update estimations by a subfamily of constraints in each iterative step. The main iteration can proceed sequentially from block to block and within each block in parallel. Traditional block iterative methods adopt a view-by-view strategy, the size of each block is $T = n_{\text{pro\_width}} * n_{\text{pro\_length}}$ and the block number is $B = n_{\text{ang}}$.

The projection step is formulated as Eq. (2.1). As mentioned, the value of matric A is decided by projection model, which describes the correspondence between projection points and voxels in object. Traditional straight-line projection model formulates projection map as a linear function. Here we adopt a quadratic curvilinear projection map. This model has been used in [2,8]. All the coefficients a_i^θ and b_i^θ are calculated by means of bundle adjustment, the details of the procedure is in [8]. The general quadratic expression is:

$$x = a_0^\theta + a_1^\theta X + a_2^\theta Y + a_3^\theta Z + a_4^\theta X^2 + a_5^\theta XY + a_6^\theta XZ + a_7^\theta Y^2 + a_8^\theta YZ + a_9^\theta Z^2$$
$$y = b_0^\theta + b_1^\theta X + b_2^\theta Y + b_3^\theta Z + b_4^\theta X^2 + b_5^\theta XY + b_6^\theta XZ + b_7^\theta Y^2 + b_8^\theta YZ + b_9^\theta Z^2$$
$$(2.3)$$

2.2 Previous Parallel Strategy on GPU

As for linear projection model, it assumes electron beams travel in straight line, so their trajectories are certainly parallel with each other and the slice perpendicular to tilting axis (in this paper we suppose the tilting axis is X) always projects in a straight line at each angle as shown in Fig. 1(a). Then 3D reconstruction problem can be decomposed into a set of independent 2D reconstruction problems [4, 13–15]. Using the reconstruction methods mentioned above, 2D slice reconstruction can be computed from a set of 1D projections (so-called sinogram [4]). But this strategy is not adaptable for curvilinear projection models because the curvilinear trajectories are not in parallel.

We need to divide data and reconstruct each part sequentially because of limited global memory on GPU and large data. TxBR [8] has proposed a GPU parallel strategy for direct reconstruction with curvilinear projection model. By regarding each projection map as a set of nonlinear transform on z-sections which sum to produce the image, it divide the object along the z-axis (see Fig. 1(b)) and reconstruct each z-section separately. Iterative algorithm ASART has implemented the parallelization on clusters [2], using similar strategy. When we use iterative algorithm, we need all the volume in one iterative step. Because of the limited storage of memory in GPUs, we need to transfer all the sections into the memory of GPUs for each iterative step, which is very time-consuming.

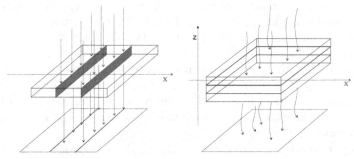

(a) Parallel strategy for straight- (b) Parallel strategy for curvilin-
line projection model   ear projection model

Fig. 1. Parallel strategy for 3D reconstruction in electron tomography

3   Block-Iterative SIRT Algorithm with Curvilinear Model

Since previous algorithms are not suitable for iterative reconstruction with curvilinear model on GPUs, we need to find a new parallel algorithm. As GPU's global memory is limited for large-field reconstruction, we can consider a scheme for data partition. We analyze the locality of curvilinear trajectory to divide data. Then by modifying traditional SIRT algorithm and using a new data-divided scheme, we can effectively reduce the time for data transfer.

3.1   Locality of Curvilinear Trajectory

As discussed above, the projection of a slice is a straight line at each view if we adopt a straight-line model. However, with a curvilinear projection model, we can get different curves as the projections of a slice at each view. Conversely, for a straight line as a projection, its corresponding voxels are formed into a curved surface (see Fig. 2), which varies with the orientation. The curved surface covers a few slices. To a certain extent, it still owns locality. We can consider to analyze the locality of the curved surface and then calculate the precise scope of X coordinates of the curved surface.

Suppose that x is constant (i.e. $x = c$, c is constant) in Eq. (2.3), we can get the expression of the curved surface mentioned above:

$$c = a_0^\theta + a_1^\theta X + a_2^\theta Y + a_3^\theta Z + a_4^\theta X^2 + a_5^\theta XY + a_6^\theta XZ + a_7^\theta Y^2 + a_8^\theta YZ + a_9^\theta Z^2 \quad (3.1)$$

For these voxels on this curved surface, the x coordinate of their projection points is c. Here we can get the range of X by calculating the global maximum and minimum of curved surface within a certain domain ($D = \{(X, Y, Z) | X \in [x_{min}, x_{max}], Y \in [y_{min}, y_{max}], Z \in [z_{min}, z_{max}]\}$). $x_{min}, x_{max} \ldots z_{max}$ are constant.

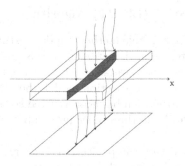

Fig. 2. Curvilinear projection model

Eq. (3.1) is an implicit function. X is dependent variable, Y and Z are independent variables. The global maximum and minimum should be a local maximum and minimum in the interior of the domain or a point on the boundary of the domain. So we can calculate these values and then choose the minimum and maximum among these values as the scope of X.

Theorem 1. *For a continuously differentiable function of several real variables, a point P is critical if all of the partial derivatives of the function are 0 at P.*

According the Theorem 1, we can use the implicit differentiation to get the partial derivatives and get the critical points in the interior of the domain.

$$\begin{cases} c = a_0^\theta + a_1^\theta X + a_2^\theta Y + a_3^\theta Z + a_4^\theta X^2 + a_5^\theta XY + a_6^\theta XZ + a_7^\theta Y^2 + a_8^\theta YZ + a_9^\theta Z^2 \\ \frac{\partial X}{\partial Z} = 0 \\ \frac{\partial X}{\partial Y} = 0 \end{cases}$$

(3.2)

Firstly, we calculate the critical points of intersecting lines. The curved surface may intersect with plane $Z = z_{min}$, $Z = z_{max}$, $X = x_{min}$, $X = x_{max}$, $Y = y_{min}$ and $Y = y_{max}$. Secondly, we consider the terminal point of the intersecting using the similar method. For example, the critical point of intersecting line (with plane $Z = z_{max}$) is calculated like:

$$\begin{cases} c = a_0^\theta + a_1^\theta X + a_2^\theta Y + a_3^\theta Z + a_4^\theta X^2 + a_5^\theta XY + a_6^\theta XZ + a_7^\theta Y^2 + a_8^\theta YZ + a_9^\theta Z^2 \\ Z = z_{max} \\ \frac{\partial X}{\partial Y} = 0 \end{cases}$$

(3.3)

According to this procedure, we can get the scope of X. Suppose the result is $\{MAX_c^\theta, MIN_c^\theta\}$. Absolutely, this result is different for different views. Next we extend the range of projection points to $\{(x, y, \theta) | x \in [c_1^\theta, c_2^\theta]\}$, c_1^θ and c_2^θ is constant, we can get the corresponding X coordinates of voxels by:

$$\begin{aligned} X_{min} = min\{MIN_i^\theta, c_1^\theta \le i \le c_2^\theta\} \\ X_{max} = max\{MAX_i^\theta, c_1^\theta \le i \le c_2^\theta\} \end{aligned}$$

(3.4)

Finally, if we consider a scope of orientations instead of one orientation, we can also get the scope of X for different views.

3.2   Block-Iterative SIRT (BSIRT) Algorithm

Different from SIRT, block iterative methods update estimations by a subfamily of constraints in each step. The iteration proceeds from block to block and usually the members of a block are selected sequentially (like SART, iterate view by view). In a straight-line projection model, the 3D object can be divided into several slices and each slice can be reconstructed separately. In a curvilinear projection model, we can divide the object into slabs according to the locality of curvilinear trajectory.

First of all, we equally divide the projection series along the tilting axis into B parts. Each part can be formulated as mathematical set S_t.

$$S_t = \{(x, y, g)|x \in (c_t, c_{t+1}], y \in [1, n_{\text{pro\_length}}], g \in [1, n_{\text{ang}}]\}$$
$$c_t = \frac{n_{\text{pro\_width}}}{B} * t \qquad t = 0, 1, 2 \ldots B - 1 \tag{3.5}$$

We can calculate the scope of X in each corresponding 3D sub-volume D_t for each S_t according to the scheme described in Section 3.1. So we can divide 3D volume into B parts. The equations for S_t is:

$$p_i = \sum_{j \in D_t} A_{ij} x_j \qquad i \in S_t \tag{3.6}$$

In each iterative step, we only need the data in sub-volume D_t instead of all the volume. As shown in Fig. 3, in the first step, we reconstruct the first sub-volume D_1 from the first part S_1 of projection images. This strategy only need to transfer data one time for one update step. The size of each sub-volume is decided by the range of projection points and the number of orientations.

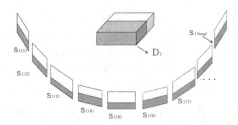

Fig. 3. Block-iterative SIRT algorithm

4   Page-Based Data Transfer Scheme

We also notice another problem: there is overlap between adjacent sub-volumes. If each sub-volume D_t is reconstructed and then transferred in each update step, the overlapping data will be transferred repeatedly. We propose a page-based data transfer scheme to eliminate the redundant data transfer. In this scheme, the overlapping data will not be transferred from the memory in the iterative steps until it is not reconstructed. Every time, we transfer useless data out of GPU and new data in GPU.

In case that the size of incoming data is bigger than the outgoing, we divide the global memory into several pages. Since the size of sub-volume D_t is different, we should calculate the size of outgoing and incoming data according to the distribution of sub-volumes. According to this, we can define the biggest number as page size.

An example is listed in Table 1. The coordinate scope of reconstruction object is $\{(X, Y, Z)|X \in [-87, 563], Y \in [-31, 652], Z \in [-62, 38]\}$. The tilting axis is X and the size of projection is $512 * 512$. We divide the projection series into 8 groups ($B = 8$). As we partition the object vertically according to the value of X, Table 1 shows that the size of X for each slab. For the first slab, after the first iterative step, we need to move out the part whose scope of X is from -78 to -12. The size of the transferred data is $(-12 - (-78) + 1) * 651 * 101 = 67 * 651 * 101$. We use 67 to indicate the size of data moved out in column "out" of Table 1. We calculated the sizes of data moved out and in for each slab and found that these values are almost the same. So we can choose the maximum ($69 * 651 * 101$) as the size of page.

Table 1. The distribution of slabs

	start position of X	end position of X	overlap range of X	out	in	total size
1st slab	-78	193	205	67	0	272
2nd slab	-11	261	206	67	68	273
3rd slab	56	329	205	69	68	274
4th slab	125	398	206	68	69	274
5th slab	193	467	207	68	69	275
6th slab	261	536	208	68	69	276
7th slab	329	563	168	67	27	235
8th slab	396	563	0	0	0	168

Next, we repartition the voxels according to the start point of each slab (shown in Table 2, the page 1 to 7). After the start points are considered, the rest of data is partitioned according to the page size (page 8,9,10).

Table 2. The distribution of pages

	start position of X	end position of X	range of X
1st page	-78	-12	67
2nd page	-11	55	67
3rd page	56	124	69
4th page	125	192	68
5th page	193	260	68
6th page	261	328	68
7th page	329	395	67
8th page	396	464	69
9th page	465	533	69
10th page	534	563	30

The whole procedure is described as follows. In the first iterative step, the first slab including several pages (from page 1 to 5) is moved in the memory of GPU. In the following iterative step, we need to move out the previous unnecessary page of data and move in the next page of data. Basically, the size of data in the memory of GPU is a little larger than that of a slab, which could ensure all the data required can be stored in GPU. When the end of data is in GPU, we can stop transfer data out until this round finish.

5   Result

In this section, we outline the experimental results. As we proposed a variant version of SIRT (called BSIRT) and studied its parallel scheme. First of all, we compare the performance of BSIRT and SIRT, show their reconstruction result. Next, we report the timing performance of SIRT and BSIRT.

The benchmark used in this section has been adopted by paper [2] and its partition has been discussed in Section 4. The thickness of sample is 350nm, the micrographs were taken in a 300kV FEI Titan TEM with a 37k magnification. The tilt series are composed of 121 micrographs, taken from $-60°$ to $+60°$. The size of each micrograph is $512 * 512$ and the size of reconstruction result is $651 * 684 * 101$.

All the experiments are carried out on machine running the Ubuntu 12.04 operating system 64-bit and the GPU card we use is NVIDIA GTX480, which owns 480 SPs and 1536M global memory.

5.1   Reconstruction Result

All the experiments are performed using quadratic projection model. Here we have considered three conditions: the first one updates the data using traditional SIRT, the second one updates the data using BSIRT parallel algorithm with 4 slabs, the third one uses BSIRT with 8 slabs. Note that BSIRT with one slab is identical to SIRT, which has been testified by experiment. In our experiment, a combination of two factors, relaxation parameter and number of iterations, need to be considered. We define the term "one iteration" as one whole sweep through all equations of the system. The relaxation parameter λ keeps constant throughout the iterations. Since the convergence of SIRT is guaranteed [16] as long as $0 < \lambda < 2$. Here we set the relaxation parameter $\lambda = 0.1$.

In our study, we applied projection error to compare the quality of the reconstructed images, which is based on the discrepancy between the experimental images and the images calculated by reprojection. Specifically, we use the mean absolute error (MAE) to calculate the projection error:

$$MAE = \frac{1}{M} \sum_{i=1}^{M} |p_i - p'_i| \tag{5.1}$$

where p_i is experimental projection images and p'_i is the calculated projection images. The line plots given in Fig.4 show these measures versus the number

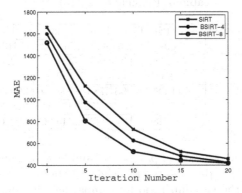

Fig. 4. Projection error

of iterations when the relaxation parameter $\lambda = 0.1$. We can see the projection error of BSIRT (8 slabs) is smaller than that of BSIRT (4 slabs) in the same iterations and the projection error of BSIRT with 4 (and 8) slabs is smaller than SIRT, especially in the first 10 iterations. It proves that BSIRT enjoys a faster convergence speed.

Fig. 5 shows the images produced after 5 iterations by SIRT and BSIRT (8 slabs). The result of BSIRT is similar to SIRT. As our algorithm is intrinsically a block-iterative algorithm, with a proper relaxation ratio, our strategy will acquire good result.

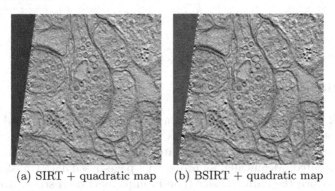

(a) SIRT + quadratic map   (b) BSIRT + quadratic map

Fig. 5. Reconstruction result comparison

5.2   Performance of Parallelization

We have timed the BSIRT code on GTX 480 using CUDA 5.0. BSIRT is test by dividing the object into 1 (SIRT), 4 and 8 slabs. We use two-dimensional blocks to accomplish parallelization. The blocks in the same row are in charge of one page. The running time is shown in Table 3. We can see it is basically in proportion to iteration number. The running time of BSIRT with 4 slabs is generally less than BSIRT with 8 slabs and its growth rate is smaller. The reason

Table 3. Running time (sec)

Iteration number	1 slab (SIRT)	4 slabs	8 slabs	CPU-only
1	24.022	31.641	57.572	1313.78
10	236.291	312.907	569.932	13224.44
20	472.321	625.321	1139.072	26396.36
30	711.686	937.785	1708.491	39653.17
40	947.617	1250.752	2277.419	52843.66
50	1183.486	1563.276	2846.961	66020.13

is less slabs means greater degree of parallelism and less communication cost. So if the hardware condition permits, the less slabs the better.

Here we calculate the speedup ratio by comparing with the CPU-only program, which is shown in Fig. 6. From this figure, we can see the speedup ratio is about 23 and 41 for BSIRT (8 slabs) and BSIRT (4 slabs).

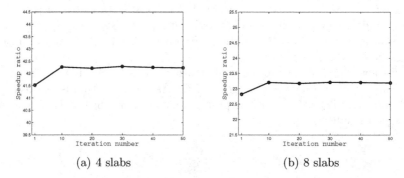

(a) 4 slabs   (b) 8 slabs

Fig. 6. Speedup Ratio

6   Conclusion

In this work, we develop a variant version of SIRT (called BSIRT) and complete its parallel scheme on GPUs. With this method, 3D reconstruction for large-field electron tomography can be calculated on GPUs even though the memory of GPUs is limited. Meanwhile, we proposed a page-based data transfer scheme, which can reduce the data transfer time in implementation stage. Now our algorithm is based on SIRT, but this parallel strategy can be applied to all kinds of block-iterative algorithms. For the future work, we are considering that this parallel scheme can be extended to other parallel platforms and the influence of relaxation parameter is still to be pending to be discussed as well.

Acknowledgments. This work is supported by grants National Natural Science Foundation for China (61232001, 61202210, 61103139 and 60921002). The authors would like to thank Albert Lawrence for his instruction on the mathematical and computational problems. The authors must also thank Sébastien Phan for providing the experimental dataset.

References

1. Phan, S., Lawrence, A., Molina, T., Lanman, J., Berlanga, M., Terada, M., Kulungowski, A., Obayashi, J., Ellisman, M.: Txbr montage reconstruction for large field electron tomography. Journal of Structural Biology 180, 154–164 (2012)
2. Wan, X., Phan, S., Lawrence, A., Zhang, F., Han, R., Liu, Z., Ellisman, M.: Iterative methods in large field electron microscope tomography. SIAM Journal on Scientific Computing 35(5), S402–S419 (2013)
3. Lawrence, A., Phan, S., Singh, R.: Parallel processing and large-field electron microscope tomography. In: World Congress on Computer Science and Information Engineering, pp. 339–343 (2009)
4. Agulleiro, J.I., Fernández, J.J.: Evaluation of a multicore-optimized implementation for tomographic reconstruction. PloS One 7(11), e48261 (2012)
5. Gilbert, P.: Iterative methods for the three-dimensional reconstruction of an object from projections. Journal of Theoretical Biology 36(1), 105–117 (1972)
6. Censor, Y., Gordon, D., Gordon, R.: Bicav: A block-iterative parallel algorithm for sparse systems with pixel-related weighting. IEEE Transactions on Medical Imaging 20(10), 1050–1060 (2001)
7. Wan, X., Zhang, F., Chu, Q., Zhang, K., Sun, F., Yuan, B., Liu, Z.: Three-dimensional reconstruction using an adaptive simultaneous algebraic reconstruction technique in electron tomography. Journal of Structural Biology 175(3), 277–287 (2011)
8. Lawrence, A., Bouwer, J.C., Perkins, G., Ellisman, M.H.: Transform-based backprojection for volume reconstruction of large format electron microscope tilt series. Journal of Structural Biology 154(2), 144–167 (2006)
9. Censor, Y.: Parallel optimization: theory, algorithms and applications. Oxford University Press (1997)
10. Aharoni, R., Censor, Y.: Block-iterative projection methods for parallel computation of solutions to convex feasibility problems. Linear Algebra and Its Applications 120, 165–175 (1989)
11. Bilbao-Castro, J.R., Carazo, J.M., Fernandez, J.J., García, I.: Parallelization and comparison of 3d iterative reconstruction algorithms. In: Proceedings of the 12th Euromicro Conference on Parallel, Distributed and Network Based Processing, pp. 96–102. IEEE (2004)
12. Herman, G.T., Meyer, L.B.: Algebraic reconstruction techniques can be made computationally efficient. IEEE Transactions on Medical Imaging 12(3), 600–609 (1993)
13. Wan, X., Zhang, F., Chu, Q., Liu, Z.: High-performance blob-based iterative reconstruction of electron tomography on multi-GPUs. In: Chen, J., Wang, J., Zelikovsky, A. (eds.) ISBRA 2011. LNCS, vol. 6674, pp. 61–72. Springer, Heidelberg (2011)
14. Fernandez, J.: High performance computing in structural determination by electron cryomicroscopy. Journal of Structural Biology 164(1), 1–6 (2008)
15. Castaño Díez, D., Mueller, H., Frangakis, A.S.: Implementation and performance evaluation of reconstruction algorithms on graphics processors. Journal of Structural Biology 157(1), 288–295 (2007)
16. Van der Sluis, A., Van der Vorst, H.: Sirt-and cg-type methods for the iterative solution of sparse linear least-squares problems. Linear Algebra and its Applications 130, 257–303 (1990)

An Improved Correlation Method
Based on Rotation Invariant Feature
for Automatic Particle Selection

Yu Chen[1,2], Fei Ren[1], Xiaohua Wan[1], Xuan Wang[3], and Fa Zhang[1]

[1] Key Lab. of Intelligent Information Processing
and Advanced Computing Research Lab., Institute of Computing Technology,
Chinese Academy of Sciences, Beijing, China
[2] University of Chinese Academy of Sciences Beijing, China
[3] Yanshan University, China
{chenyu,renfei,wanxiaohua,zhangfa}@ict.ac.cn,
wangxuan@ysu.edu.cn

Abstract. Particle selection from cryo-electron microscopy (cryo-EM) images is very important for high-resolution reconstruction of macromolecular structure. However, the accuracy of existing selection methods are normally restricted to noise and low contrast of cryo-EM images. In this paper, we presented an improved correlation method based on rotation invariant features for automatic, fast particle selection. We first selected a preliminary particle set applying rotation invariant features, then filtered the preliminary particle set using correlation to reduce the interference of high noise background and improve the precision of correlation method. We used Divide and Conquer technique and cascade strategy to improve the recognition ability of features and reduce processing time. Experimental results on the benchmark of cryo-EM images show that our method can improve the accuracy of particle selection significantly.

Keywords: Particle selection, Rotation invariant feature, Correlation, Divide and Conquer, Cascade strategy.

1 Introduction

Single particle cryo-electron microscopy (cryo-EM) has been widely applied to the study of macromolecular three dimensional (3D) reconstruction [1]. In cryo-EM, each biological sample must be prepared in a relatively homogeneous form, and then this sample is rapidly frozen as a thin film, transferred to the electron microscope and imaged, final, sufficiently sampled angular view of the 2D projections need to be aligned, combined to retrieve the samples 3D structure. However, to minimize radiation damage, micrographs showing projections of particles must be recorded at very low election dose, resulting in a high level of noise (the typical single-to-noise rate (SNR) is < 1) and very low contrast [2,3]. To obtain atomic level reconstructed resolution, hundreds of thousands of particles

M. Basu, Y. Pan, and J. Wang (Eds.): ISBRA 2014, LNBI 8492, pp. 114–125, 2014.

may be necessary, which makes it impractical to manually pick the particles. In addition, particle detection by visual observation may be inaccurate and fairly subjective.

Several methods have been developed for automatic or semi-automatic particle detection [4]. Those algorithms,can be roughly grouped into two classes,template matching approach and feature-based approach.Feature-based approach usually relies on recognizing local or global salient features of particle images, such as statistical features [5,6], geometric features [7,8], cross-correlation features [9] or discriminative shape-related features [10].However, the main weakness of these methods is that it may be difficult to extract distinctive features from very low-contrast images. Template matching is a basic technique used in many signal processing and image analysis applications for detection and localization of patterns in signal corrupted by noise. In particle detection, this type of approach is based on cross-correlation using some templates (references) which are generated from either a 3D reference structure or the average of a few particles picked manually. It detects position of particle by comparing correlation coefficients of templates and target images.

Although higher accuracy results have been achieved by template-matching methods, comparing to feature-based approaches [11,12] , two key issues remain to be resolved. The first is how to improve precision by reducing the interference of noisy background when comparing correlation coefficients. As mentioned above, cryo-EM images usually have extremely low SNR, which will reduce the difference of correlation coefficients with templates between true particles and false particles(background or other type particles).The second is how to deal with the random orientation of particles more quickly. The orientation of particles is randomly in micrographs. To discriminate the orientation, templates must be rotated to generate a template set in different direction, and correlation with a template set is time-consuming especially when the number of target images is large.

To overcome these problems, we propose an improved correlation method based on rotation invariant feature to implement automatic and fast selection of particles. Rotation invariance of image means, for arbitrary rotation, function parameters may be changed, but the function value remained constant, and rotation invariant feature is a kind of feature that is shared by the same type particle in any orientation. Compared to conventional correlation-based methods, our method has several advantages.First,a preliminary particle set is selected using rotation invariant features, which contains about 98% target particles.Second, we apply correlation to filter the preliminary particle set instead of a whole cryo-EM image. Since the preliminary particle set contains less false particles compared to a whole image, filtering by correlation can further reduce the interference of noisy background and improve the precision of correlation method.Third,Divide and Conquer technique is applied in the extraction of rotation invariant feature to improve the recognition ability of features.Also a cascade strategy is used into the preliminary set generation to reduce processing time.

The remainder of the paper is organized as follows. In section 2, we introduce the framework of our method. First, we introduce the usage of rotation invariant features to generate a preliminary particle set and how the Divide and Conquer and cascade techniques are used to improve the processing. Then we introduce how to filter the preliminary set with correlation method. In section 3, we present our experimental results and analysis. Section 4 is focused on potential improvements.

2   Method

Our method consists of three steps(See Fig.1). The first step is to extract rotation invariant features to generate a strong classifier (discriminating true particles and false particles) by training. In this step, in order to improve the discrimination of classifiers, Divide and Conquer technique is used. The second step is to scan over the whole image with a particle size window and generate a preliminary particle set with classifiers. In this step, cascade strategy is used to speed up the recognition process. The third step is to filter the preliminary particle set with correlation function. In this step, mask is applied to every template and candidate particle before correlation to reduce the interference of background noise.

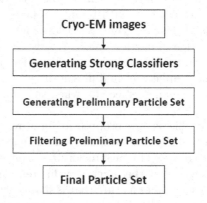

Fig. 1. Flow diagram of the improved correlation method

2.1   Rotation Invariant Features

Because of the random orientation of particles, it is difficult to generate some shared features in traditional way. However, if the center of a particle is determined, even if the orientation of the particle may be random, all the pixels of the particle are limited into a series of circles. And no matter how the orientation of the particle changes, every pixel just have a different position in the same circle. According to this, we generate some new features (see Fig.2(a)), which are invariant under rotation and well reflect the distribution of image gray, as follows. Let $sum(r_0, r_1)$ is the sum of each pixel whose radius is more than r_0 but less than r_1, and SUM is the sum of all pixels of a particle size image. A feature is defined as Eq.(2.1).

$$f(r_0, r_1) = \frac{sum(r_0, r_1)}{SUM}$$ (2.1)

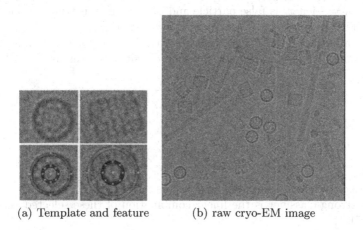

(a) Template and feature   (b) raw cryo-EM image

Fig. 2. (a) templates and features of end and side-view (left and right,respectively); (b) raw cryo-EM image

2.2   Classifier Generation

After the first step discussed above, we generates a number of features. There is more than one reason to reduce the number of features to a sufficient minimum. Computational complexity is the obvious one. A related reason is that although two features may carry good classification information when treated separately, there is little gain if they are combined together in a feature vector. In this section, we describe a feature selection method which can select a set of the most important features from this huge feature space.

The Weak Classifier. For each feature f in feature space, we learn the training dataset and generate a classifier as Eq.(2.2).

$$C(f) = \begin{cases} 1 & , \quad \text{if} \quad (leftThr \le f \le rightThr) \\ 0 & , \quad \text{else} \end{cases}$$ (2.2)

Where $C(f)$ is the weak classifier for feature f. And $(leftThr, rightThr)$ is set to make $C(f)$ discriminate true particles and false particles correctly in most cases.

The Strong Classifier. We learn a set of training images and select only a few of optimal features for some weak classifiers. These selected features should have minimum mistakes to discriminate true particles and false particles in the training dataset. And we use a feature selection method to generate the strong classifier.

The feature selection method is described below:

1. Let $(x_1, y_1)...(x_m, y_m)$ where $x_i \in X = \{training \quad dataset\}$, $y_i \in Y = \{-1, +1\}$, $y_i = -1$ if x_i is false particle and $y_i = +1$ if x_i is true particle.
2. Initialize $D(i) = \frac{1}{m}, i = 1, ..., m$ as the weight of x_i.
3. Normalize as Eq.(2.3). And then for each feature $f_n, n = 1, .., N$, train a weak classifier $C(f_n)$. And the error of each classifier is given as Eq.(2.4).

$$D(i) = D(i)/\sum_{i=1}^{m} D(i) \tag{2.3}$$

$$e_n = \sum_{i=1}^{m} D(i)|C(f_n) - y_i| \tag{2.4}$$

4. Select T weak classifiers of lowest errors.
5. The strong classifier is the linear combination of these selected classifiers, as Eq.(2.5).

$$F(I) = \begin{cases} 1 & , & \sum_{t=1}^{T} a_t * C_t(I) \geq \tau * a_t \\ 0 & , & \text{else} \end{cases} \tag{2.5}$$

$$a_t = log(1 - e_t)/log(e_t) \tag{2.6}$$

Where a_t is the weight; τ is the threshold; I is a particle size image; top T features are used.

Divide and Conquer. The weight of each weak classifier is Eq.(2.6) in the strong classifier, but it might be not the optimal weight for each weak classifier. Thus Divide and Conquer technique is employed to optimize their weights and improve the recognition precision.

Firstly, we break the training dataset into several parts, and then, generate a strong classifier in each part respectively. Finally, we combine all strong classifiers into an integrated classifier with different weights as Eq.(2.7).The combination gives a larger weight to a weak classifier shared by two or more strong classifiers.

$$IntF(I) = \sum_{d=1}^{D} W_d * F_d(I), W_d = \frac{S_d}{S} \tag{2.7}$$

Where S_d is the image number of training dataset in dth part, W_d is the weight of the strong classifier F_d, S is the total image number of training dataset.

The Divide and Conquer technique is evaluated according to the performance of features which are trained with a training dataset containing 210 true particles and 210 false particles under different segmentation strategies. In the early stage,the recognition precision improves as the number of segments increases. However, too much division reduces training data in each divided part leading to an obvious deterioration.In this paper, we employ the features trained with a three-segment training dataset, of which each divided part contains at least 70 true particles and 70 false particles.

2.3   Preliminary Particle Set Generation

Threshold. As mentioned above, we need a threshold to control the performance of classifiers, which will influence the quality of the preliminary set. Generally, we evaluate a particle set with two attributes: the false positive rate(FPR)(See Eq.2.8) and the false negative rate(FNR)(See Eq.2.9).

If a proper threshold is used, one can get a preliminary particle set with acceptable FPR/FNR. Normally, we can obtain a proper threshold of one image by analysing the generated particle set, and then apply it to the whole image set captured under similar conditions. However, difference between proper thresholds of different images always exits, which is small but affects the accuracy, leading to a tradeoff between accuracy and labour.

If a low threshold is employed, one can get a preliminary set with about 98% target particles ($\sim$ 2% FNR) but also with some false particles (high FPR), and then correlation can be used to filter the preliminary set to achieve an improved result.

$$FPR = \frac{the\ number\ of\ false\ selected\ particles}{total\ number\ of\ selected\ particles} \qquad (2.8)$$

$$FNR = \frac{the\ number\ of\ missed\ true\ particles}{total\ number\ of\ true\ particles} \qquad (2.9)$$

Cascade Strategy. As explained in the previous section, the strong classifier is a linear combination of several weak classifiers. Because most locations do not contain true particles, it is time-consuming to reject those locations by considering every weak classifier, and it is possible to reject those locations by only a few weak classifiers out of the selected ones [6]. Hence, we can employ a cascade of classifiers to speed up recognition. A cascade of classifiers groups all the selected weak classifiers in one strong classifier into several stages as shown in Fig.3. Each stage contains a few classifiers out of the selected weak classifiers. Most locations in the micrograph contain false particles are rejected at the early stage. The reason why the cascade of classifiers can speed up the detection process is that it removes the evaluation of other weak classifiers.

In this paper, we use three stages with 5, 10 and the other weak classifiers (about 20) respectively. The cryo-EM image is resized as $1024 * 1024$, and the processing time can be brought down to around 42 s from roughly 1254 s on a Intel(R) 2.33 GHz processor with threshold = 0.7.

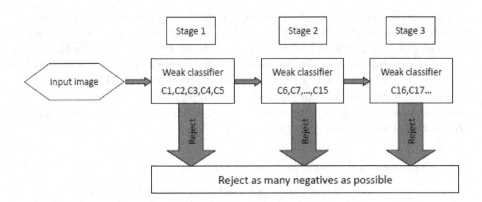

Fig. 3. A three stage cascade of classifiers

2.4   Filter the Preliminary Particle Set with Correlation

Template. Since dealing with a preliminary particle set instead of a whole cryo-EM image, we can simply use raw particle images as templates (see Fig.2(a)) and no image preprocessing is needed. However, because of the low SNR of cryo-EM images, some other views of particles, which should be considered as false particles, will also be selected into the preliminary particle set to become a disturbance especially when a low threshold is used. To deal with that, we should select some abtemplates those represent interfering views of particles.

Because of the random orientation of particles, all templates should be rotated, 5 degree per rotation, to generate a template set.

Correlation and Classify Candidate Particles. First, every template or candidate particle is masked with Eq.(2.10) to reduce the interference of background noise ,and then a fast Fourier based implementation of the correlation function is used to get correlation coefficients of candidate particles and templates/abtemplates.

$$I(x,y) = \begin{cases} I(x,y) & , \quad \sqrt{x^2 + y^2} \leq R \\ 0 & , \quad \text{else} \end{cases} \tag{2.10}$$

Where I is a template or candidate particle, R is the radius of a particle.

According to the correlation coefficients of one candidate particle with all templates and abtemplates, we assign the candidate particle to the class represented by the template/abtemplate of highest correlation coefficient, which can provide orientation information for further research. And then we mark the score of this candidate particle as Eq.(2.11).

$$S(p_i) = \begin{cases} \max\{|coef(p_i, t_j)|\} & , \quad \text{assigned to template} \\ -1 * \max\{|coef(p_i, abt_j)|\} & , \quad \text{assigned to abtemplate} \end{cases} \tag{2.11}$$

Where $coef(p_i, t_j)$ is the correlation coefficient of the ith candidate particle and the jth template, and $coef(p_i, abt_j)$ is the correlation coefficient of the particle and the jth abtemplate.

After scoring all candidate particles, we can remove the candidate particles with low scores, which are usually false particles.

3   Results and Discussion

3.1   Test Dataset

Our method is tested with the keyhole limpet hemocyanin (KLH) dataset (available from the AMI group, The Scripps Research Institute, CA USA, http://ami.scripps.edu/ptrl_data/). KLH is a cylindrically shaped $\sim 8MD$ particle, a homo-oligomeric didecamer with D5 point group symmetry [13]. KLH particles are preferentially oriented into end-view and side-view (see Fig.2(a)). And only the selection of side-view is discussed in this paper.

82 digital micrographs of keyhole limpet hemocyanin (KLH) (see Fig.2(b)) are used. These images were acquired in a Phillips CM200 transmission electron microscope at a magnification of 66,000x and a voltage of 120kV. And they were recorded by a Tietz CCD camera of size $2048 * 2048$.

3.2   Preliminary Particle Set Generation with different Thresholds

As mentioned in section 2, threshold influences the quality of the preliminary particle set. Next, we discuss the results of two representative cases.

Preliminary Set Generation with Proper Thresholds. Fig.4(a) shows the preliminary particle set of one image with a proper threshold. Although most true particles are selected with a proper threshold, there are some shortcomings of rotation invariant features should be improved. The selection of particle 2, 3 shows rotation invariant features are not sensitive to details of particles and easy to select some unqualified particles, increasing FPR; The selection of particle 4 and the miss of particle 5, 6 shows some noisy background area may satisfy rotation invariant features better than true particles.

For 82 images, with a proper threshold, rotation invariant features generate a preliminary particle set of 25.3% FPR and 22.2% FNR.

Preliminary Set Generation with Low Thresholds. Fig.4(b) shows the preliminary particle set of one image with a low threshold. All true particles are selected when some false particles are accepted wrongly.

For 82 images, with a low threshold, rotation invariant features generate a preliminary particle set of 69.4% FPR and 2.1% FNR; $\sim 98\%$ of true particles are picked out.

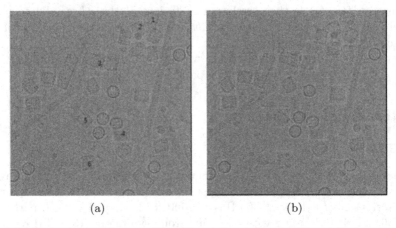

(a)   (b)

Fig. 4. Preliminary particle sets with different thresholds. (a) The preliminary particle set with a proper threshold. Threshold = 0.79, 14 candidate particles are selected containing 11 true particles. Particle 1 is rejected because it is too close to image edge; particle 2 has a small amount of contamination but still recognized as a true particle; particle 3 is part of a long particle but still accepted; particle 4 is noisy background but still satisfy high threshold; particle 5, 6 are true particles but missed. (b) The preliminary particle set with a low threshold. Threshold = 0.5, 23 candidate particles are selected containing 14 true particles.

3.3 Improved Correlation Method Based on Rotation Invariant Features

Fig.5(a) shows the final particle set of one image after filtering the preliminary set generated with a low threshold(see Fig.4(b)). There is only one false particle (particle 1) is left. Fig.5(b) shows the classification of 160 rectangular particles selected from 15 images.

For 82 images, our method generates a final particle set of 12.89% FPR and 7.5% FNR.

Table.1 shows the comparison of our method and some other methods. The FPR of our method is 12.89% and higher than Sorzano's algorithm(9.3%), Volkman's algorithm(12.2%) and Malick's algorithm(11.7%), but compared to these three algorithms, our method has a lower FNR. The FNR of our method is 7.5% and only higher than Yu's algorithm(7.3%), but our method has a lower FPR. In summary, our method improves the accuracy significantly and generates one of the best results.

4   Potential Improvements

4.1 Accuracy Improvement

With correlation function, we filter a preliminary particle set to remove false particles and cut down the FPR (69.4% down to 12.89%). However, the FNR increases to 7.5%, which means we remove some true particles by error. Although

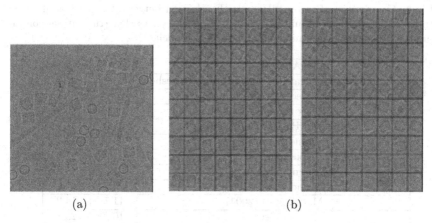

Fig. 5. Results of improved correlation method. (a) : The final particle set of one image after filtering the preliminary set generated with a low threshold. Particle 1 is part of a long particle but left wrongly. Some edge-detection methods may be used to solve such problems (b) : The classification of 160 rectangular particles selected from 15 images.

$\sim 10\%$ FNR is acceptable and will not have a big impact on the accuracy of 3D reconstruction if we process more images to get enough particles, extra calculation is needed and slows down the method.

We can improve the FNR from two aspects. On one side, filtering methods can be applied to improve the SNR of cryo-EM images [14], as the quality of images improved, rotation invariant features will be improved to generate a better preliminary set, and better candidate particle images also means better performance of correlation function. On the other side, we can use traditional methods (like averaging templates to improve SNR) to generate better templates instead of raw particle images to reduce the increase of FNR.

However, both optimization methods mentioned above come with time consumption. It is important to balance speed and accuracy, especially in practical application.

4.2   Speed Up

As the requirement for precision increases, extra processing, like filtering, becomes necessary. Thus we should try to speed up the improved correlation method from other aspect.

After further evaluation, one can easily find out good parallel characteristics of our method. The method applies exactly the same process mode to every target image of particle size when generating the preliminary set, and for every candidate particle in the preliminary set, the same thing happens. Thus, it is possible to process each target image or candidate particle parallel on GPU, which will speed up the method in an amazing way.

Table 1. Comparison of rectangular particle selection between our method and other methods. It shows the FNR and FPR of detecting side views. The true dataset contains about 1042 particles picked manually by Fabrice Mouche.

Algorithm	FPR(%)	FNR(%)
Rotation invariant features with a proper threshold	25.3	22.2
Our improved correlation method	12.89	7.5
C.O.S. Sorzano et al.(2009)	9.3	30.9
Hall and Patwardhan (2004)	22.0	27.4
Ludtke(1999)	23.7	17.7
Volkman(2004)	12.2	27.4
Penczek(2004)	38.8	48.8
Yu and Bajaj (2004)	24.7	7.3
Malick et al.(2004)	11.7	14.2
Bern	16.2	23.8

5 Conclusion

In this paper, we propose an improved correlation method based on rotation invariant feature. Our focus is to solve two key issues of conventional correlation-based particle selection algorithms and improve the accuracy by introducing new rotation invariant features. The experimental results demonstrate that our method improves the accuracy of the particle selection significantly and the acceleration strategy used in the method is effective. We also discuss the potential improvements and good parallel characteristics of our method.

Our method is developed into software called Picker. Picker is freely available at our team's homepage http://ear.ict.ac.cn .

Acknowledgments. This work is supported by grants National Natural Science Foundation of China (61232001, 61202210, 61103139 and 60921002).The authors would like to thank Yangguang Shi,Fan Xu, Renmin Han,Jingrong Zhang for their helps about this work.

References

1. Joachim, F.: Three-dimensional electron microscopy of macromolecular assemblies. Academic Press (1996)
2. Henderson, R.: The potential and limitations of neutrons, electrons and x-rays for atomic resolution microscopy of unstained biological molecules. Quarterly Reviews of Biophysics 28(02), 171–193 (1995)
3. Sali, A., Glaeser, R., Earnest, T., Baumeister, W.: From words to literature in structural proteomics. Nature 422(6928), 216–225 (2003)
4. Zhu, Y., Carragher, B., Glaeser, R.M., Fellmann, D., Bajaj, C., Bern, M., Mouche, F., de Haas, F., Hall, R.J., Kriegman, D.J., et al.: Automatic particle selection: results of a comparative study. Journal of Structural Biology 145(1), 3–14 (2004)
5. Hall, R.J., Patwardhan, A.: A two step approach for semi-automated particle selection from low contrast cryo-electron micrographs. Journal of Structural Biology 145(1), 19–28 (2004)

6. Mallick, S.P., Zhu, Y., Kriegman, D.: Detecting particles in cryo-em micrographs using learned features. Journal of Structural Biology 145(1), 52–62 (2004)
7. Zhu, Y., Carragher, B., Mouche, F., Potter, C.S.: Automatic particle detection through efficient hough transforms. IEEE Transactions on Medical Imaging 22(9), 1053–1062 (2003)
8. Yu, Z., Bajaj, C.: Detecting circular and rectangular particles based on geometric feature detection in electron micrographs. Journal of Structural Biology 145(1), 168–180 (2004)
9. Sorzano, C., Recarte, E., Alcorlo, M., Bilbao-Castro, J., San-Martín, C., Marabini, R., Carazo, J.: Automatic particle selection from electron micrographs using machine learning techniques. Journal of Structural Biology 167(3), 252–260 (2009)
10. Abrishami, V., Zaldívar-Peraza, A., de la Rosa-Trevín, J., Vargas, J., Otón, J., Marabini, R., Shkolnisky, Y., Carazo, J., Sorzano, C.: A pattern matching approach to the automatic selection of particles from low-contrast electron micrographs. Bioinformatics 29(19), 2460–2468 (2013)
11. Roseman, A.: Findema fast, efficient program for automatic selection of particles from electron micrographs. Journal of Structural Biology 145(1), 91–99 (2004)
12. Sigworth, F.J.: Classical detection theory and the cryo-em particle selection problem. Journal of Structural Biology 145(1), 111–122 (2004)
13. Orlova, E.V., Dube, P., Harris, J.R., Beckman, E., Zemlin, F., Markl, J., van Heel, M.: Structure of keyhole limpet hemocyanin type 1 (klh1) at 15 å resolution by electron cryomicroscopy and angular reconstitution. Journal of Molecular Biology 271(3), 417–437 (1997)
14. Kumar, V., Heikkonen, J., Engelhardt, P., Kaski, K.: Robust filtering and particle picking in micrograph images towards 3d reconstruction of purified proteins with cryo-electron microscopy. Journal of Structural Biology 145(1), 41–51 (2004)

An Effective Algorithm for Peptide *de novo* Sequencing from Mixture MS/MS Spectra

Yi Liu[1], Bin Ma[2], Kaizhong Zhang[1], and Gilles Lajoie[3]

[1] Department of Computer Science, The University of Western Ontario, London,
Ontario, Canada, N6A 5B7
`{yliu766,kzhang}@csd.uwo.ca`

[2] David R. Cheriton School of Computer Science, University of Waterloo, Waterloo,
Ontario, Canada N2L 3G1
`binma@uwaterloo.ca`

[3] Department of Biochemistry, The University of Western Ontario, London, Ontario,
Canada, N6A 5B8
`glajoie@uwo.ca`

Abstract. In the past decade, extensive research has been conducted for
the computational analysis of mass spectrometry based proteomics data.
Yet, there are still remaining challenges, among which, one particular
challenge is that the identification rate of the MS/MS spectra collected
is rather low. One significant reason that contributes to this situation is
the concurrent fragmentation of multiple precursors in a single MS/MS
spectrum. Nearly all the mainstream computational methods take the
assumption that the acquired spectra come from a single precursor, thus
they are not suitable for the identification of mixture spectra. In this
research, we formulated the mixture spectra *de novo* sequencing problem
mathematically, and proposed a dynamic programming algorithm for the
problem. Experiment shows that our proposed algorithm can serve as a
complimentary method for the identification of mixture spectra.

Keywords: Mass Spectrometry, Mixture Spectra, Peptide *de novo* Sequencing, Computational Proteomics.

1 Introduction

Over the past decade, mass spectrometry has gradually become a standard technique for the identification and quantification of large biomolecules, including peptides and proteins[1]. In a typical LC-MS/MS experiment, protein sample is digested into peptides, and the resulting peptide mixtures are ionized first, and then the charged peptides are fragmented and measured by mass spectrometers in a high-throughput manner[2]. The large amount of data collected in an MS experiment requires highly effective computational approaches to automate the process of spectra interpretation. Generally, the mainstream computational methods fall into two categories: database search and *de novo* sequencing. Nowadays, lots of algorithms and software packages have been designed for both categories. In a database search method[3,4], the identification of MS/MS spectra is

M. Basu, Y. Pan, and J. Wang (Eds.): ISBRA 2014, LNBI 8492, pp. 126–137, 2014.

assisted with a protein sequence database, and the primary task is to correctly correlate MS/MS spectra with amino acid sequences in the protein database. While in a *de novo* method[5,6], the computation of peptide sequence doesn't rely on the protein database, the algorithm directly constructs the peptide sequence that best matches the spectra.

Although much effort has been made to develop new computational approaches to analyse mass spectrometry data, there are still several unsatisfactory areas remaining challenging. One specific challenge is that in a high throughput MS/MS experiment, usually only a fraction of the acquired spectra can be confidently interpreted by the existent computational methods. Many factors may contribute to this situation which include: low precursor intensity, poor fragmentation of selected precursor, existence of modified residues, and particularly the concurrent fragmentation of co-eluting peptides. Some preliminary research confirmed that peptides with similar mass and chromatographic properties being sequenced together can happen quite frequently and result in mixture spectra containing ion fragments from multiple precursors[7,8]. In addition, some innovative experimental modes also rendered the increasing necessity for peptide identification using mixture spectra, for instance the data-independent acquisition(DIA) strategy[9].

Classical computational methods often assume that each MS/MS spectrum is generated from a single precursor, therefore they are not suitable for effective identification of mixture spectra. There have been several attempts to address such an issue. Zhang *et al.* introduced a database search engine, ProbIDtree to identify co-eluting peptides from mixture spectra[10]. This method works in an iterative process of database searching in which ions assigned to a tentative peptide are subtracted from the acquired spectrum, and the remaining spectrum is used to detect another matched peptide. The software then organized the tentative matched peptides in a tree structure and calculated an adjusted probability score to determine the correct identifications. Wang *et al.* proposed M-SPLIT, an MS/MS spectral library search software and demonstrated its potential to identify peptides from mixture spectra by matching the acquired spectra with previously identified spectra[11]. This method is limited in application due to the fact that it can only be used to identify peptide that has been observed before. Recently, Wang *et al.* provided a new database search tool, MixDB, which enables to interpret mixture spectra by using a specifically designed scoring function for matching mixture spectra with a pair of peptide sequences filtered from the protein database[12].

Before this research, no algorithm has been reported for peptide *de novo* sequencing from mixture spectra. Such study is necessary because the *de novo* sequencing can be regarded as a complimentary method for spectra interpretation when the targeted peptide is not included in the database. Moreover, researchers have showed an increasing interest in combining *de novo* sequencing with database search to efficiently interpret MS/MS spectra lately[13,14]. In this paper, we formulate the problem of mixture spectra *de novo* sequencing mathematically, and propose a dynamic algorithm to solve the problem.

2   Mathematical Model of the Problem

Assume that a mixture spectrum $\mathcal{M}$ is generated by co-fragmentation of peptides P_1 and P_2, and $\mathcal{M}$ is represented by a peak list $\mathcal{M} = \{(x_i, h_i)|i = 1, 2, ..., n\}$. Assume that $\mathcal{M}$ is preprocessed and only contains ions of charge one, consequently the mass to charge ratio of an ion is equal to its mass value. And we use two molecular weight MW_1 and MW_2 to denote the precursor mass values of the two peptides which satisfy $|MW_1 - MW_2| \leq \Delta$. Δ is a small value predefined by the width of the mass spectrometer selection window.[1] In addition, we use Σ to denote the alphabet of 20 different types of amino acids. For an amino acid $a \in \Sigma$, we use $\|a\|$ to symbolize the mass of the amino acid residue. Here we will have $\max_{a \in \Sigma} \|a\| = 186.08$ and $\min_{a \in \Sigma} \|a\| = 57.02$. Intuitively, a peak in $\mathcal{M}$ whose position matches the mass of a fragment ion of peptide P is a positive evidence that $\mathcal{M}$ resulted from P. Therefore for a mixture spectrum, the more and higher peaks are matched to ions of P_1 and P_2, the more likely that P_1 and P_2 are correct. In this section we will formulate this intuition and model the mixture spectra *de novo* sequencing problem.

2.1   Mass Representation of the Ion Fragments

Let $P = a_1 a_2 ... a_k$ be the string of amino acids, we define the residue mass of the peptide as $\|P\| = \Sigma_{1 \leq j \leq k} \|a_i\|$ and the actual mass of the peptide as $\|P\| + \|H_2O\|$. Denote b_i and y_i to be the mass of the b-ion and y-ion of P with i amino acids respectively, therefore we have $b_i + y_{k-i} = \|P\| + 20$.[2]

Let $\Pi = \{y, b, a, c, x, z, y^*, y^o, b^*, b^o\}$ be all the ion types that we consider throughout the paper. Assume x is the mass value of a b-ion, we use $\mathcal{B}(x)$ to denote the set of all the ion masses corresponding to this b-ion, then we have:

$$\mathcal{B}(x) = \{x, x - 17, x - 18, x + 17, x - 28\}. \tag{1}$$

Similarly, for each y-ion with mass x, we will have the following notation,

$$\mathcal{Y}(x) = \{x, x - 18, x - 17, x + 26\}. \tag{2}$$

that represent all the ion masses related to this y-ion.[3]

Theoretically, the spectrum of the co-fragmented peptides $P_1 = a_1 a_2 ... a_n$ and $P_2 = b_1 b_2 ... b_m$ should contain a peak at each of the following mass values:

$$S(P_1, P_2) = \bigcup_{i=1}^{n-1} [\mathcal{B}(b_i^1) \cup \mathcal{Y}(y_i^1))] \bigcup_{j=1}^{m-1} [\mathcal{B}(b_j^2) \cup \mathcal{Y}(y_j^2)] \tag{3}$$

, in which b_i^1 and y_i^1 are ions generated from P_1, b_j^2 and y_j^2 are those from P_2.

---

[1] In the typical data-dependent acquisition mode, the selection window of the mass spectrometer is usually a few Daltons wide.

[2] The summation contains the mass values of all amino acid residues and the mass of a water and the mass of two protons added to the peptide in the ionization process.

[3] Reason that $\mathcal{Y}(x)$ has one fewer element than $\mathcal{B}(x)$ is because y-ion losing an ammonia and its corresponding z-ion are both $\|y\| - 17$.

2.2   Problem Statement

Because that the mass values acquired from mass spectrometers are not accurate, we use $\delta > 0$ to represent the maximum error of the mass spectrometers. Moreover, we denote $\bar{S} = \{(x_i, h_i) \in \mathcal{M} | \exists y \in S,\ \text{s.t.}\ |y - x_i| \leq \delta\}$, in which $\bar{S}$ is the subset of $\mathcal{M}$ containing all the peaks *explained* by the mass values in S.[4]

Let $S(P_1, P_2)$ be all the possible ion masses of P_1 and P_2, $S(P_1, P_2)$ can be computed by formula (3). Then $\overline{S(P_1, P_2)}$ contains all the peaks in $\mathcal{M}$ that can be explained by the ions of P_1 and P_2. Intuitively, the more and higher peaks included in $\overline{S(P_1, P_2)}$ indicates the more likely that $\mathcal{M}$ is generated from P_1 and P_2. Thus, the MIXTURE SPECTRA DE NOVO SEQUENCING problem can be formulated as follows: Given a mixture spectrum $\mathcal{M}$, and two precursor mass values MW_1 and MW_2, $|MW_1 - MW_2| \leq \Delta$ and a predefined error bound δ, we want to construct two peptides P_1 and P_2, such that $|\|P_1\| + 20 - MW_1| \leq \delta$, $|\|P_2\| + 20 - MW_2| \leq \delta$, and the following summation is maximized:

$$H(P_1, P_2) = \sum_{(x,h) \in \overline{S(P_1,P_2)}} h \tag{4}$$

The equation above is the summation of all the intensity values of the included peaks. It's worthy to notice that the scoring function mentioned above is totally replaceable. In practice, we can apply a more sophisticated scoring function which involves more influential factors than just the height of peaks.

3   Algorithm for Mixture Spectra *de novo* Sequencing

Besides the *de novo* methods suffer due to the imperfect data, there will be new complications when dealing with mixture spectra. Firstly, the difficulty to design a scoring function to evaluate the similarity between the mixture spectra and a pair of constructed peptides simultaneously. The Scoring function in (4) considers the peak intensity of ions from both peptides, providing us an easy way for such a task. Secondly, the difficulty to establish an efficient method to address the overlapping peaks. Overlapping peaks occur more frequently for co-sequenced peptides, which makes the interpretation more complicated and less accurate. Our proposed algorithm below can handle it effectively by gradually constructing prefix-suffix pairs for both peptides in a specially designated pathway.

Assume that $MW_1 = \|P_1\| + 20$ and $MW_2 = \|P_2\| + 20$, and let $A = a_1 a_2 ... a_k$ be a prefix(N-terminus) of peptide $P_1 = a_1 a_2 ... a_n$, the mass of the b-ion produced by a cleavage between a_i and a_{i+1} is represented as $\|a_1 a_2 ... a_i\|_b = \Sigma_{1 \leq x \leq i} \|a_x\| + 1$, thus the mass of the y-ion from the same cleavage site is $MW_1 - \|a_1 a_2 ... a_i\|_b$. Let $B = a_n a_{n-1} ... a_{n-(t-1)}$ be the *reverse* string of a suffix(C-terminus) of peptide P_1, and we denote the mass of the y-ion by a cut between a_{n-i} and $a_{n-(i+1)}$ as $\|a_n a_{n-1} ... a_{n-i}\|_y = \Sigma_{n-i \leq x \leq n} \|a_x\| + 19$, thus the mass of the corresponding

---

[4] The term *explained* means that a theoretical mass value matches with a peak in the acquired spectrum.

b-ion is $MW_1 - \|a_n a_{n-1}...a_{n-i}\|_y$. Let $C = b_1 b_2...b_u$ and $D = b_m b_{m-1}...b_{m-(v-1)}$ be the prefix and *reverse* suffix of peptide $P_2 = b_1 b_2...b_m$ respectively.

We use $S_N(A)$ to denote the values of all the ions caused from a cleavage in A, then we have

$$S_N(A) = \bigcup_{i=1}^{k} [\mathcal{B}(\|a_1 a_2...a_i\|_b) \cup \mathcal{Y}(MW_1 - \|a_1 a_2...a_i\|_b)]$$

Similarly, we use $S_C(B)$ to denote all the mass values induced by a cut in the suffix string B. Therefore we have the following equation:

$$S_C(B) = \bigcup_{i=1}^{t} [\mathcal{Y}(\|a_n a_{n-1}...a_{n-(i-1)}\|_y) \cup \mathcal{B}(MW_1 - \|a_n a_{n-1}...a_{n-(i-1)}\|_y)]$$

Accordingly, we have the following two equations for C and D of peptide P_2:

$$S_N(C) = \bigcup_{i=1}^{u} [\mathcal{B}(\|b_1 b_2...b_i\|_b) \cup \mathcal{Y}(MW_2 - \|b_1 b_2...b_i\|_b)]$$

and

$$S_C(D) = \bigcup_{i=1}^{v} [\mathcal{Y}(\|b_m b_{m-1}...b_{m-(i-1)}\|_y) \cup \mathcal{B}(MW_2 - \|b_m b_{m-1}...b_{m-(i-1)}\|_y)]$$

According to the four equations above, if it satisifes that $P_1 = A\alpha \overleftarrow{B}$, and $P_2 = C\beta \overleftarrow{D}$, $\alpha \in \Sigma, \beta \in \Sigma$, here $\overleftarrow{B}$ and $\overleftarrow{D}$ are the *reverse* strings of B and D. Then the following equation is easy to obtain:

$$S(P_1, P_2) = S_N(A) \cup S_C(B) \cup S_N(C) \cup S_C(D) \tag{5}$$

This formula indicates that the MIXTURE SPECTRA DE NOVO SEQUENCING can be achieved by constructing appropriate prefixes and suffixes of P_1 and P_2. In the following, we will describe a method that constructs the appropriate prefix-suffix pairs for both P_1 and P_2.

Given an amino acid string $s = s_1 s_2..s_n$, we call $\|s\| = \Sigma_1^n \|s_i\|$ the weight of s, and we use $s_> = s_1 s_2...s_{n-1}$ to denote the string with the the nth amino acid removed. We assume $MW = \min\{MW_1, MW_2\}$ in this paper. For four strings A, B, C, D, we additionally define three useful sets $\mathcal{Q} = \{A, B, C, D\}$, $\mathcal{R} = \{A, B\}$, and $\mathcal{T} = \{C, D\}$. For simplicity, we denote $\|A\| = \|A\|_b$, $\|B\| = \|B\|_y$, $\|C\| = \|C\|_b$ and $\|D\| = \|D\|_y$ in the following section.

Definition 1. *The string quadruple $\mathcal{Q} = \{A, B, C, D\}$ is called a Rigid Quartet if it satisfies that: $\|A\| + \|B\| \leq MW_1$ and $\|C\| + \|D\| \leq MW_2$, and it also satisfies that for any pair of strings $\mathcal{E}, \mathcal{F} \in Q$, either of the following inequations holds: $\|\mathcal{F}_>\| < \|\mathcal{E}\| \leq \|\mathcal{F}\|$ or $\|\mathcal{E}_>\| \leq \|\mathcal{F}\| < \|\mathcal{E}\|$, and it also satisfies that for any string $\mathcal{G} \in Q$, the following inequation holds: $\|\mathcal{G}\| < MW - \|\mathcal{G}_>\| - 54$.*

Definition 2. *The string quadruple* $Q = \{A, B, C, D\}$ *is called a General Quartet if it satisfies that:* $\|A\| + \|B\| \leq MW_1$ *and* $\|C\| + \|D\| \leq MW_2$, *and for the string pair* $\mathcal{R} = \{A, B\}$, *either of the following inequations holds:* $\|B_>\| < \|A\| \leq \|B\|$ *or* $\|A_>\| \leq \|B\| < \|A\|$, *and for the string pair* $\mathcal{T} = \{C, D\}$, *either of the following inequations holds:* $\|D_>\| < \|C\| \leq \|D\|$ *or* $\|C_>\| \leq \|D\| < \|C\|$, *and it also satisfies that for any string* $\mathcal{G} \in Q$, *the following inequation holds:* $\|\mathcal{G}\| < MW - \|\mathcal{G}_>\| - 54$.

Corollary 1. *If* $Q = \{A, B, C, D\}$ *is a Rigid Quartet, then it is also a General Quartet.*

Proof. It is easy to get this conclusion from Definition 1 and Definition 2.□

Lemma 1. *Let* $Q = \{A, B, C, D\}$ *be a Rigid Quartet and let* $a \in \Sigma$ *be an amino acid. If A has the smallest weight among the quadruple and* $\|A\| + \|a\| + \|B\| \leq MW_1$ *and* $\|A\| + \|a\| < MW - \|A\| - 54$, *then the quadruple* $\{Aa, B, C, D\}$ *is also a Rigid Quartet.*

Proof. We have the following inequations hold: $\|B_>\| \leq \|A\| \leq \|B\|$, $\|C_>\| \leq \|A\| \leq \|C\|$, and $\|D_>\| \leq \|A\| \leq \|D\|$.

We know that $\|Aa\| > \|A\|$, so if $\|Aa\| \leq \|B\|$, then we have $\|B_>\| < \|Aa\| \leq \|B\|$. If $\|Aa\| > \|B\|$, then we have $\|A\| \leq \|B\| < \|Aa\|$. So we know that string pair $\{Aa, B\}$ always satisfies the restrictions in Definition 1.

Accordingly, we can easily prove that string pairs $\{Aa, C\}$ and $\{Aa, D\}$ also satisfies the restrictions in Definition 1, and we already have $\|A\| + \|a\| + \|B\| \leq MW_1$. Together with the precondition $\|A\| + \|a\| < MW - \|A\| - 54$, we will have the conclusion that $Q(Aa, B, C, D)$ is a Rigid Quartet.□

It also works for the case that B, C or D is the smallest. Lemma 1 tells that for a given Rigid Quartet, if we extend the smallest weighted string with one amino acid, the new quadruple we get is also a Rigid Quartet.

Lemma 2. *Let* $Q(A, B, C, D)$ *be a Rigid Quartet, and* $Q(Aa, B, C, D)$, $a \in \Sigma$ *also be a Rigid Quartet, then A has the smallest weight among the quadruple.*

Proof. If $\|A\|$ is not the smallest weight, without losing generosity, we assume $\|A\| > \|B\|$. Because $\{A, B, C, D\}$ is a Rigid Quartet, then the following inequation holds: $\|B\| < \|A\| < \|Aa\|$. it means for the string pair $\{Aa, B\}$, neither of the inequations $\|B_>\| < \|A\| \leq \|B\|$ and $\|A_>\| \leq \|B\| < \|A\|$ holds, then $Q(Aa, B, C, D)$ is not a Rigid Quartet. This conflicts with the condition.

Similarly, we can prove that if $\|A\| > \|C\|$ or if $\|A\| > \|D\|$, $Q(Aa, B, C, D)$ will not satisfy the constraints of Rigid Quartet.□

Lemma 3. *Let* $Q(A, B, C, D)$ *be a General Quartet, and letters* $a \in \Sigma$, $b \in \Sigma$, *let* $\|A\|$ *be the smaller weighted one in* $\mathcal{R}$, *and* $\|C\|$ *be the smaller weighted one in* $\mathcal{T}$. *If* $\|A\| + \|a\| + \|B\| \leq MW_1$ *and* $\|C\| + \|b\| + \|D\| \leq MW_2$ *then both* $Q(Aa, B, C, D)$ *and* $Q(A, B, Cb, D)$ *are General Quartet.*

Proof. The proof is very similar to those in Lemma 1, therefore omitted here.□

Lemma 4. *Let $Q(A, B, C, D)$ and $Q(Aa, B, C, D)$ both be Rigid Quartet, and $|MW_1 - MW_2| \leq \Delta$, then we have:*

$$(1)\ \overline{\mathcal{B}(\|Aa\|) \cap \overline{S_N(A_>) \cup S_C(B_>) \cup S_N(C_>) \cup S_C(D_>)}} = \phi$$

and

$$(2)\ \overline{\mathcal{Y}(MW_1 - \|Aa\|) \cap \overline{S_N(A_>) \cup S_C(B_>) \cup S_N(C_>) \cup S_C(D_>)}} = \phi$$

Proof. From Lemma 2. We know that $\|A\| \leq \|B\|$, $\|A\| \leq \|C\|$, and $\|A\| \leq \|D\|$. Without losing generosity, we assume $MW = MW_1$ and $MW_2 = MW_1 + \Delta$.

Let Z be any prefix of A, and $Z \neq A$. We know that $\mathcal{B}(\|Aa\|)$ is apart from $\mathcal{B}(\|Z\|)$. Indeed we have $[\mathcal{B}(\|Aa\|) - \mathcal{B}(\|Z\|)]_{\min} = 2 \times \min_{a \in \Sigma} \|a\| - (28 + 17) > 69$. This means that $\overline{\mathcal{B}(\|Aa\|) \cap \mathcal{B}(\|Z\|)} = \phi$. We have $\|Aa\| < MW - \|A\| - 54 \leq MW_1 - \|Z\| - (54 + \min \|a\|_{a \in \Sigma})$. this means that $\overline{\mathcal{B}(\|Aa\|) \cap \mathcal{Y}(MW_1 - \|Z\|)} = \phi$. Therefore, we have the following conclusion: $\overline{\mathcal{B}(\|Aa\|) \cap S_N(A_>)} = \phi$.

Similarly, assume Z be any prefix of B, and $Z \neq B$, we have the following inequations: $\|Aa\| \geq \|A\| + \min \|a\|_{a \in \Sigma} \geq \|B_>\| + \min \|a\|_{a \in \Sigma} \geq \|Z\| + \min \|a\|_{a \in \Sigma}$, this means that $\overline{\mathcal{B}(\|Aa\|) \cap \mathcal{Y}(\|Z\|)} = \phi$. We also know that $\|Aa\| < MW - \|A\| - 54 \leq MW_1 - \|B_>\| - 54$, this indicates that $\overline{\mathcal{B}(\|Aa\|) \cap \mathcal{B}(MW_1 - \|Z\|)} = \phi$. Therefore we have the following equation: $\overline{\mathcal{B}(\|Aa\|) \cap S_C(B_>)} = \phi$.

Similarly, assume Z be any prefix of C, and $Z \neq C$, we have the following holds: $\|Aa\| \geq \|C_>\| + \min \|a\|_{a \in \Sigma} \geq \|Z\| + \min \|a\|_{a \in \Sigma}$. This tells that $\overline{\mathcal{B}(\|Aa\|) \cap \mathcal{B}(\|Z\|)} = \phi$. We have $\|Aa\| < MW - \|A\| - 54 \leq MW_2 - \|Z\| - (54 + \Delta)$. This inequation indicates that $\|Aa\|$ is apart from $\|MW_2\| - \|Z\|$, $\overline{\mathcal{B}(\|Aa\|) \cap \mathcal{B}(MW_2 - \|Z\|)} = \phi$. Therefore we have $\overline{\mathcal{B}(\|Aa\|) \cap S_N(C_>)} = \phi$.

Similarly we can prove that $\overline{\mathcal{B}(\|Aa\|) \cap S_C(D_>)} = \phi$. Therefore (1) is proved. Similar arguments can be applied to $\mathcal{Y}(MW_1 - \|Aa\|)$ to prove (2). $\square$

Lemma 5. *Let $\mathcal{Q}(A, B, C, D)$ be a Rigid Quartet, and a letter $a \in \Sigma$, and define function*

$$\Psi(v, w, m, n) = \overline{\mathcal{B}(v) \cup \mathcal{Y}(MW_1 - v) \cup \mathcal{Y}(w) \cup \mathcal{B}(MW_1 - w)}$$
$$\overline{\cup\, \mathcal{B}(m) \cup \mathcal{Y}(MW_2 - m) \cup \mathcal{Y}(n) \cup \mathcal{B}(MW_2 - n)}$$

(1) If $\mathcal{Q}(Aa, B, C, D)$ is a Rigid Quartet, and

$$f_1(u, v, w, m, n) = H(\overline{\mathcal{B}(u) \cup \mathcal{Y}(MW_1 - u)} \backslash \Psi(v, w, m, n)),\ then$$

$$H(\overline{S_N(Aa) \cup S_C(B) \cup S_N(C) \cup S_C(D)}) = f_1(\|Aa\|, \|A\|, \|B\|, \|C\|, \|D\|)$$
$$+ H(\overline{S_N(A) \cup S_C(B) \cup S_N(C) \cup S_C(D)})$$

(2) If $\mathcal{Q}(A, Ba, C, D)$ is a Rigid Quartet, and

$$f_2(u, v, w, m, n) = H(\overline{\mathcal{B}(MW_1 - u) \cup \mathcal{Y}(u)} \backslash \Psi(v, w, m, n)),\ then$$

$$H(\overline{S_N(A) \cup S_C(Ba) \cup S_N(C) \cup S_C(D)}) = f_2(\|Ba\|, \|A\|, \|B\|, \|C\|, \|D\|)$$
$$+ H(\overline{S_N(A) \cup S_C(B) \cup S_N(C) \cup S_C(D)})$$

(3) If $\mathcal{Q}(A, B, Ca, D)$ is a Rigid Quartet, and

$$f_3(u, v, w, m, n) = H(\overline{\mathcal{B}(u) \cup \mathcal{Y}(MW_2 - u)} \backslash \Psi(v, w, m, n)), \text{ then}$$

$$H(\overline{S_N(A) \cup S_C(B) \cup S_N(Ca) \cup S_C(D)}) = f_3(\|Ca\|, \|A\|, \|B\|, \|C\|, \|D\|)$$
$$+ H(\overline{S_N(A) \cup S_C(B) \cup S_N(C) \cup S_C(D)})$$

(4) If $\mathcal{Q}(A, B, C, Da)$ is a Rigid Quartet, and

$$f_4(u, v, w, m, n) = H(\overline{\mathcal{B}(MW_2 - u) \cup \mathcal{Y}(u)} \backslash \Psi(v, w, m, n)), \text{ then}$$

$$H(\overline{S_N(A) \cup S_C(B) \cup S_N(C) \cup S_C(Da)}) = f_4(\|Da\|, \|A\|, \|B\|, \|C\|, \|D\|)$$
$$+ H(\overline{S_N(A) \cup S_C(B) \cup S_N(C) \cup S_C(D)})$$

Proof. Due to limited space, the related proof is omitted. Similar proof for a different problem can be found in [5].$\square$

Lemma 5 indicates that the summation H in Formula-(4) for a newly generated quadruple can be gradually calculated from previously quadruple by adding some new peaks contained in f.

For string pair $\mathcal{R} = (A, B)$, and string pair $\mathcal{T} = (C, D)$, we define three relations between $\mathcal{R}$ and $\mathcal{T}$: *Neighbouring, Crossing,* and *Nesting.* Assuming that $\|A\| \leq \|B\|$, and $\|C\| \leq \|D\|$. If it satisfies that $\|A\| \leq \|B\| \leq \|C\| \leq \|D\|$, then we say the two string pairs $\mathcal{R}$ and $\mathcal{T}$ are *Neighbouring.* If it satisfies that $\|A\| \leq \|C\| \leq \|B\| \leq \|D\|$, then we say the two string pairs $\mathcal{R}$ and $\mathcal{T}$ are *Crossing.* If it satisfies that $\|A\| \leq \|C\| \leq \|D\| \leq \|B\|$, then we say the two string pairs $\mathcal{R}$ and $\mathcal{T}$ are *Nesting.*

Suppose that for a *Rigid Quartet* $\mathcal{Q} = \{A, B, C, D\}$ in which $\|A\|$ is the smallest weighted. When adding letter $a \in \Sigma$, if it satisfies the following conditions: $\|A\| + \|a\| < MW - \|A\| - 54$, but $\|A\| + \|a\| + \|B\| = MW_1$, then the first peptide is already fully constructed. Meanwhile, if it satisfies $\|A\| + \|a\| + \|B\| < MW_1$, but $\|A\| + \|a\| \geq MW - \|A\| - 54$. In this case, we will have the following conclusion: $2 \times \|A\| + \|a\| \geq MW - 54$ and $2 \times \|B\| + \|a\| \geq MW - 54$, together we have $\|A\| + \|B\| + \|a\| \geq MW - 54 \geq MW_1 - \Delta - 54$. It tells that if we add letter a to string A, the remaining weight will not fit in any other letter. We only consider the case that for some letter $b \in \Sigma$, that $\|A\| + \|B\| + \|b\| = MW_1$ exactly. In both cases, we will continue to construct the other peptide. Lemma 6 below illustrates the details of the extension of the other peptide.

Lemma 6. *Let $\mathcal{Q} = (A, B, C, D)$ be a Rigid Quartet, and let $\|A\|$ be the smaller weighted one in $\mathcal{R}$, and $\|C\|$ be the smaller weighted one in $\mathcal{T}$, and $\|A\| \leq \|C\|$. If for an amino acid $b \in \Sigma$, the following condition holds: $\|A\| + \|b\| + \|B\| = MW_1$. Denote $\Lambda = \overline{S_N(A_>) \cup S_C(B_>) \cup S_N(C_>) \cup S_C(D_>)}$ and assume $a \in \Sigma$ is the letter to be added to $\mathcal{T}$, then we have:*
(1) If $\mathcal{R}$ and $\mathcal{T}$ are neighbouring or crossing, then $\mathcal{T}$ can be extended at most 2

times and we have $\overline{\mathcal{B}(\|Ca\|) \cup \mathcal{Y}(MW_2 - \|Ca\|)} \cap \Lambda = \phi.$ [5]

(2) If $\mathcal{R}$ and $\mathcal{T}$ are nesting, then $\mathcal{T}$ can be extended at most 5 times, and we have $\overline{\mathcal{B}(\|Ca\|) \cup \mathcal{Y}(MW_2 - \|Ca\|)} \cap \Lambda = \phi.$

Proof. (1) For the case of *Neighbouring*, we know that $\|A\| \leq \|B\| \leq \|C\| \leq \|D\|$ and we also have $\|A\| + \|a_1\| + \|B\| = MW_1$. Without losing generosity, we assume $MW = MW_1$ and $MW_2 = MW_1 + \Delta$. Then we know that $\|C\| + \|D\| + \|a_1\| \geq \|A\| + \|B\| + \|a_1\| = MW_1 = MW_2 - \Delta$. From this inequation, we can infer $MW_2 - (\|C\| + \|D\|) \leq \|a_1\| + \Delta \leq \max \|a\|_{a \in \Sigma} + \Delta$. For string pair $\mathcal{T}$ can be at most extended $\lfloor (\max \|a\|_{a \in \Sigma} + \Delta)/ \min \|a\|_{a \in \Sigma} \rfloor - 1 = 2$ times. The proof for both $\overline{\mathcal{B}(\|Ca\|)} \cup \Lambda = \phi$ and $\overline{\mathcal{Y}(MW_2 - \|Ca\|)} \cup \Lambda = \phi$ is similar to Lemma 4.

For the case of *Crossing*, the proof is similar and therefore omitted here.

(2) From the relation of $\mathcal{R}$ and $\mathcal{T}$, we know that $\|C\| + \|D\| \geq 2 \times \|A\|$. Meanwhile we know that $\|A\| + \|a_1\| + \|B\| = MW_1$ and $\|B\| \leq \|A\| + \max \|a\|_{a \in \Sigma}$. Together, we have $2 \times \|A\| + \max \|a\|_{a \in \Sigma} + \|a_1\| \geq MW_2 - \Delta$. It means $MW_2 - (\|C\| + \|D\|) \leq \max \|a\|_{a \in \Sigma} + \|a_1\| + \Delta \leq 2 \times \max \|a\|_{a \in \Sigma} + \Delta$, then we know that for $\mathcal{T}$, it can be extended at most $\lfloor (2 \times \max \|a\|_{a \in \Sigma} + \Delta)/ \min \|a\|_{a \in \Sigma} \rfloor - 1 = 5$ times.

It is easy to get that $\overline{\mathcal{B}(\|Ca\|)} \cup S_N(A_>) = \phi$ and $\overline{\mathcal{B}(\|Ca\|)} \cup S_N(C_>) = \phi$ and $\overline{\mathcal{B}(\|Ca\|)} \cup S_C(D_>) = \phi$. From the conditions, we know $\|A\| \leq \|C\| \leq \|D\| \leq \|B\|$ and $\|Ca\| < MW - \|C\| - 54$ and $\|B_>\| \leq \|A\|$, so we have $\|Ca\| < MW_1 - \|A\| - 54 < MW_1 - \|B_>\| - 54$. We also have $\|Ca\| \geq \|C\| + \min \|a\|_{a \in \Sigma} \geq \|B_>\| + \min \|a\|_{a \in \Sigma}$. Together, we get $\overline{\mathcal{B}(\|Ca\|)} \cup S_C(B_>) = \phi$. Therefore, we prove $\overline{\mathcal{B}(\|Ca\|)} \cup \Lambda = \phi$. Similarly, we can prove $\overline{\mathcal{Y}(MW_2 - \|Ca\|)} \cup \Lambda = \phi$. □

Our basic idea of PEPTIDE DE NOVO SEQUENCING FROM MIXTURE SPECTRA is based on our mathematical model and proof above. We gradually and carefully construct four strings $\{A, B, C, D\}$. They are the N-terimus prefix and the reverse C-terminus suffix of the targeted peptide P_1 and P_2 respectively. The algorithm starts from growing a *Rigid Quartet*, according to Lemma 1, when we add one amino acid to the smallest weighted string, the newly acquired string quadruple is still a *Rigid Quartet*. At some point in the procedure, it will occur that the extension of the quadruple can't be constrained by the defintion of *Rigid Quartet*. In such condition, we consider the quadruple as a *General Quartet* and continue to extend the other prefix-suffix pair until we obtain the whole sequence. At the same time, from Lemma 6, we know that when this situation happens, the following computation for extending the *General Quartet* can be completed in constant time, and we call this procedure PROCEEDTOEND in the following algorithm. In each step of extending a *Rigid Quartet*, the score for the string quadruple is calculated based on Lemma 5. After all the possible amino acid extension is considered, we then rank all the string quadruples that satisfy the given conditions and output the highest scored candidate pair in the end.

---

[5] We say $\mathcal{R}$ or $\mathcal{T}$ is *extended*, if we add a letter $a \in \Sigma$ to the smaller weighted string in $\mathcal{R}$ or $\mathcal{T}$, the new quadruple is at least a *General Quartet*.

Algorithm 1. Mixture Spectra *de novo* Sequencing

INPUT: Given mixture spectrum $\mathcal{M}$, and two precursor mass values MW_1 and MW_2, $|MW_1 - MW_2| \leq \Delta$, and a predefined error bound δ, and the finest mass spectrometer calibration β

OUTPUT: Constructing two peptides P_1 and P_2, which satifies $|\|P_1\|+20-MW_1| \leq \delta$, $|\|P_2\| + 20 - MW_2| \leq \delta$, and $H(P_1, P_2)$ is maximized.

1: Initializing all $DP[i, j, k, l] = -\infty$; Let $DP[1, 19, 1, 19] = 0$
2: Let $MW = \min\{MW_1, MW_2\}$ and $\mathcal{W} = MW - \min \|a\|_{a \in \Sigma} - 54$
3: **for** x from 1 to $\mathcal{W}/2$, by step β **do**
4:     **for** y from $\max(19, x - \max \|a\|)$ to $\min(x + \max \|a\|, MW_1 - x)$ **do**
5:         **for** m from $\max(1, x - \max \|a\|)$ to $\min(x + \max \|a\|, \mathcal{W}/2)$ **do**
6:             **for** n from $\max(19, x - \max \|a\|$ to $\min(x + \max \|a\|, MW_2 - m)$ **do**
7:                 **if** x is the smallest value **then**
8:                     **for** $a \in \Sigma$ **do**
9:                         **if** $|x + y + \|a\| - MW_1| \leq \delta$ **then**
10:                             PROCEEDTOEND
11:                         **else if** $x + \|a\| < MW - x - 54$ & $x + y + \|a\| < MW_1$ **then**
12: $$DP[x + \|a\|, y, m, n] = \max \begin{cases} DP[x + \|a\|, y, m, n] \\ DP[x, y, m, n] + f_1 \end{cases}$$
13:                 **else if** y is the smallest value **then**
14:                     **for** $a \in \Sigma$ **do**
15:                         **if** $|x + y + \|a\| - MW_1| \leq \delta$ **then**
16:                             PROCEEDTOEND
17:                         **else if** $y + \|a\| < MW - y - 54$ & $x + y + \|a\| < MW_1$ **then**
18: $$DP[x, y + \|a\|, m, n] = \max \begin{cases} DP[x, y + \|a\|, m, n] \\ DP[x, y, m, n] + f_2 \end{cases}$$
19:                 **else if** m is the smallest value **then**
20:                     **for** $a \in \Sigma$ **do**
21:                         **if** $|m + n + \|a\| - MW_2| \leq \delta$ **then**
22:                             PROCEEDTOEND
23:                         **else if** $m + \|a\| < MW - m - 54$ & $m + n + \|a\| < MW_2$ **then**
24: $$DP[x, y, m + \|a\|, n] = \max \begin{cases} DP[x, y, m + \|a\|, n] \\ DP[x, y, m, n] + f_3 \end{cases}$$
25:                 **else**
26:                     **for** $a \in \Sigma$ **do**
27:                         **if** $|m + n + \|a\| - MW_2| \leq \delta$ **then**
28:                             PROCEEDTOEND
29:                         **else if** $n + \|a\| < MW - n - 54$ & $m + n + \|a\| < MW_2$ **then**
30: $$DP[x, y, m, n + \|a\|] = \max \begin{cases} DP[x, y, m, n + \|a\|] \\ DP[x, y, m, n] + f_4 \end{cases}$$
31: Compute the best $DP[x, y, m, n]$ for all x, y, m, n and a_1 and a_2 satisfying $|x + y + \|a_1\| - MW_1| \leq \delta$ and $|m + n + \|a_2\| - MW_2| \leq \delta$
32: Backtrack and Output the best peptide pairs $Aa_1 \overleftarrow{B}$ and $Ca_2 \overleftarrow{D}$

Theorem 1. *Algorithm computes the optimal solution of the* Mixture Spectra De Novo Sequencing Problem *in time bounded by*

$$O((\frac{max\|a\|_{a\in\Sigma}}{\beta})^3 \times \frac{\delta}{\beta} \times |\Sigma|^5 \times \frac{MW}{\beta})$$

Proof. The procedure ProceedToEnd in Algorithm 1 is a recursive procedure. According to Lemma 6, we know the height of the recursive tree is bounded by a small number. Therefore the number of nodes in the recursive tree is bounded by $O(|\Sigma|^5)$ which indicates that the computation in this recursive procedure is constant time. Lemma 5 tells us that the function $f(u,v,w,m,n)$ is to find the possible overlaping peaks in two sets and sum up the height of all the explaned peaks, it can be computed in $O(\frac{\delta}{\beta})$. Overall, the time complexity is indeed linear to the mass of the peptide, MW here is the larger mass of P_1 and P_2. $\square$

4   Experiment Results and Discussion

Because there is currently no publicly available data with validated identifications of mixture MS/MS spectra, thus similar to the methods in [12], we created a dataset of 253 simulated mixture spectra to justify the performance of our algorithm. Firstly, we selected some of the confidently identified MS/MS spectra from the PRIDE library[15]. Secondly, the two spectrums we chose to merge have precursor m/z values with difference less than 3 Th. And we use a coefficient to simulate the fact that two co-sequenced precursors may have different intensities. $\mathcal{M} = A + \alpha B$, in which α is the mixture coefficient. Thirdly, raw spectra with the unexplained peaks are used to create the test dataset, such that we can simulate the noises that always exist in MS/MS spectra. We compare our algorithm with the PEAKS [4] *de novo* online version on the simulated dataset. We use PEAKS software to identify two peptides in an iterative manner, first identify the peptide with largr precursor intensity and then identify the second peptide from the spectra with the explained peaks of the first peptide subtracted. The experimental results are show in the following table.

Table 1. Two basic aspects are compared in this experiment. A) Number of reported pairs in which both peptides have less than 4 incorrect characters. B)Number of reported pairs that both peptides have longer than 3 consecutive correct characters.

Coefficient α	Both with $\leq$ 4 incorrect chars		Both with $\geq$ 3 tags		Total
	Our Algorithm	PEAKS	Our Algorithm	PEAKS	
1.0	100(39.5%)	89(35.2%)	172(68.0%)	140(55.3%)	253
0.8	95(37.5%)	94(37.1%)	168(66.4%)	153(60.5%)	253
0.6	89(35.2%)	106(41.9%)	161(63.6%)	164(64.8%)	253

From the results, we can see that our method works better when the two co-sequenced peptides both have relatively high intensity, this is reasonable because our method formulates two peptides together and consider the mixture spectra *de novo* problem as a whole, while for PEAKS method, in case that two peptides with high intensity being fragmented together, the relatively higher peaks of one peptide will impede the constructing of another peptide. To sum up, our algorithm can serve as a complimentary method to the mainstream *de novo* sequencing approaches for efficient mixture spectra identification.

Acknowledgments. KZ was partially supported by an NSERC Discovery Grant and a Discovery Accelerator Supplements Grant. YL was partially supported by CSC(China Scholarship Council) Scholarship.

References

1. Fenn, J.B., Mann, M., Meng, C.K., Wong, S.F., et al.: Electrospray Ionization for Mass Spectrometry of Large Biomolecules. Science 246(4926), 64–71 (1989)
2. Peng, J., Elias, J.E., et al.: Evaluation of Multidimensional Chromatography Coupled with Tandem Mass Spectrometry(LC/LC-MS/MS) for Large-scale Protein Analysis: the Yeast Proteome. J. Proteome Res. 2(1), 43–50 (2003)
3. Cottrell, J., et al.: Probability-based Protein Identification by Searching Sequence Database using Mass Spectrometry Data. Electrophoresis 20(18), 3551–3567 (1999)
4. Ma, B., et al.: PEAKS: Powerful Software for Peptide De Novo Sequencing by Tandem Mass Spectrometry. Rapid Commun. Mass Spectrom. 17(20), 2337–2342 (2003)
5. Ma, B., Zhang, K., Liang, C.: An Effective Algorithm for Peptide De Novo Sequencing from MS/MS Spectra. J. Comput. Syst. Sci. 70(3), 418–430 (2005)
6. Frank, A., Pevzner, P.: Pepnovo: De Novo Peptide Sequencing via Probabilistic Network Modeling. Anal. Chem. 77(4), 964–973 (2005)
7. Alves, G., Ogurtsov, A.Y., Kwok, S., Wu, W.W., Wang, G., et al.: Detection of Co-eluted Peptides using Database Search Methods. Biol. Direct. 3(27) (2008)
8. Houel, S., Abernathy, R., Rengariathan, K., Meyer-Arendt, K., et al.: Quantifying the Impact of Chimera MS/MS Spectra on Peptide Identification in Large-scale Proteomics Studies. J. Proteome Res. 9(8), 4152–4160 (2010)
9. Venable, J., et al.: Automated Approach for Quantitative Analysis of Complex Peptide Mixtures from Tandem Mass Spectra. Nat. Methods 1(1), 39–45 (2004)
10. Zhang, N., et al.: ProbIDtree: An Automated Software Program Capable of Identifiying Multiple Peptides from a Single Collision-induced Dissociation Spectrum Collected by a Tandem Mass Spectrometer. Proteomics 5(16), 4096–4106 (2005)
11. Wang, J., Perez-Santiago, J., Katz, J.E., et al.: Peptide Identification from Mixture Tandem Mass Spectra. Mol. Cell. Proteomics 9(7), 1476–1485 (2010)
12. Wang, J., Bourne, P.E., Bandeira, N.: Peptide Identification by Database Search of Mixture Tandem Mass Spectra. Mol. Cell. Proteomics 10(12) (2011)
13. Zhang, J., et al.: PEAKS DB: De Novo Sequencing Assisted Database Search for Sensitive and Accurate Peptide Identification. Mol. Cell. Proteomics 11(4) (2012)
14. Frank, A., Savitski, M.M., et al.: De Novo Peptide Sequencing and Identification with Precision Mass Spectrometry. J. Proteome Res. 6(1), 114–123 (2007)
15. Vizcaino, J.A., et al.: The Proteomics Identification(PRIDE) Database and Associated Tools: Status in 2013. Nucleic Acids Res. 41(D1), D1063–D1069 (2013)

Identifying Spurious Interactions
in the Protein-Protein Interaction Networks
Using Local Similarity Preserving Embedding

Lin Zhu[1], Zhu-Hong You[2], and De-Shuang Huang[1,*]

[1] School of Electronics and Information Engineering, Tongji University
4800 Caoan Road, Shanghai 201804, China
[2] College of Computer Science and Software Engineering, Shenzhen University, Shenzhen, Guangdong, China
dshuang@tongji.edu.cn

Abstract. Over the last decade, the development of high-throughput techniques has resulted in a rapid accumulation of protein-protein interaction (PPI) data. However, the high-throughput experimental interaction data is prone to exhibit high level of noise. In this paper, we propose a new approach called Local Similarity Preserving Embedding(LSPE) for assessing the reliability of interactions. Unlike previous approaches which seek to preserve a global predefined distance matrix in the embedding space, LSPE tries to adaptively and locally learn a Euclidean embedding under the simple geometric assumption of PPI networks. The experimental results show that our approach substantially outperforms previous methods on PPI assessment problems. LSPE could thus facilitate further graph-based studies of PPIs and may help infer their hidden underlying biological knowledge.

Keywords: bioinformatics, protein-protein interaction (PPI), denoising.

1   Introduction

In the past decades, due to the progress in large-scale experimental technologies such as yeast two-hybrid (Y2H) screens [1, 2], tandem affinity purification (TAP) [3], mass spectrometric protein complex identification (MS-PCI) [4] and other high-throughput biological techniques, a large amount of protein-protein interaction(PPI) data for different species has been accumulated [1-6]. These PPI data can be modeled by networks, where nodes in networks represent proteins and edges between the nodes represent physical interactions between proteins. PPI networks provide a comprehensive view of the global interaction structure of an organism's proteome, as well as detailed information on specific interactions[7]. Analyzing the structure of PPI networks could lead

---

* Corresponding author.

M. Basu, Y. Pan, and J. Wang (Eds.): ISBRA 2014, LNBI 8492, pp. 138–148, 2014.
© Springer International Publishing Switzerland 2014

to new knowledge about complex biological mechanisms. For example, Lapp et al. used the human PPI network to optimize their biological experiments which would detect up to 90% of the human interactome with less than one-third of the proteome used as bait in large-scale pull down experiments [8]. Therefore, PPI networks occupy a central position in cellular systems biology.

Although the expectation is that the PPI data will be as useful as other biological data (e.g., sequence data) in unveiling new biology, the PPI network researches face numerous changes. For example, due to the limitations of the associated experimental techniques and the dynamic nature of protein interaction maps, the high-throughput methods are prone to a high rate of false-positives, i.e. protein interactions which are identified by the experiment do not take place in the cell. For example, the false positive rates for Y2H could be as high as 64% and for TAP experiments they could be as high as 77% [9]. These false positives may lead to the connection between the unrelated proteins, forming huge interaction clusters, which complicate elucidation of the biological importance of these interactions [1]. Incorrect biological conclusions may also be derived from these interactions. Therefore, the mathematical and computational analysis techniques for assessing and ranking reliability of PPIs are highly desirable.

A large number of approaches have been introduced for eliminating unreliable interactions and increasing the reliability of protein interactome. Among them, the network-topology-based methods attracted extensive attention. Those approaches are promising as they only require the input from the PPI network topology [10-14]. The representative algorithms include interaction generality [13, 14], interaction reliability by alternative path (IRAP) [15], Czekanowski-Dice distance (CD-Dist) [16], and functional similarity weight (FSWeight) [17]. The main idea of these approaches is to rank the reliability of an interacting protein pair based on the topology of the interactions between the protein pair and their neighbors within a short radius [18]. The major shortcoming of these indices is that their performance will deteriorate rapidly when they are applied to sparse PPI networks [18]. Motivated by this issue, a number of researchers have demonstrated that the performance of PPIs assessment can be improved by exploiting the entire connectivity information of the network simultaneously [19-23]. Several approaches use random walk as the theoretical basis [21, 24]. Performance of this type of methods may be significantly affected by hub nodes that are connected to many nodes in the network. To circumvent this problem, Lei et al. recently proposed the Random Walk with Resistance (RWR) approach which can reduce the influence of hub nodes[23]. On the other hand, Zhu et al. proposed to indentify false PPI links via a novel Generative Network Model (RIGNM) where the scale-free property of the PPI network is considered[22].

Additionally, several methods were proposed based on the geometric assumption of PPI networks[20, 25], i.e., in a PPI network nodes correspond to points in a metric space and edges are created between pairs of nodes if the corresponding points are close enough according to some distance norm [25, 26]. Based on this assumption, Przulj's method(denoted as MDS-GEO henceforth) firstly constructs a distance matrix between the proteins which satisfy the geometric assumptions, then the proteins are embedded in a low-dimensional space using multidimensional scaling (MDS), i.e., the spectral decomposition of the distance matrix[27]. After the embedding is learned, a pair of proteins that is connected in the original PPI network will be assigned an interaction if and only if they are close to each other in an embedded space.

In our previous work [20], we utilized manifold learning theory [28] to develop an effective tool for assessing the reliability of protein interactions. The isometric feature mapping (ISOMAP) algorithm is firstly applied to seek a manifold embedding of the PPI network, then we assign an interaction reliability score based on FSWeight to each protein pair according to the similarity between the points in the embedded space. Experimental results show that our method (denoted as ISOMAP-FSWeight hence-forth) is able to achieve a satisfying performance, especially for large sparse PPI net-work on which the traditional algorithms fail.

Despite the advantages of MDS-GEO and ISOMAP-FSWeight, their performance is limited by some drawbacks: (1) The original geometric assumption simply assumes that points connected in the network are close and points which are unconnected are more distant[19]. However, MDS-GEO and ISOMAP-FSWeight both seek to preserve a predefined metric. Clearly, they enforce more structural assumptions on the embed-ding, which may deteriorate the fitting performance. Furthermore, both ISOMAP and MDS-GEO use the shortest path-lengths on the graph to define the global similarity between nodes, which is also sensitive to the false links in the graph [21, 29].

With these considerations in mind, in this paper we propose a novel approach called Local Similarity Preserving Embedding(LSPE) for assessing the reliability of interac-tions. As in MDS-GEO and ISOMAP-FSWeight, LSPE is based on the geometric assumption of PPI networks and requires only the connectivity information as input. Unlike learning in MDS-GEO and ISOMAP-FSWeight, however, LSPE does not seek to preserve a pre-specified global metric in the embedding space. Instead, it adaptively learns a metric embedding under the criterion that it can better satisfy the geometric assumption. During the iterations, the embedding algorithm iteratively pushes local pairs of real disconnected proteins apart and pulls pairs of similar proteins together. Thus, at the end of training, proteins that are near each other in the embedding space are likely to be semantically related. The experimental results show that our approach substantially outperforms previous methods on PPI assessment problems.

The remainder of this paper is organized as follows. Section 2 outlines the metho-dologies used in this paper. A variety of experimental results are presented in Section 3. Finally, we provide some concluding remarks in Section 4.

2   Methodology

A PPI network can be naturally represented as a neighborhood graph $G \sim (V, E)$, where the set of vertices $V = \{v_1, v_2, \cdots, v_n\}$ are the proteins, and the set of edges $E = \{e_{ij}\}$ indicate interaction relationships between the proteins. The main idea of our approach is to learn a mapping $g : v \to \Phi(v) \in \mathbb{R}^{1 \times d}$ which maps the nodes of V into a vector space that captures their "semantic similarity", i.e., we would like the Euclidean distance between node pairs that is known to interact to be smaller than the distances corresponding to non-interacting pairs to be larger than ε, and obtain a probabilistic estimation of whether two nodes interact.

For each protein, v_i, and each potential neighbor in the network, v_j, we start by computing the symmetric probability, P_{ij}, that v_i would pick v_j as its neighbor:

$$P_{ij} = \begin{cases} \dfrac{1}{n_i} & e_{ij} \in E \\ 0 & e_{ij} \notin E \end{cases} \tag{1}$$

where n_i is the vertex degree of node i. It is easy to verify that $\sum_j P_{ij} = 1$.

On the other hand, in the low-dimensional space we model the similarity of map point $\Phi(v_j)$ to map point $\Phi(v_i)$ by the following decreasing function of their distance:

$$Q_{ij} = \frac{q_{ij}}{Z_i} \tag{2}$$

where $q_{ij} = \left\| \phi(v_i) - \phi(v_j) \right\|_2^{-2\alpha_i}$, $Z_i = \sum_{k \neq i} q_{ik}$ is the normalization term that enforces $\sum_j Q_{ij} = 1$, α_i is the parameter of the local similarity function.

If the map points $\phi(v_i)$ and $\phi(v_j)$ correctly model the similarity between the proteins v_i and v_j, the conditional probabilities P_{ij} and Q_{ij} will be equal. Motivated by this observation, the aim of the embedding is to match these two distributions defined in (1) and (2) for every i as well as possible. More specifically, we minimize a cost function which is a sum of cross entropy between the original $\{P_{ij}\}$ and $\{Q_{ij}\}$ distributions for each protein:

$$L\big(\Phi(V), A\big) = -\sum_i \sum_j P_{ij} \log \frac{Q_{ij}}{\sum_{k \neq i} Q_{ik}} \tag{3}$$

where $A = \{\alpha_i\}$.

Minimizing $L\big(\Phi(V), A\big)$ in Equation (3) with respect to Y is equivalent to the optimization problem:

$$\underset{\Phi(V), q}{\text{minimize}} \; L\big(\Phi(V), A, q\big) = -\sum_{ij} P_{ij} \log \frac{q_{ij}}{\sum_k q_{ik}} \tag{4}$$

$$\text{s.t.} \; q_{ij} = H_i\left(\left\| \Phi(v_i) - \Phi(v_j) \right\|_2^2\right) = \left\| \phi(v_i) - \phi(v_j) \right\|_2^{-2\alpha_i}$$

The extended objective using the Lagrangian technique is given by

$$\tilde{L}\big(\Phi(V),A,q\big) = -\sum_{ij} P_{ij} \log \frac{q_{ij}}{\sum_k q_{ik}} + \sum_{ij} \lambda_{ij} \left(q_{ij} - H_i \left(\left\| \Phi(v_i) - \Phi(v_j) \right\|_2^2 \right) \right) \qquad (5)$$

Setting $\dfrac{\partial \tilde{L}\big(\Phi(V),A,q\big)}{\partial q_{ij}} = 0$ yields

$$\lambda_{ij} = -\frac{1}{\sum_k q_{ik}} + \frac{P_{ij}}{q_{ij}} \qquad (6)$$

Inserting these Lagrangian multipliers to the gradient of $\tilde{L}\big(\Phi(V),A,q\big)$ with respect to $\Phi(v_i)$, we have

$$
\begin{aligned}
\tilde{L}\big(\Phi(V),A,q\big) &= -\sum_{ij} P_{ij} \log \frac{q_{ij}}{\sum_k q_{ik}} - \sum_{ij} \left(\frac{1}{\sum_k q_{ik}} - \frac{P_{ij}}{q_{ij}} \right) \left(q_{ij} - H_i \left(\left\| \Phi(v_i) - \Phi(v_j) \right\|_2^2 \right) \right) \\
&= -\sum_{ij} P_{ij} \log \frac{q_{ij}}{\sum_k q_{ik}} + \sum_{ij} \left(\frac{1}{\sum_k q_{ik}} - \frac{P_{ij}}{q_{ij}} \right) H_i \left(\left\| \Phi(v_i) - \Phi(v_j) \right\|_2^2 \right) \\
&= -\sum_{ij} P_{ij} \log \frac{q_{ij}}{\sum_k q_{ik}} + \sum_{ij} \left(\frac{q_{ij}}{\sum_k q_{ik}} - P_{ij} \right) \frac{1}{q_{ij}} H_i \left(\left\| \Phi(v_i) - \Phi(v_j) \right\|_2^2 \right)
\end{aligned} \qquad (7)
$$

Define $\dfrac{\delta \log H_i(\tau)}{\delta \tau} = G_i(\tau)$, we have

$$
\begin{aligned}
\frac{\partial L\big(\Phi(V),A\big)}{\partial \Phi(v_i)} &= \frac{\partial \tilde{L}\big(\Phi(V),A,q\big)}{\partial \Phi(v_i)} \\
&= 2\sum_{ij} \left[(P_{ij} - Q_{ij}) G_i \left(\left\| \Phi(v_i) - \Phi(v_j) \right\|_2^2 \right) \left(\Phi(v_i) - \Phi(v_j) \right) \right] \\
&\quad + 2\sum_{ij} \left[(P_{ji} - Q_{ji}) G_j \left(\left\| \Phi(v_i) - \Phi(v_j) \right\|_2^2 \right) \left(\Phi(v_i) - \Phi(v_j) \right) \right] \\
&= 2\sum_{ij} \left[\alpha_i (P_{ij} - Q_{ij}) + \alpha_j (P_{ji} - Q_{ji}) \right] \frac{1}{\left\| \Phi(v_i) - \Phi(v_j) \right\|_2^2} \left(\Phi(v_i) - \Phi(v_j) \right)
\end{aligned} \qquad (8)
$$

Similarly, we have

$$\frac{\partial L\big(\Phi(V),A\big)}{\partial \alpha_i} = \frac{\partial \tilde{L}\big(\Phi(V),A,q\big)}{\partial \alpha_i} = \sum_{j \neq i} (Q_{ij} - P_{ij}) \ln \left(\left\| \Phi(v_i) - \Phi(v_j) \right\|_2 \right) \qquad (9)$$

We adopt an alternating projection strategy to minimize $L(\Phi(V),\varepsilon)$ until convergence. More specifically, each time we optimize one parameter, such as $\Phi(V)$, with the other parameters fixed.

The learning of $\alpha_i, 1 \leq i \leq n$ with $\Phi(V)$ fixed can be straightforwardly decomposed into n single variant optimization problems and we solve them using gradient descent method, which works well in practice.

Then we learn $\Phi(V)$ with $\alpha_i, 1 \leq i \leq n$ fixed. The partial derivative (8) can be further written as the following compact form:

$$\frac{\partial L(\Phi(V),\varepsilon)}{\partial \Phi(V)} = (L_W - L_D)\Phi(V) \tag{10}$$

where $L_W = \text{diag}\left(\sum_j w_{1j}, \cdots, \sum_j w_{nj}\right) - W$, $L_D = \text{diag}\left(\sum_j d_{1j}, \cdots, \sum_j d_{nj}\right) - D$, the matrices W and D are defined as

$$\begin{aligned} w_{ij} &= 2\alpha_i P_{ij} + 2\alpha_j P_{ji} \\ D_{ij} &= 2\alpha_i Q_{ij} + 2\alpha_j Q_{ji} \end{aligned} \tag{11}$$

During the experiments, we have noticed that learning $\Phi(V)$ with the standard gradient descent direction (10) is very slow and requires many tiny steps to converge. Letting the gradient (10) to zero, we instead investigate several splits in an attempt to identify a fixed point iteration method. For instance, we can consider

$$(L_P - L_Q)\Phi(V) = 0 \Rightarrow \Phi(V) = (L_P)^{-1} L_Q \Phi(V) \tag{12}$$

Although this iteration is not fixed point iteration and does not always converge, it does suggest using a new search direction $\Delta = (L_P)^{-1} L_Q \Phi(V) - \Phi(V)$ along which we can decrease $L(\Phi(V),\varepsilon)$ with a line search $\Phi(V) \leftarrow \Phi(V) + \alpha\Delta$ for $\alpha > 0$. As is proven in the appendix, Δ is a descent direction, i.e., the directional derivative of the search direction always remains negative. Hence, as a result of Zoutendijk's theorem, we are guaranteed to converge to a local optimum of $L(\Phi(V),\varepsilon)$ if we use the search direction in combination with a line-search that satisfies the Wolfe conditions[30], i.e., a line-search step that simultaneous satisfies the Armijo condition and the curvature condition:

$$L(\Phi(V) + \alpha\Delta, \varepsilon) \leq L(\Phi(V),\varepsilon) + c_1 \alpha \left\langle \Delta, \frac{\partial L(\Phi(V),\varepsilon)}{\partial \Phi(V)} \right\rangle \tag{13}$$

$$\left\langle \Delta, \frac{\partial L(\Phi(V) + \alpha\Delta, \varepsilon)}{\partial \Phi(V)} \right\rangle \geq c_2 \left\langle \Delta, \frac{\partial L(\Phi(V),\varepsilon)}{\partial \Phi(V)} \right\rangle \tag{14}$$

where c_1 and c_2 are free parameters.

3   Experiments

In this section we provide the experiments to evaluate the performance of our method on several PPI networks described below. Throughout the numerical evaluation in this work, our algorithm was implemented in Matlab 7.12, and the experiments were run on a hardware environment with Intel i5-750 CPU and 8G RAM.

3.1   Data Sources and Evaluation Metric

We used two PPI datasets of S.Cerevisiae to evaluate the effectiveness of our methodology. Specifically, the first dataset was obatained from [31], and contains 7622 non-redundant interactions between 2171 of the yeast proteins. The second PPI dataset was downloaded from the Database of Interacting Proteins (DIP) [32], which is a database that documents experimentally determined protein-protein interactions. This data set is composed of 17491 interactions between 4934 yeast proteins. We removed from the dataset any protein with self-interactions and repeated interactions, and the final dataset consists of 17173 interaction pairs involving 4875 proteins. The characteristics of the datasets are summarized in Table 1.

Table 1. Characteristics of three protein interaction networks

PPI Dataset	Number of proteins	Number of interactions	Data source
Tong	2171	7622	[31]
Ho	4875	17173	[32]

3.2   Increasing the Reliability of PPI Networks Using LSPE

To validate the usefulness of the proposed method for assessing the reliability of protein interactions, in this section, we systematically compare it with ISOMAP-FSweight and MDS-GEO approaches.

The topology-based methods like CD-Dist and the recently proposed RWR [23] and RIGNM[22] are also included for comparison. As in [18, 20], we utilize the degree of functional homogeneity and localization coherence of protein pairs as the measure to evaluate the performance.

It is well known that the strategy of 'guilt by association' provides the evidence that interacting proteins are likely to share a common function and cellular localization [33], which means true interacting protein pairs should share at least a common functional role or they should at least be at a common cellular localization if a pair of proteins to be interacting in vivo. Since LSPE assumes that the distance between two proteins in the embedding space is a monotonically decreasing function of the probability that they interact, it is expected that if we only consider protein pairs with smaller distance in the latent space to be have true positive interaction, the proportion of interacting proteins with functional homogeneity and localization coherence should increase correspondingly.

In the study, the Gene Ontology (GO)[1] based annotations is used to evaluate the functional homogeneity and localization coherence. GO is one of the most important ontology within the bioinformatics community. The three organizing principles of Gene Ontology are cellular component, biological process, and molecular function. Here we used the first taxonomies of the GO terms for localization coherence calculation, and the last two taxonomies of the GO terms for functional homogeneity calculation. The GO terms are organized hierarchically into functional subfamilies. Two different GO terms may have a common parent or a common child in the hierarchy. GO terms at high levels may occur in many genes (or proteins), while GO terms at low levels appear in very few proteins. In our experiment, we just choose those GO terms at middle levels. More specifically, we choose the GO terms which occur in at least 30 proteins, but none of its children appears in at least 30 proteins.

We rank interactions of proteins according to their distance in the embedding space from the lowest to highest, and measure the functional homogeneity and localization coherence by computing the rate of interacting protein pairs with common functional roles and cellular localization. The experimental results on the three datasets are respectively showed in Fig. 1-2. The vertical axis is the proportion of interacting protein pairs which share a common function or cellular localization. The horizontal axis is the coverage of the PPI network comparing the original network.

As can be seen in Fig.1, LSPE is the best in assessing false positive interactions in the Tong network: as more interactions which were detected as potential false positive interaction were removed from the interactions, the degree of functional homogeneity and localization coherence in the resulting interactome increases at a faster rate than using other methods. 93.1% of the top 40% of interacting protein pairs ranked by LSPE have a common functional role and 90.2% of them have a common subcellular localization, while the corresponding performance of the best competing method(CD-Dist) are 86.9% and 83.1%

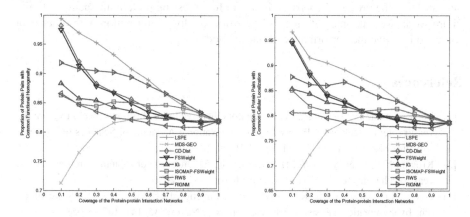

Fig. 1. Comparison of various algorithms on Tong network for assessing the reliability of interactions in term of functional homogeneity and localization coherence

---

[1] http://www.geneontology.org/

For DIP network, the conclusions are similar. On the whole, LSPE achieves the best performance as compared to the other approaches for increasing the reliability of protein interactomes, which confirms the usefulness of our method.

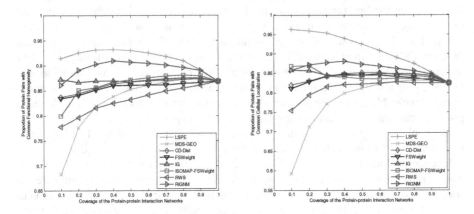

Fig. 2. Comparison of various algorithms on DIP network for assessing the reliability of interactions in term of functional homogeneity and localization coherence

4   Conclusions

In this paper, we have developed a novel technique to assess protein interactions from high-throughput experimental data. The experimental results show our method consistently performs better than previously methods on the two PPI networks, which indicates that the proposed approach is useful for assessing protein interactions. In the future work, we will integrate the PPI network topology information with other data source to improve the performance of our method.

References

[1] Ito, T., Chiba, T., Ozawa, R., et al.: A comprehensive two-hybrid analysis to explore the yeast protein interactome. Proceedings of the National Academy of Sciences of the United States of America 98, 4569–4574 (2001)

[2] Krogan, N.J., Cagney, G., Yu, H.Y., et al.: Global landscape of protein complexes in the yeast saccharomyces cerevisiae. Nature 440, 637–643 (2006)

[3] Gavin, A.C., Bosche, M., Krause, R., et al.: Functional organization of the yeast proteome by systematic analysis of protein complexes. Nature 415, 141–147 (2002)

[4] Ho, Y., Gruhler, A., Heilbut, A., et al.: Systematic identification of protein complexes in saccharomyces cerevisiae by mass spectrometry. Nature 415, 180–183 (2002)

[5] Uetz, P., Giot, L., Cagney, G., et al.: A comprehensive analysis of protein-protein interactions in saccharomyces cerevisiae. Nature 403, 623–627 (2000)

[6] Giot, L., Bader, J.S., Brouwer, C., et al.: A protein interaction map of drosophila mela-nogaster. Science 302, 1727–1736 (2003)

[7] Morrison, J.L., Breitling, R., Higham, D.J., et al.: A lock-and-key model for pro-tein-protein interactions. Bioinformatics 22, 2012–2019 (2006)

[8] Lappe, M., Holm, L.: Unraveling protein interaction networks with near-optimal effi-ciency. Nature Biotechnology 22, 98–103 (2004)

[9] Edwards, A.M., Kus, B., Jansen, R., et al.: Bridging structural biology and genomics: as-sessing protein interaction data with known complexes. TRENDS in Genetics 18, 529–536 (2002)

[10] Albert, I., Albert, R.: Conserved network motifs allow protein–protein interaction predic-tion. Bioinformatics 20, 3346–3352 (2004)

[11] Deng, M., Zhang, K., Mehta, S., et al.: Prediction of protein function using pro-tein-protein interaction data. Journal of Computational Biology 10, 947–960 (2003)

[12] Liu, G., Li, J., Wong, L.: Assessing and predicting protein interactions using both local and global network topological metrics. Genome Informatics 22, 138–149 (2008)

[13] Saito, R., Suzuki, H., Hayashizaki, Y.: Interaction generality, a measurement to assess the reliability of a protein–protein interaction. Nucleic Acids Research 30, 1163–1168 (2002)

[14] Saito, R., Suzuki, H., Hayashizaki, Y.: Construction of reliable protein–protein interac-tion networks with a new interaction generality measure. Bioinformatics 19, 756–763 (2003)

[15] Chen, J., Hsu, W., Lee, M.L., et al.: Discovering reliable protein interactions from high-throughput experimental data using network topology. Artificial Intelligence in Medicine 35, 37–47 (2005)

[16] Brun, C., Chevenet, F., Martin, D., et al.: Functional classification of proteins for the pre-diction of cellular function from a protein-protein interaction network. Genome Biolo-gy 5, 6 (2003)

[17] Chua, H.N., Sung, W.K., Wong, L.: Exploiting indirect neighbours and topological weight to predict protein function from protein-protein interactions. Bioinformatics 22, 1623–1630 (2006)

[18] Chua, H.N., Wong, L.: Increasing the reliability of protein interactomes. Drug Discovery Today 13, 652–658 (2008)

[19] Kuchaiev, O., Rasajski, M., Higham, D.J., et al.: Geometric de-noising of protein-protein interaction networks. Plos Computational Biology 5, e1000454 (2009)

[20] You, Z.H., Lei, Y.K., Gui, J., et al.: Using manifold embedding for assessing and predict-ing protein interactions from high-throughput experimental data. Bioinformatics 26, 2744–2751 (2010)

[21] Fang, Y., Benjamin, W., Sun, M.T., et al.: Global geometric affinity for revealing high fi-delity protein interaction network. Plos One 6, e19349 (2011)

[22] Zhu, Y., Zhang, X.-F., Dai, D.-Q., et al.: Identifying Spurious Interactions and Predicting Missing Interactions in the Protein-Protein Interaction Networks via a Generative Net-work Model. IEEE/ACM Transactions on Computational Biology and Bioinformatics 10, 219–225 (2013)

[23] Lei, C., Ruan, J.: A novel link prediction algorithm for reconstructing protein–protein in-teraction networks by topological similarity. Bioinformatics 29, 355–364 (2013)

[24] Fouss, F., Pirotte, A., Renders, J.M., et al.: Random-walk computation of similarities between nodes of a graph with application to collaborative recommendation. IEEE Transactions on Knowledge and Data Engineering 19, 355–369 (2007)

[25] Przulj, N., Corneil, D.G., Jurisica, I.: Modeling interactome: scale-free or geometric? Bioinformatics 20, 3508–3515 (2004)

[26] Milenkovic, T., Lai, J., Przulj, N.: Graphcrunch: a tool for large network analyses. BMC Bioinformatics 9, 70 (2008)

[27] Higham, D.J., Rasajski, M., Przulji, N.: Fitting a geometric graph to a protein-protein interaction network. Bioinformatics 24, 1093–1099 (2008)

[28] Tenenbaum, J.B., De Silva, V., Langford, J.C.: A global geometric framework for nonlinear dimensionality reduction. Science 290, 2319–2323 (2000)

[29] Gomez-Rodriguez, M., Leskovec, J., Krause, A.: Inferring networks of diffusion and influence. ACM Transactions on Knowledge Discovery from Data (TKDD) 5, 21 (2012)

[30] Nocedal, J., Wright, S.J.: Numerical optimization, 2nd edn. Springer (2006)

[31] Tong, A.H.Y., Lesage, G., Bader, G.D., et al.: Global mapping of the yeast genetic interaction network. Science's STKE 303, 808 (2004)

[32] Xenarios, I., Rice, D.W., Salwinski, L., et al.: DIP: the database of interacting proteins. Nucleic Acids Research 28, 289–291 (2000)

[33] Oliver, S.: Guilt-by-association goes global. Nature 403, 601–603 (2000)

Multiple RNA Interaction
with Sub-optimal Solutions

Syed Ali Ahmed* and Saad Mneimneh**

The Graduate Center and Hunter College, City University of New York,
New York, USA
sahmed3@gc.cuny.edu, saad@hunter.cuny.edu

Abstract. The interaction of two RNA molecules involves a complex interplay between folding and binding that warranted recent developments in RNA-RNA interaction algorithms. However, biological mechanisms in which more than two RNAs take part in an interaction exist. It is reasonable to believe that interactions involving multiple RNAs are generally more complex to be treated pairwise. In addition, given a pool of RNAs, it is not trivial to predict which RNAs are interacting without sufficient biological knowledge. Therefore, structures resulting from multiple RNA interactions often cannot be predicted by the existing algorithms.

We recently proposed a system for multiple RNA interaction that overcomes the difficulties mentioned above by formulating a combinatorial optimization problem called *Pegs and Rubber Bands*. A solution to this problem encodes a structure of interacting RNAs. In general, however, the optimal solution obtained does not necessarily correspond to the actual structure observed experimentally. Moreover, a structure produced by interacting RNAs may not be unique. In this work, we extend our previous approach to generate multiple sub-optimal solutions. By clustering these solutions, we are able to reveal representatives that correspond to realistic structures. Specifically, our results on the U2-U6 complex in the spliceosome of yeast and the CopA-CopT complex in E. Coli are consistent with published biological structures.

1   Introduction

The interaction of two RNA molecules has been independently formulated as a computational problem in several works, e.g. [1–3]. In their most general form, these formulations lead to NP-hard problems (which means computationaly intractable, i.e. the running time of the algorithm that produces an optimal solution increases exponentially with the problem size). To overcome this hurdle, researchers have been either reverting to approximation algorithms, or imposing algorithmic restrictions; for instance, the avoidance of the formation of certain structures.

---

* Supported by NSF Award CCF-AF 1049902 and a CUNY GC Science Fellowship.
** Corresponding author. Supported by NSF Award CCF-AF 1049902.

M. Basu, Y. Pan, and J. Wang (Eds.): ISBRA 2014, LNBI 8492, pp. 149–162, 2014.
© Springer International Publishing Switzerland 2014

While these algorithms had limited use in the beginning, they became important venues for (and in fact popularized) an interesting biological fact: RNAs interact. For instance, micro-RNAs (miRNAs) bind to a complementary part of messenger RNAs (mRNAs) and inhibit their translation [4]. But more complex forms of RNA-RNA interaction exist. In E. Coli, CopA binds to the ribosome binding site of CopT, also as a regulation mechanism to prevent translation [5]; so does OxyS to fhlA [6]. In both of these structures, the simultaneous folding (within the RNA) and binding (to the other RNA) are non-trivial to be predicted as separate events. To account for this, most of the RNA-RNA interaction algorithms calculate the probability for a pair of subsequences (one of each RNA) to participate in the interaction, and in doing so they generalize the energy model used for the partition function of a single RNA to the case of two RNAs [7–12]. This generalization takes into consideration the simultaneous aspect of folding and binding.

Not surprisingly, there exist other mechanisms in which more than two RNA molecules take part in an interaction. Typical scenarios involve the interaction of multiple small nucleolar RNAs (snoRNAs) with ribosomal RNAs (rRNAs) in guiding the methylation of the rRNAs [4], and multiple small nuclear RNAs (snRNA) with mRNAs in the splicing of introns [13]. Even with the existence of a computational framework for a single RNA-RNA interaction, it is reasonable to believe that interactions involving multiple RNAs are generally more complex to be treated pairwise. In addition, given a pool of RNAs, it is not trivial to predict which RNAs interact without some prior biological information. Some attempts for multiple RNA interaction have been considered, e.g. [14, 15], but they only generalize the partition function algorithm of [16] by concatenation of all RNAs into one, and so can only produce restricted structures, e.g. no kissing loops. Even though algorithms for kissing loops exist, e.g. [17], advances in pairwise interaction of RNAs suggest that the concatenation model is less suitable.

We recently proposed a new computational approach for handling multiple RNA interaction based on a combinatorial optimization problem that we call Pegs and Rubber Bands [18]. In this work, we extend this approach to generate multiple sub-optimal solutions, and show that these solutions correspond to realistic structure.

2   Background and Approach

2.1   Pegs and Rubber Bands: A Formulation

We now present the problem of Pegs and Rubber Bands as a framework for multiple RNA interaction. The link between the two will be made shortly following a formal description of Pegs and Rubber Bands.

Consider m levels numbered 1 to m with n_l pegs in level l numbered 1 to n_l. There is an infinite supply of rubber bands that can be placed around two pegs in consecutive levels. For instance, we can choose to place a rubber band around peg i in level l and peg j in level $l+1$; we call it a rubber band at $[l, i, j]$. Every such pair of pegs $[l, i]$ and $[l+1, j]$ contribute their own weight $w(l, i, j)$. The Pegs and Rubber Bands problem is to maximize the total weight by placing rubber bands around pegs in such a way that no two rubber bands intersect. In other words, each peg can have at most one rubber band around it, and if a rubber band is placed at $[l, i_1, j_1]$ and another at $[l, i_2, j_2]$, then $i_1 < i_2 \Leftrightarrow j_1 < j_2$. We assume without loss of generality that $w(l, i, j) \neq 0$ to avoid the unnecessary

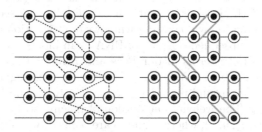

Fig. 1. Pegs and Rubber Bands. All positive weights are equal to 1 and are represented by dashed lines. The optimal solution achieves a total weight of 8.

placement of rubber bands and, therefore, either $w(l, i, j) > 0$ or $w(l, i, j) = -\infty$. Figure 1 shows an example.

Given an optimal solution, it can always be reconstructed from left to right by repeatedly placing some rubber band at $[l, i, j]$ such that, at the time of this placement, no rubber band is around peg $[l, k]$ for $k > i$ and no rubber band is around peg $[l + 1, k]$ for $k > j$. This process can be carried out by a dynamic programming algorithm to compute the maximum weight (Section 3.1).

2.2   Multiple RNA Interaction as Pegs and Rubber Bands

To provide some initial context we now describe how the formulation of Pegs and Rubber Bands, though in a primitive way, captures the problem of multiple RNA interaction. We think of each level as an RNA and each peg as one base of the RNA. The weight $w(l, i, j)$ corresponds to the negative of the energy contributed by the binding of the ith base of RNA l to the jth base of RNA $l + 1$. This can be obtained using existing algorithms for RNA-RNA interaction that act on pairs of RNAs. It should be clear, therefore, that an optimal solution for Pegs and Rubber Bands represents the lowest energy conformation in a base-pair energy model, when a pseudoknot-like restriction is imposed on the RNA interaction (rubber bands cannot intersect). In doing so, we obviously assume that an order on the RNAs is given with alternating sense and antisense, and that the first RNA interacts with the second RNA, which in turn interacts with the third RNA, and so on. We later relax this ordering and the stringency of the interaction pattern of the RNAs. While a simple base-pairing model is not likely to give realistic results, our goal here was simply to establish a correspondence between the two problems.

2.3   Windows and Gaps: A Better Formulation for RNA Interaction

In the previous section, we described our initial attempt to view the interaction of m RNAs as a Pegs and Rubber Bands problem with m levels, where the first RNA interacts with the second RNA, and the second with the third, and so on (so they alternate in sense and antisense). This used a simple base-pair energy model, which is not too realistic. We now address this issue (and leave the issues of the ordering and the interaction pattern to the following section). A better model for RNA interaction will consider windows of interaction instead of single bases. For instance, subsequence $[i_1, i_2]$ of RNA l can interact with subsequence $[j_1, j_2]$ of RNA $l + 1$. In terms of our Pegs and Rubber Bands problem, this translates to placing rubber bands around a stretch of contiguous pegs in two consecutive levels, e.g. around pegs $[l, i_1]$, $[l, i_2]$, $[l + 1, j_1]$, and $[l + 1, j_2]$, where $i_2 \geq i_1$ and $j_2 \geq j_1$. The weight contribution of placing such a rubber band is now given by $w(l, i_2, j_2, u, v)$, where i_2 and j_2 are the last two pegs covered by the rubber band in level l and level $l+1$, and $u = i_2 - i_1 + 1$ and $v = j_2 - j_1 + 1$ represent the length of the two windows covered in level l and level $l + 1$, respectively.

As a *heuristic*, we also allow for the possibility of imposing a gap $g \geq 0$ between windows as a way to ensure that windows are energetically independent. This gap is also taken into consideration when we perform the gap filling procedure described in Section 3.1.

We use windows satisfying $2 \leq u, v \leq w = 26$ and a gap $g = 0$. The weights $w(l, i, j, u, v)$ are obtained from RNAup, a tool to compute energies of pairwise interactions [7], as (negative of energy values):

$$w(l, i, j, u, v) \propto \log p_l(i - u + 1, i) + \log p_{l+1}(j - v + 1, j)$$

$$+ \log Z_l^I(i - u + 1, i, j - v + 1, j)$$

where $p_l(i_1, i_2)$ is the probability that subsequence $[i_1, i_2]$ is free (does not fold) in RNA l, and $Z_l^I(i_1, i_2, j_1, j_2)$ is the partition function (as computed in [7]) of the interaction of subsequences $[i_1, i_2]$ in RNA l and $[j_1, j_2]$ in RNA $l+1$ (subject to no folding within RNAs). As such, the weight considers intra-molecular and inter-molecular energies. The windows are filtered for sub-additivity as described in Section 3.1.

2.4   Order and Interaction Pattern via Permutations

We now describe how to relax the ordering and the stringency of the interaction pattern of the RNAs. We first identify each RNA as being *even* (sense) or *odd* (antisense), but this convention can obviously be switched. Given m RNAs and a permutation on the set $\{1, \ldots, m\}$, we map the RNAs onto the levels of a Pegs and Rubber Bands problem as follows: We place the RNAs in the order in which they appear in the permutation on the same level as long as they have the same parity (they are either all even or all odd). We then increase the number of levels by one, and repeat. RNAs that end up on the same level are *virtually* considered as one RNA that is the concatenation of all. However, in the corresponding Pegs and Rubber Bands problem, we do not allow windows to span multiple RNAs, nor do we enforce a gap between two windows in different RNAs. We describe in Section 3.2 a greedy algorithm that searches heuristically for the best permutation.

3 Algorithms

3.1 Complexity of the Problem and Approximations

We proved that Pegs and Rubber Bands is NP-hard [18]. Therefore, any algorithm that finds an optimal solution generally requires exponential time. However, while our problem is NP-hard, we also proved that the same formulation can be adapted to obtain a polynomial time approximation. A maximization problem admits a polynomial time approximation scheme (PTAS) iff for every **fixed** $\epsilon > 0$ there is an algorithm with a running time polynomial in the size of the input that finds a solution within $(1 - \epsilon)$ of optimal [19].

Let OPT be the weight of the optimal solution and denote by $W[i \ldots j]$ the weight of the optimal solution when the problem is restricted to levels $i, i + 1, \ldots, j$ (a sub-problem). For a given $\epsilon > 0$, let $k = \lceil \frac{1}{\epsilon} \rceil$. Consider the following k solutions (weights), each obtained by a concatenation of optimal solutions for sub-problems consisting of at most k levels.

$$W_1 = W[1 \ldots 1] + W[2 \ldots k + 1] + W[k + 2 \ldots 2k + 1] + \ldots$$
$$W_2 = W[1 \ldots 2] + W[3 \ldots k + 2] + W[k + 3 \ldots 2k + 2] + \ldots$$

$$\vdots$$

$$W_k = W[1 \ldots k] + W[k + 1 \ldots 2k] + W[2k + 1 \ldots 3k] + \ldots$$

The best of these solutions is a $(1 - \epsilon)$ approximation [18], i.e.

$$\max_i W_i \geq \frac{k - 1}{k} OPT \geq (1 - \epsilon) OPT$$

Therefore, for a given integer k, the $(1 - 1/k)$-factor approximation algorithm is to simply choose the best $W_i = W[1 \ldots i] + W[i+1 \ldots i+k] + W[i+k+1 \ldots i+2k] + \ldots$ as a solution, where $W[i \ldots j]$ denotes the weight of the optimal solution for the sub-problem consisting of levels $i, i+1 \ldots, j$. Some more theoretical results on approximation based on our formulation were obtained in [20].

As a practical step, and instead of using the W_i's for the comparison, we can fill in for each W_i some additional rubber bands (interactions) between (RNAs) level i and level $i + 1$, between level $i + k$ and level $i + k + 1$, and so on, by identifying the pegs of these levels (regions of RNAs) that are not part of the solution. This does not affect the theoretical guarantee but gives a larger weight to the solution. We call it *gap filling*.

Figure 2 describes an algorithm for m levels based on dynamic programming by defining $W(i_1, i_2, \ldots, i_m)$ to be the maximum weight when we truncate the levels at pegs $[1, i_1], [2, i_2], \ldots, [m, i_m]$. The maximum weight is given by $W(n_1, n_2, \ldots, n_m)$ and the optimal solution can be obtained by standard backtracking.

$$W(i_1, i_2, \ldots, i_m) = \max \begin{cases} W(i_1 - 1, i_2, \ldots, i_m) \\ W(i_1, i_2 - 1, i_3, \ldots, i_m) \\ \vdots \\ W(i_1, \ldots, i_{m-1}, i_m - 1) \\ W((i_1 - u - g)^+, (i_2 - v - g)^+, i_3, \ldots, i_m) + w(1, i_1, i_2, u, v) \\ W(i_1, (i_2 - u - g)^+, (i_3 - v - g)^+, i_4, \ldots, i_m) + w(2, i_2, i_3, u, v) \\ \vdots \\ W(i_1, \ldots, i_{m-2}, (i_{m-1} - u - g)^+, (i_m - v - g)^+) + w(m - 1, i_{m-1}, i_m, u, v) \end{cases}$$

where x^+ denotes $\max(0, x)$, $w(l, i, j, u, v) = -\infty$ if $u > i$ or $v > j$, $0 < u, v \leq w$ (the maximum window size), $g \geq 0$ (the gap), and $W(0, 0, \ldots, 0) = 0$.

Fig. 2. Dynamic programming algorithm for Pegs and Rubber Bands with the windows and gaps formulation

The running time of the algorithm is $O(mw^2 n^m)$ (exponential) and $O(mw^2 \lceil \frac{1}{\epsilon} \rceil n^{\lceil \frac{1}{\epsilon} \rceil})$ for the approximation scheme (polynomial), where w is the maximum window length. If we impose that $u = v$ in $w(l, i, j, u, v)$, then those running times become $O(mwn^m)$ and $O(mw \lceil \frac{1}{\epsilon} \rceil n^{\lceil \frac{1}{\epsilon} \rceil})$ respectively.

For the correctness of the algorithm, we have to assume that windows are *sub-additive*. In other words, we require the following condition (otherwise, the algorithm may compute an incorrect optimum due to the possibility of achieving the same window by two or more smaller ones with higher total weight):

$$w(l, i, j, u_1, v_1) + w(l, i - u_1, j - v_1, u_2, v_2)$$

$$\leq w(l, i, j, u_1 + u_2, v_1 + v_2)$$

In our experience, most existing RNA-RNA interaction algorithms produce weights (the negative of the energy values) of RNA interaction windows that mostly conform to the above condition. In rare cases, we filter the windows to eliminate those that are not sub-additive. For instance, if the above condition is not met, we set $w(l, i, j, u_1, v_1) = w(l, i - u_1, j - v_1, u_2, v_2) = -\infty$.

3.2   Heuristic for a Single Solution

A heuristic for resolving the ordering and the interaction pattern of the RNAs is shown in Figure 3. As described in Section 2.4, the order and interaction pattern are determined by a permutation. The main idea of this heuristic is to first start with an arbitrary permutation, and then iteratively change it by moving along neighboring permutations with better solutions (larger weights). This is repeated until no more improvement can be achieved. Using the PTAS, this algorithm finds **one** solution within a $(1 - \epsilon)$-factor of optimal (which could itself be the optimal).

```
Given ε = 1/k and m RNAs
    produce a random permutation π on {1, ..., m}
    let W be the weight of the (1 − ε)-optimal solution given π
    repeat
        better←false
        generate a set Π of neighboring permutations for π
        for every π' ∈ Π (in any order)
            let W' be the weight of the (1 − ε)-optimal solution given π'
            if W' > W
                then W ← W'
                     π ← π'
                     better←true
    until not better
```

Fig. 3. A heuristic for multiple RNA interaction using the PTAS algorithm

To generate neighboring permutations for this heuristic algorithm one could adapt a standard 2-opt method used in the Traveling Salesman Problem (or other techniques). For instance, given permutation π, a neighboring permutation π' can be obtained by dividing π into three parts and making π' the concatenation of the first part, the reverse of the second part, and the third part. In other words, if $\pi = (\alpha, \beta, \gamma)$, then $\pi' = (\alpha, \beta^R, \gamma)$ is a neighbor of π, where β^R is the reverse of β.

3.3   Multiple Sub-optimal Solutions

We now describe how to generate (all) solutions with a weight of at least some threshold T.

Generation: RNAs often interact in more than one way. To explore this, we assume that the order and interaction pattern have been already determined, e.g. by the algorithm of Section 3.2. We then seek sub-optimal solutions. Denote by $S(i_1, \ldots, i_m)$ a solution where i_l is the smallest index at level l such that peg $[l, i_l]$ is covered by a window, $l = 1 \ldots m$. We will also use $S(i_1, \ldots, i_m)$ interchangeably to represent the weight of that solution. Similarly, we will use $w(l, i, j, u, v)$ interchangeably to denote a window and its weight. We denote by $S(i_1, \ldots, i_m) + w(l, i, j, u, v)$ an extension of solution S by the addition of window w.

We say that a window $w(l, i, j, u, v)$ in $S(i_1, \ldots, i_m)$ is a *terminal* window iff:

- $i - u + 1 = i_l$,
- $j - v + 1 = i_{l+1}$, and
- no other window $w(l', i', j', u', v')$ in $S(i_1, \ldots, i_m)$ satisfies $i' - u' + 1 = i_{l'}$, $j' - v' + 1 = i_{l'+1}$, and $l' > l$.

This imposes some order on the windows to prevent generating the same solution in multiple ways. To that end, we can only extend a solution by adding to it a terminal window (a window that becomes the terminal for the extended solution). Observe that whenever $W(i_1 - g - 1, \ldots, i_m - g - 1) + S(i_1, \ldots, i_m) < T$, where g is the gap parameter as described in Section 2.3, S cannot be extended in anyway to meet the threshold.

Let $\phi = S(n_1 + g + 1, \ldots, n_m + g + 1)$ represent the empty solution (with zero weight). We have the following algorithm (Figure 4) for generating every solution with weight at least T, starting with Process(ϕ). Because windows are considered in order, the running time of the algorithm is linear in the size of its output plus a crude $O(2^{|\mathbb{W}|})$ bound (all possible solutions), where $\mathbb{W}$ is the set of windows.

```
Process(S(i₁,...,iₘ))
    if W(i₁ − g − 1,...,iₘ − g − 1) + S(i₁,...,iₘ) < T
       then return
       else for every window w(l, i, j, u, v) that is
                terminal in S(i₁,...,iₘ) + w(l, i, j, u, v)
                with iₗ − i > g and i_{l+1} − j > g
                Process(S(i₁,...,iₘ) + w(l, i, j, u, v))
    if S(i₁,...,iₘ) ≥ T
       then output S
```

Fig. 4. Generating multiple sub-optimal solutions

Clustering: The sub-optimal solutions generated above may be a lot more than what we need. We use a pseudo-clustering algorithm to identify a small set of representative solutions. We use the term pseudo-clustering because our algorithm does not attempt to optimize clusters in any way. Let $d(S, C)$ be the distance between a solution S and a cluster C (Section 3.3.3), and assume a threshold D. The idea is to add a solution S to a cluster C if $d(S, C) < D$. Figure 5 shows an algorithm that clusters solutions until all solutions are in clusters or the maximum number of clusters c has been reached. The running time of this algorithm is, therefore, $O(|\mathbb{S}|cf(m, n))$, where $\mathbb{S}$ is the set of solutions, and $f(m, n)$ is the time needed to compute the distance on instances with m RNAs of length n.

```
Cluster
    r = 0
    for every solution S in decreasing order of weight
        if there exists a cluster Cᵢ such that d(S, Cᵢ) < D
           then Cᵢ ← Cᵢ ∪ {S}
           else r ← r + 1
                Cᵣ ← {S}
                output S (the best in its cluster)
                if r = maximum number of clusters c
                   return
```

Fig. 5. An algorithm for pseudo-clustering the solutions

Distance. Recall that peg $[l, i]$ represents the ith base of RNA l. Therefore, if peg $[l, i]$ is covered by a window in some solution for Pegs and Rubber Bands, we say that base i is interacting. Otherwise, we distinguish between two cases: base i is free, or there is a base j of RNA l that is **not** interacting such that base

i folds onto base j (makes a bond) in the optimal folding of RNA l. Therefore, given a solution, RNA l can be represented by a string s_l where $s_l[i]$, the ith character in s_l, is one of three letters: I for interacting (with another RNA), F for free, and B for bonding (to the same RNA).

We define a distance function $d(S_1, S_2)$ between two solutions, and set $d(S, C)$ as the distance between S and the representative solution of cluster C. The last paragraph in this section describes how such a representative is determined.

Jaccard: Given a solution S, convert s_l for every $l = 1 \ldots m$ into a binary vector v_l by replacing I with 10, B with 01, and F with 00. Concatenate all such vectors into one vector $v = v_1 v_2 \ldots v_m$. If u is the vector corresponding to solution S_1 and v is the vector corresponding to solution S_2, then:

$$d(S_1, S_2) = \frac{\sum_i u[i] \otimes v[i]}{\sum_i u[i] \oplus v[i]}$$

where $v[i]$ is the ith bit of vector v, and $\otimes$ and $\oplus$ stand for the binary operators XOR (exclusive OR) and OR, respectively. Intuitively, this reflects a Hamming distance scaled by the number of entries that can potentially differ [21]. We also define a coarser version of this distance below.

Levenshtein: Given a solution S, collapse s_l for every $l = 1 \ldots m$ by replacing repeated consecutive letters by one occurrence of the given letter, e.g. replace BBBBB by B. With this modification, if $s_1, \ldots, s_m$ correspond to solution S_1 and $t_1, \ldots, t_m$ correspond to solution S_2, then:

$$d(S_1, S_2) = \frac{\sum_{l=1}^{m} \mathrm{Lev}(s_l, t_l)}{\sum_{l=1}^{m} \max(|s_l|, |t_l|)}$$

where $\mathrm{Lev}(s, t)$ is the Levenshtein distance (in modern terms, an edit distance where each mismatch and deletion contributes a 1 [22]), and $|\ \ |$ denotes the length of a string.

We either use the Jaccard distance, or the average of Jaccard and Levenshtein when the Jaccard distance is not sensitive to small variations. In computing $d(S, C)$, the representative of cluster C is either the best solution in the cluster (i.e. the one with the largest weight, which is also the one that started the cluster), or the consensus of the cluster. The consensus can be obtained in terms of the vector v, where $v[i] = 1$ for the consensus solution if and only if a strict majority of the solutions in C have the ith bit equal to 1.

4   Results

For all of our experiments, we only show the interaction pattern (no folding within the individual RNAs).

4.1   Single Solutions

We use the algorithm for Section 3.2 We pick the largest weight solution among several runs of the algorithm. The value of k and the gap filling criterion depend on the scenario, as described below.

Fishing for Pairs. Six RNAs in E. Coli of which three pairs are known to interact are used [8]. The interest here is to see whether the algorithm can identify the three pairs. For this purpose, it will suffice to set $k = 2$ and to ignore gap filling. Furthermore, we only consider solutions in which each RNA interacts with at most one other RNA. The solution with the largest weight identifies the three pairs correctly (Figure 6). In addition, the interacting sites in each pair are consistent with the predictions of existing RNA-RNA interaction algorithms, e.g. [10].

```
OxyS  5' ...CCCUUG...GUG...UCCAG... 3'        MicA  5' ...GCGCA...CUGUUUUC...CGU... 3'
              ||||||   |||   |||||                     |||||   ||||||||   |||
fhlA  3' ...GGGAAC...CAC...AGGUC... 5'        lamB  3' ...CGCGU...GAUAGAGG...GCA... 5'


     CopA  5' CGGUUUAAGUGGG...UCGUACUCGCCAAAGUUGA...UUUUGCUU 3'
              |||||||||||||   |||||||||||||||||||   ||||||||
     CopT  3' GCCAAAUUCACCC...AGCAUGAGCGGUUUCAACU...AAAACGAA 5'
```

Fig. 6. Known pairs of interacting RNAs

Structural Separation. The yeast snRNA complex U2-U6 is necessary for the splicing of a specific mRNA intron [23]. Only the preserved regions of the intron are considered, which consist of two structurally autonomous parts, resulting in an instance with a total of four RNAs, U2, U6, I1, and I2. Six RNAs are used: CopA, CopT, and the four mentioned RNAs. The interest here is to see whether the algorithm can separate the CopA-CopT complex from that of yeast. The algorithm is performed with $k = 3$ and gap filling. The solution with the largest weight successfully predicts and separates the RNA complex CopA-CopT of Figure 6 from the RNA structure shown in Figure 7a for the U2-U6 complex in the splicing of its intron.

4.2   Multiple Sub-optimal Solutions

We now use the algorithm of Section 3.3 with an appropriate threshold T to generate enough solutions. Additional parameters for this algorithms are: the distance function $d(S, C)$, the threshold D for adding a solution to a cluster, and whether the cluster representative is the best solution in the cluster or the consensus of the cluster (refer to Section 3.3.3 for computing distances). The choice of these parameters will be given for each scenario. Only the best solution in each cluster is reported (see Figure 5 for detail).

Helices and Co-axial Stacking. The U2-U6 complex in yeast has been reported to have two distinct experimental structures, e.g. [24]. In one conformation, U2 and U6 interact to form helix Ia (interaction as in Figure 7c). In another conformation, the interaction reveals a structure containing an additional helix, helix Ib. It has been conjectured in [25] that co-axial stacking is essential for the stabilization of helix Ia in U2-U6 and, therefore, inhibition of the co-axial stacking, possibly by protein binding, may activate the second conformation. Regardless of what underlying mechanisms are responsible for this conformational switch, our sub-optimal solutions cluster in a way that reveal the two conformations (Figure 7).

```
(a) cluster 1                              (b) cluster 2

    I1 3' NNNNGUAUGUNNNN 5'                    I1 3' NNNNGUAUGUNNNN 5'
          |||                                        |||
    U6 5' ACAGAGAUGAUC--AGC 3'                 U6 5' ACAGAGAUGAUC--AGC 3'
          |||||  |||                                 |||||  |||
    U2 3' AUGAU-GUGAACUAGAUUCG 5'              U2 3' AUGAU-GUGAACUAGAUUCG 5'
          |||||  |||                                 |||||  |||
    I2 5' NNNNUACUAACACCNNNN 3'                I2 5' NNNNUACUAACACCNNNN 3'


(c) cluster 3                              (d) cluster 4

    I1 3' NNNNGUAUGUNNNN 5'                    I1 3' NNNNGUAUGUNNNN 5'
          |||
    U6 5' ACAGAGAUGAUC--AGC 3'                 U6 5' ACAGAGAUGAUC--AGC 3'
          |||||                                      |||||
    U2 3' AUGAU-GUGAACUAGAUUCG 5'              U2 3' AUGAU-GUGAACUAGAUUCG 5'
          |||||  |||                                 |||||  |||
    I2 5' NNNNUACUAACACCNNNN 3'                I2 5' NNNNUACUAACACCNNNN 3'
```

Fig. 7. U2 and U6 truncated up to helix Ib. Algorithm performed with the following parameters: distance is Jaccard, threshold=0.15, representative is consensus. (a) Helices Ia and Ib with correct binding of introns. (b) same as (a) with I1 not binding. (c) Helix Ia only with correct binding of introns. (d) Same as (c) with I1 not binding.

Artifact Interactions and Reversible Kissing Loops. Due to the optimization nature of the problem, it is sometimes easy to pick up interactions that are biologically unreal. This is because dropping these interactions from the solution would make it less optimal. The third interaction window of CopA-CopT in Figure 6 is an example of such an artifact. As shown in Figure 8 on the Left, our second cluster of sub-optimal solutions succeeds in dropping this window.

Reversible kissing loops represent an even harder mechanism to capture by optimization. With this mechanism, the initial kissing complex occurs between a subset of loop bases in both RNAs, but this interaction is fully reversible and very unstable [26]. Therefore, in the final interaction, the kissing loop will be missing few bases towards its center. An example of this scenario is the middle interaction window of CopA-CopT in Figure 6 and Figure 8 on the Left (considering the folding pattern of CopA and CopT reveals that this interaction window is a kissing loop). By isolating this window and generating sub-optimal solutions, our third cluster starts to reveal a separation of the interaction close to the center, as shown in Figure 8 on the Right.

(a) cluster 1

```
(a) cluster 1

CopA 5' CGGUUUAAGUGGG...UCGUACUCGCCAAAGUUGA...UUUUGCUU 3'
        ||||||||||||    ||||||||||||||||||    ||||||||
CopT 3' GCCAAAUUCACCC...AGCAUGAGCGGUUUCAACU...AAAACGAA 5'


(b) cluster 2

CopA 5' CGGUUUAAGUGG...UCGUACUCGCCAAAGUUGAA... 3'
        |||||||||||    |||||||||||||||||||||
CopT 3' GCCAAAUUCACC...AGCAUGAGCGGUUUCAACUU... 5'
```

```
(a) cluster 1

CopA 5' ...UCGUACUCGCCAAAGUUG... 3'
           ||||||||||||||||||
CopT 3' ...AGCAUGAGCGGUUUCAAC... 5'


(b) cluster 2

CopA 5' ...UCGUACUCGCCAAAGUUG... 3'
           |||||
CopT 3' ...AGCAUGAGCGGUUUCAAC... 5'


(c) cluster 3

CopA 5' ...UCGUACUCGCCAAAGUUG... 3'
           ||||||  |||||||||||
CopT 3' ...AGCAUGAGCGGUUUCAAC... 5'
```

Fig. 8. Left: Algorithm performed with the following parameters: distance is average of Jaccard and Levenshtein, threshold=0.25, representative is best. (a) As in Figure 6. (b) A less optimal but realistic structure in which the third interaction window is dropped. Right: CopA-CopT complex truncated to its middle window. Algorithm performed with the following parameters: distance is average of Jaccard and Levenshtein, threshold=0.3, representative is best. The third cluster starts to reveal a separation (reversible kissing loop) in the middle interaction window.

5   Conclusion

While RNA-RNA interaction algorithms exist, they are not suitable for predicting RNA structures with more than two RNAs; for instance, treating the RNAs pairwise may not lead to the best global structure. Moreover, the best structure may not be the real structure, and the real structure may not be unique. In this work, we build on our recent formulation for multiple RNA interaction as a combinatorial optimization problem, and extend it to produce multiple sub-optimal solutions. Our experiments reveal that such an approach can provide several candidate structures when they exist, e.g. the U2-U6 complex in the spliceosome of yeast, and find realistic structures that are not necessarily optimal in the computational sense, e.g. CopA-CopT in E. Coli.

References

1. Pervouchine, D.D.: Iris: Intermolecular RNA interaction search. In: 15th International Conference on Genome Informatics (2004)
2. Alkan, C., Karakoc, E., Nadeau, J.H., Sahinalp, S.C., Zhang, K.: RNA-RNA interaction prediction and antisense RNA target search. Journal of Computational Biology 13(2) (2006)
3. Mneimneh, S.: On the approximation of optimal structures for RNA-RNA interaction. IEEE/ACM Transactions on Computational Biology and Bioinformatics (2009)

4. Meyer, I.M.: Predicting novel RNA-RNA interactions. Current Opinions in Structural Biology 18 (2008)

5. Kolb, F.A., Malmgren, C., Westhof, E., Ehresmann, C., Ehresmann, B., Wagner, E.G.H., Romby, P.: An unusual structure formed by antisense-target RNA binding involves an extended kissing complex with a four-way junction and a side-by-side helical alignment. RNA Society (2000)

6. Argaman, L., Altuvia, S.: fhla repression by oxys: Kissing complex formation at two sites results in a stable antisense-target RNA complex. Journal of Molecular Biology 300 (2000)

7. Muckstein, U., Tafer, H., Hackermuller, J., Bernhart, S.H., Stadler, P.F., Hofacker, I.L.: Thermodynamics of RNA-RNA binding. In: Journal of Bioinformatics (2006)

8. Chitsaz, H., Backofen, R., Sahinalp, S.C.: biRNA: Fast RNA-RNA binding sites prediction. In: Salzberg, S.L., Warnow, T. (eds.) WABI 2009. LNCS, vol. 5724, pp. 25–36. Springer, Heidelberg (2009)

9. Chitsaz, H., Salari, R., Sahinalp, S.C., Backofen, R.: A partition function algorithm for interacting nucleic acid strands. Journal of Bioinformatics (2009)

10. Salari, R., Backofen, R., Sahinalp, S.C.: Fast prediction of RNA-RNA interaction. Algorithms for Molecular Biology 5(5) (2010)

11. Huang, F.W.D., Qin, J., Reidys, C.M., Stadler, P.F.: Partition function and base pairing probabilities for RNA-RNA interaction prediction. Journal of Bioinformatics 25(20) (2009)

12. Li, A.X., Marz, M., Qin, J., Reidys, C.M.: RNA-RNA interaction prediction based on multiple sequence alignments. In: Journal of Bioinformatics (2010)

13. Sun, J.S., Manley, J.L.: A novel u2-u6 snrna structure is necessary for mammalian MRNA splicing. Genes and Development 9 (1995)

14. Andronescu, M., Chuan, Z.Z., Codon, A.: Secondary structure prediction of interacting RNA molecules. Journal of Molecular Biology 345(5) (2005)

15. Dirks, R.M., Bois, J.S., Schaffer, J.M., Winfree, E., Pierce, N.A.: Thermodynamic analysis of interacting nucleic acid strands. SIAM Review 49(1) (2007)

16. McCaskill, J.S.: The equilibrium partition function and base pair binding probabilities for RNA secondary structure. Journal of Biopolymers 29(6-7) (1990)

17. Chen, H.L., Codon, A., Jabbari, H.: An $o(n^5)$ algorithm for mfa prediction of kissing hairpins and 4-chains in nucleic acids. Journal of Computational Biology 16(6) (2009)

18. Ahmed, S.A., Mneimneh, S., Greenbaum, N.L.: A combinatorial approach for multiple RNA interaction: Formulations, approximations, and heuristics. In: Du, D.-Z., Zhang, G. (eds.) COCOON 2013. LNCS, vol. 7936, pp. 421–433. Springer, Heidelberg (2013)

19. Cormen, T., Leiserson, C.E., Rivest, R.L., Stein, C.: Approximation Algorithms in Introduction to Algorithms. MIT Press (2010)

20. Tong, W., Goebel, R., Liu, T., Lin, G.: Approximation algorithms for the maximum multiple RNA interaction problem. In: Widmayer, P., Xu, Y., Zhu, B. (eds.) COCOA 2013. LNCS, vol. 8287, pp. 49–59. Springer, Heidelberg (2013)

21. Jaccard, P.: Bulletin de la Societe Vaudoise des Sciences Naturelles 38(69) (1902)

22. Levenshtein, V.: Binary codes capable of correcting deletions, insertions, and reversals. Soviet Physics Doklady (10), 70710 (1966)

23. Newby, M.I., Greenbaum, N.L.: A conserved pseudouridine modification in eukaryotic u2 snrna induces a change in branch-site architecture. RNA 7(6), 833–845 (2001)

24. Sashital, D.G., Cornilescu, G., Butcher, S.E.: U2-u6 RNA folding reveals a group ii intron-like domain and a four-helix junction. Nature Structural and Molecular Biology 11(12) (2004)
25. Cao, S., Chen, S.J.: Free energy landscapes of RNA/RNA complexes. Journal of Molecular Biology (357), 292–312 (2006)
26. Kolb, F.A., Slagter-Jager, J.G., Ehresmann, B., Ehressmann, C., Westhof, E., Gerhart, E., Wagner, H., Romby, P.: Progression of a loop-loop complex to a four-way junction is crucial for the activity of a regulatory antisense RNA. The EMBO Journal 19(21), 5905–5915 (2000)

Application of Consensus String Matching in the Diagnosis of Allelic Heterogeneity[*]
(Extended Abstract)

Fatema Tuz Zohora and M. Sohel Rahman

AℓEDA Group
Department of CSE, Bangladesh University of Engineering and Technology,
Dhaka-1000, Bangladesh
{anne.06.cse,sohel.kcl}@gmail.com

Abstract. In this paper, an algorithm is proposed that detects the existence of a common ancestor gene sequence for non-overlapping inversion (reversed complement) metric given two input DNA sequences. Theoretical average and worst case time complexity of the algorithm is proven to be $O(n^3)$ and $O(n^4)$ respectively, where n is length of input sequences. However, practically those are found to be $O(n^2)$ and $O(n^3)$ respectively, where the worst case occurs when both input sequences have the similarity of around 90%. Similarly, theoretical worst case space complexity is $O(n^3)$, whereas it is $O(n^2)$ practically. The work is motivated by the purpose of diagnosing unknown genetic disease that shows *allelic heterogeneity*, a case where a normal gene mutates in different orders resulting in two different gene sequences causing two different genetic diseases. The algorithm can be useful as well in the study of breed-related hereditary conditions to determine the genetic spread of a defective gene in the population.

1 Introduction

Computer Alignment of molecular sequences is widely used for biological sequence comparisons. It is very interesting to know that different mutations in the same gene result in different phenotypes which may lead to diseases with entirely different clinical features [1]. For example, mutations in the *RET* gene have been implicated in the etiology of *Hirshprung* disease as well as *Multiple Endocrine Neoplasia (MEN) Type 2*. Allelic heterogeneity is considered as the greatest challenge for molecular genetic diagnosis as stated in the book by Meisenberg et al. [2]. It makes the use of usual clinical diagnostic approaches like allele-specific oligonucleotide probes impractical and needs different approaches like mismatch scanning, gene sequencing, linkage analysis etc., which all are highly expensive solutions. Allelic heterogeneity motivates us with its importance in the field of medical science. It also causes autism and rigid-compulsive behaviors [3]. Recently Castellani et al. [4] have presented CFTR2 for the clinical

---

[*] This research work has been conducted at CSE, BUET as part of the M.Sc. Engg. Thesis of Zohora under the supervision of Rahman.

M. Basu, Y. Pan, and J. Wang (Eds.): ISBRA 2014, LNBI 8492, pp. 163–175, 2014.
© Springer International Publishing Switzerland 2014

diagnosis of genetic disorders emphasizing especially allelic heterogeneity. But as the clinical diagnosis incurs high cost so we derive an algorithm for detecting the possibility of allelic heterogeneity in polynomial time. In this paper we use the term *common ancestor* to indicate the same gene sequence from which different mutations give different gene sequences x and y, resulting different diseases. Our aim is to find the common ancestor given x and y as input.

The Consensus problem in strings is motivated by the requirement of finding commonality of a large number of strings and has a variety of applications in bioinformatics [5]. We can map the problem of determining allelic heterogeneity to the well known *consensus string problem* defined as follows: given a set of strings $S = \{s_1, \ldots, s_N\}$ and a constant d, find, if it exists, a string s^* such that the distance of s^* from each of the strings in S does not exceed d, for some suitable and meaningful definition of the term 'distance'. In this paper, we propose an algorithm for deciding whether two input DNA sequences x and y are mutated (by non-overlapping inversions only) from the same sequence. This is equivalent to determining the *existence* of consensus string (s^*), given two strings x and y, of length n, alphabet of size $k = 4$ under the distance metric called *non overlapping inversion* i.e., reversed complements. Since the minimum distance d is not present as parameter, this problem can be thought of as a relaxed version of the original consensus string problem. This problem has been introduced very recently by Cho et al. [6]. In particular, they have provided an $O(n^3)$ algorithm using $O(n^2)$ space, where n is the size of the two input strings. But hypothetically we have found by experimentation that their experiment fails in returning the correct answers in some cases because of not tracking the prefix of the common ancestors. For example, there can never exists any common ancestors between $x = GTGGC$ and $y = CTGGT$, as the number of complement bases ($A - T$ and $C - G$) is different in x and y. But their algorithm returns YES. Erroneous results occur also with the same number of complement bases in x and y. In this paper we present a new algorithm which correctly solves this problem with the same time and space complexity. We further present experimental evidence that our algorithm in practice runs in quadratic time rather than cubic time. Again, the authors in [6] could not identify specific domain of application which is done by us for the first time to our knowledge.

In general, alignment with inversions does not have a known polynomial time algorithm and a simplification to the problem considers only non-overlapping inversions. In previous works of Schoniger et al. [7] and Vellozo et al. [8], a non-overlapping inversion occurs only in one string and transforms the string to the other string. On the other hand, we and Cho et al. [6] consider the more difficult version where non overlapping inversions are allowed in both the strings simultaneously. From now on, whenever we refer to the term 'inversion', we assume non-overlapping inversions.

In our approach a special matrix is built in $O(n^2)$ time, for the input gene sequence x, which gives the possible gene bases {A, T, C, G} at each index considering all possible inversions on x. Some successor links among the matrix cells are considered which let us build a DFS tree producing all possible inversed sequences. But there are exponential number of inversed sequences. So instead of doing that, we iteratively compare two matrices (defined for x and y) to produce the result in polynomial time. There exist previous works where iterative tree matching is applied using some hypothesis. For example, Charnoz et at. [9] propose an original tree matching algorithm for intra-patient hepatic vascular system registration, where starting from the tree root, edges and nodes are iteratively matched based on a set of matching hypotheses which is updated to keep best matches. However, in our case the aim is just to identify whether there exists any common path of length n from root to leaf, but not to return those paths. Also enough number of overlapped branches exist in the solution space. So this helps us proposing an exact algorithm using matrix representation. If we need retrieving the paths also, then after matches are found, we just need to run a DFS over the existing links in matrix cells from left to right. It incurs less time complexity as most of the links are discarded at the time of matching phase.

The rest of the paper is organized as follows. In Section 2, we provide some definitions and observations necessary for presenting the algorithm. Then in Section 3 we discuss the main algorithm. We discusses its correctness, complexity and experimental results in section 4. In Section 5, we explain the steps for detecting Allelic Heterogeneity using our algorithm. Finally we conclude in Section 6 discussing some future research directions.

2   Preliminaries

We consider the biological operation *Inversion* which is the *reverse* and *complement* of a string x. *Inversion Sequence*, θ is defined as a set of the non overlapping inversions. So the *Inversed Sequence*, $\theta(x)$ is the resultant DNA sequence after applying the set of inversions, θ over x. Let $x = AGGC$ is a DNA sequence and $\theta' = \{(1,2),(4,4)\}$, then $\theta'(x) = CTGG$. Again, $\theta'(x)$ upto index 3 is CTG.

Given two sequences x and y of length n, We follow the notation of [6] and use $T_x[n+1][n]$ and $T_y[n+1][n]$ to present the sets of all possible inversions of x and y respectively. In Figure 1, each $T_x[j][i]$ is called a *Inversion fragment*. It represents a tuple $\langle (p,q), \alpha \rangle$ where α is the base $\{A, C, T, G\}$ yielded at index i after applying the inversion (p,q) over x as mentioned below:

$$\langle (p,q), \alpha \rangle = \begin{cases} \langle (i,i)', x[i] \rangle & \text{if } j = i \text{ (no inversion at index } i) \\ \langle (i,j-1), \theta(x[j-1]) \rangle & \text{if } j > i \\ \langle (j,i), \theta(x[j]) \rangle & \text{if } j < i \end{cases}$$

The $\theta(x)$ can be constructed by connecting the inversion fragments in a path specified by θ and concatenating their yielded base letters in that order. In Figure 1, for a given $\theta' = \{(1,4),(5,5),(6,8)\}$, the $\theta'(x) = GGCTTAGC$ is presented by the path shown by shaded cells in T_x. The $\theta'(x)$ up to index $i = 3$ is GGC. Note that the same inversion fragment can belong to different inversion sequences and present a totally different inversed sequences. For θ', $T_x[4][2]$ presents a fragment that belongs to the inversion $(1,4)$ according to the path: $T_x[5][1] \rightarrow \mathbf{T_x[4][2]} \rightarrow T_x[2][3] \rightarrow T_x[1][4] \equiv (1,4), G \rightarrow \mathbf{(2,3)}, \mathbf{G} \rightarrow (2,3), C \rightarrow$

i \ j	1	2	3	4	5	6	7	8
1	(1,1)',A	(1,2),T	(1,3),T	(1,4),T	(1,5),T	(1,6),T	(1,7),T	(1,8),T
2	(1,1),T	(2,2)',G	(2,3),C	(2,4),C	(2,5),C	(2,6),C	(2,7),C	(2,8),C
3	(1,2),C	(2,2),C	(3,3)',C	(3,4),G	(3,5),G	(3,6),G	(3,7),G	(3,8),G
4	(1,3),G	(2,3),G	(3,3),G	(4,4)',C	(4,5),G	(4,6),G	(4,7),G	(4,8),G
5	(1,4),G	(2,4),G	(3,4),G	(4,4),G	(5,5)',A	(5,6),T	(5,7),T	(5,8),T
6	(1,5),T	(2,5),T	(3,5),T	(4,5),T	(5,5),T	(6,6)',G	(6,7),C	(6,8),C
7	(1,6),C	(2,6),C	(3,6),C	(4,6),C	(5,6),C	(6,6),C	(7,7)',C	(7,8),G
8	(1,7),G	(2,7),G	(3,7),G	(4,7),G	(5,7),G	(6,7),G	(7,7),G	(8,8)',T
9	(1,8),A	(2,8),A	(3,8),A	(4,8),A	(5,8),A	(6,8),A	(7,8),A	(8,8),A

i \ j	1	2	3	4	5	6	7	8
1	(1,1)',T	(1,2),A	(1,3),A	(1,4),A	(1,5),A	(1,6),A	(1,7),A	(1,8),A
2	(1,1),A	(2,2)',C	(2,3),G	(2,4),G	(2,5),G	(2,6),G	(2,7),G	(2,8),G
3	(1,2),G	(2,2),G	(3,3)',G	(3,4),C	(3,5),C	(3,6),C	(3,7),C	(3,8),C
4	(1,3),C	(2,3),C	(3,3),C	(4,4)',G	(4,5),C	(4,6),C	(4,7),C	(4,8),C
5	(1,4),C	(2,4),C	(3,4),C	(4,4),C	(5,5)',G	(5,6),C	(5,7),C	(5,8),C
6	(1,5),C	(2,5),C	(3,5),C	(4,5),C	(5,5),C	(6,6)',G	(6,7),G	(6,8),G
7	(1,6),G	(2,6),G	(3,6),G	(4,6),G	(5,6),G	(6,6),G	(7,7)',T	(7,8),A
8	(1,7),A	(2,7),A	(3,7),A	(4,7),A	(5,7),A	(6,7),A	(7,7),A	(8,8)',T
9	(1,8),A	(2,8),A	(3,8),A	(4,8),A	(5,8),A	(6,8),A	(7,8),A	(8,8),A

i   ii

Fig. 1. (i)$T_x[][]$ for $x = AGCCAGCT$; (ii)$T_y[][]$ for $y = TCGGGCTT$

$(1, 4), T$. For $\theta'' = \{(1, 1), (2, 3), (4, 4)', (5, 8)\}$, the same fragment belongs to the inversion $(2, 3)$ according to the path: $\mathbf{T_x}[4][2] \to T_x[2][3] \equiv (2, 3), \mathbf{G} \to (2, 3), C$. These $\theta'(x)$ and $\theta''(x)$ are two different inversed sequences of the same x. Note that, for a fixed θ, there is only one choice as we move from i to $i + 1$. For example, only one path, i.e., one inversed sequence can be derived for θ' (shaded cells) and θ'' (arrow). In this way we can generate all possible inversed sequences of x (though not necessary in our algorithm). In what follows for the select of notational ease we will drop x or y from $T[][]$ when it is clear from the content or applicable independent of x & y.

Two inversion fragments $T[j'][i] = \langle (p_1, p_2), \alpha_1 \rangle$ and $T[j''][i+1] = \langle (q_1, q_2), \alpha_2 \rangle$ are called *Agreed Fragments* if one of the following two conditions holds.

Condition 1. $p_1 + p_2 = q_1 + q_2$ and $j' > j''$: *The condition holds when we move from $T_x[5][1] \to T_x[4][2] \equiv (1, 4), G \to (2, 3), G$ for the inversion $(1, 4)$.*

Condition 2. $q_1 = p_2 + 1$ and $j'' \geq j'$: *This condition holds when we move from $T_x[1][4] \to T_x[6][5] \equiv (1, 4), T \to (5, 5), T$, i.e., the inversion $(1, 4)$ finishes and next inversion $(5, 5)$ starts.*

Otherwise, we call it disagreed fragments. For example, $T_x[5][1] \equiv (1, 4), G$ and $T_x[5][2] \equiv (2, 4), G$ are disagreed as none of the conditions holds. Again, for two pairs of agreed fragments $(\langle (p_1, p_2), \alpha_1 \rangle, \langle (q_1, q_2), \alpha_2 \rangle)$ and $(\langle (q_1, q_2), \alpha_2 \rangle, \langle (r_1, r_2), \alpha_3 \rangle)$, we say these two pairs are connected by $\langle (q_1, q_2), \alpha_2 \rangle$ and thus all these three inversion fragments are agreed fragments. The *Agreed Sequence* is formed by taking an *inversion fragment* from each column $i = 1, 2, \ldots, n$, such that, for any two consecutive fragments $T[j'][i] = \langle (p_1, p_2), \alpha_1 \rangle$ and $T[j''][i+1] = \langle (q_1, q_2), \alpha_2 \rangle$, they are *agreed fragments*. So one *agreed sequence* actually presents an *inversed sequence*, $\theta(x)$. In an *agreed sequence*, we have the following two cases.

Case 1 - Upward movement at index i. It happens in an *agreed sequence* at index i when, $T[j'][i]$ and $T[j''][i+1]$ are agreed fragments based on Condition 1. It implies that an inversion (p, q) is continuing from some index $i' \leq i$. Inversion fragment $T[j'][i]$ involved in such a scenario is called a *continuing_inversion* *fragment* for the corresponding θ'. In Figure 1, for θ', $T_x[4][2]$ is following Case 1 for the inversion $(1, 4)$, and thus is a continuing inversion fragment.

Case 2 - Horizontal or Downward movement at index i. This happens when $T[j'][i]$ and $T[j''][i+1]$ constitute an agreed fragments based on Condition 2. It implies that an inversion (p, q) has started at some index $i' \leq i$, ends at index i (having $j' = p = i'$ and $j'' = q = i$), and next inversion starts at index $i + 1$. Involved $T[j'][i]$, is called the *ending_inversion fragment* for the corresponding θ. In Figure 1, for θ', $T_x[1][4]$ belongs to Case 2 for the inversion $(1, 4)$ and thus is the ending inversion fragment.

Observation 1. *In Table $T[][]$, an inversion (p, q) starts at inversion fragment $T[q + 1][p]$, continue moving upward up to $T[p][q]$. Then it moves horizontal or downward indicating no change, i.e., $(p, p)'$ or start of a new inversion.*

The $Pair(t, r)$ is defined for any inversion fragment $T[j][i]$, where t is the starting index $i' \leq i$ of the last inversion in the inversion sequence θ it belongs to, and r is the current row j. If the same inversion fragment belongs to multiple inversion sequences then multiple $Pair(t, r)$ can exist for it. In such case the value of t would be different for different inversion sequences. *Pairs* corresponding to the agreed fragments are called *Agreed Pairs*. Similarly, a Pair corresponding to *continuing_inversion fragment* is called *continuing_inversion pair* which has $t \neq -1$ and $t \leq i$. Also the *ending_inversion pair* is defined for an *ending_inversion fragment* and has $t = -1$, and r = starting index of the last inversion (by Observation 1). We define another important set, *cont_inv_i'*, containing only the *continuing_inversion* pairs presenting inversions started at index i' and still exist as *continuing_inversion* pair, at index i, $i \geq i'$. Subset of any *cont_inv_i'* is denoted as *cont_inv*, that contains one or multiple *continuing_inversion* pairs (each having the same value for t) presenting all those inversions which produce the same prefix from index $t \leq i$, up to i.

We define S_x and S_y to be the sets of all possible inversion sets θ over x and y respectively. In general, $\theta_x \in S_x$ and $\theta_y \in S_y$ are used to present the matching phase. Deciding whether any consensus sequence exists between two given DNA sequences x and y having the same length n, involves finding out the existence of common agreed sequences of x and y. For this purpose we track the matched pairs between $T_x[n+1][n]$ and $T_y[n+1][n]$ for each index or column $i = 1, 2, \ldots, n$. For the same index i, if an inversion fragment in T_x mapped by the Pair (t', r') and another in T_y mapped by Pair (t'', r''), yield the same α, and the respective inversed sequences $\theta_x(x)$ and $\theta_y(y)$ up to i is the same, then those two pairs are called *Matched Pairs* and corresponding inversion fragments are called *Matched Fragments*. The matched pairs are denoted as $\langle X sibling \rangle$ - $\langle Y sibling \rangle$ for the ease of representation. Both of $X sibling$ and $Y sibling$ may contain one or more Pairs. In the rest of the section, we define some table like data structures that will be used in our algorithm. Each table will record some information of the matched pairs and will be named based on the type of $\theta(x)$ at each column i. Column i of each table presents some alignment of $\theta_x(x)$ and $\theta_y(y)$ up to index i.

ICA_table[i] - Inversions Completed at i. This table holds rows of $\langle X sibling \rangle$-$\langle Y sibling \rangle \equiv \langle (t', r') \rangle$-$\langle (t'', r'') \rangle \equiv \langle (-1, r') \rangle$-$\langle (-1, r'') \rangle$ presenting an alignment of $\theta_x(x)$ with $\theta_y(y)$ up to i, where the last inversion in θ_x and θ_y are $(p', q') \equiv (r', i)$ and $(p'', q'') \equiv (r'', i)$ respectively by Observation 1. That is, the last inversion in θ_x and θ_y are running from index r' and r'' respectively, both ending at i.

ISA_table_x[ ][i] - Inversions Started At i. This table presents an alignment of $\theta_x(x)$ (upto i), having the last inversion ended at $i - 1$, and a new inversion starting from i, with the $\theta_y(y)$ (upto i), having the last inversion started before or at i, still continuing or ended at i. It contains the pairs $\langle (t', r_1), \ldots, (t', r_s) \rangle$ of x, presenting the **Inversions Started at** i and ended at i or later. These pairs map to the inversion fragments $T_x[j'][i]$, where $i \leq j' \leq n + 1$.

$ISA\_table\_x[][i]$ holds $k = 4$ rows, one for each of the base letters $\alpha \in \{A, T, C, G\}$ such that the row $ISA\_table\_x[\alpha][i]$, holds pairs yielding base letter α at index i. Each row consists of two fields: $X sibling$ and $Y sibling$.

The $X sibling$ consists of a x_cont_inv set (having type $cont\_inv$) and a x_end_inv (having type $ending\_inversion$) pair. The x_cont_inv holds the $continuing\_$
$inversion$ pairs starting from index i, and thus have $t = i$ and $r = j$, $j \geq i + 2$. The x_end_inv is the $ending\_inversion$ pair having $t = -1$ and $r = i$ or $i + 1$, representing inversion $(i, i)'$ (no change) or (i, i) (flip) respectively.

Initially $Y sibling$ is empty. In the matching phase, $Y sibling$ maintains a list of pointers to the matched pairs of $X sibling$ in T_y, and categorized into two types, namely, single $ending\_inversion$ pair named as y_end_inv (Type 1) and set $cont\_inv$ named as y_cont_inv (Type 2) where all pairs have the same t, $t \leq i$.

Now we explain the intuition behind keeping these records. Both types of pointers ($Y sibling$) mentioned above are considered as matched pairs of x_cont_inv set. But for x_end_inv, only Type 2 is considered as the matched pair in this table. For each Type 1, i.e., y_end_inv in $Y sibling$ list, we keep a separate record $\langle X sibling \rangle$-$\langle Y sibling \rangle \equiv \langle x\_end\_inv \rangle$-$\langle y\_end\_inv \rangle$ in the $ICA\_table[i]$. Though this creates redundancy but this separation makes the data structure conceptually simpler and keeps the final decision checking simple at the end of the algorithm. Please refer to the Figure 2 for an illustration.

Observation 2. *At any* i, $\sum_{\alpha \in \{A, T, C, G\}} |X sibling| = n - i + 2$, *where* $X sibling \in ISA\_table\_x[\alpha][i]$. *Here the total number of continuing_inversion pairs is* $n - i$ *and the number of ending_inversion pairs is 2.*

ISB_table[i] - Inversions Started Before i. It holds rows $\langle X sibling \rangle$-$\langle Y sibling \rangle$ just as before presenting alignments of $\theta_x(x)$ yielding α at index i (but having the last inversion started **before** i, still continuing or has ended at i), with $\theta_y(y)$ yielding the same base letter α at index i (having the last inversion started before or at i, still continuing, or has ended at i). Here the x_cont_inv set of $X sibling$ has $t = i' < i$ and the x_end_inv holds some upper diagonal $ending\_inversion$ pair. Structure of $Y sibling$ and the intuition behind the records are the same as that in $ISA\_table[][i]$ (refer to the Figure 3).

ISA_table_y[][i]. It contains $\langle Y sibling \rangle \equiv \langle y\_cont\_inv, y\_end\_inv \rangle$ just like the $X sibling$ in $ISA\_table\_x[][i]$. This $Y sibling$ is actually get pointed by the $Y sibling$ lists of $X siblings$, at $ISA\_table\_x[][i]$, $ISB\_table[i]$ and $ICA\_table[i]$.

3   Algorithm

Common inversed sequences between x and y are computed by tracking the matched pairs between $T_x[][]$ and $T_y[][]$ from column $i = 1$ to n. The following procedures will be used in our algorithm.

Procedure 1. *Next_Calculation((t', r'), i, T_x):* If the input pair (t', r') is of type *continuing_inversion*, it returns the pointer to one unique next agreed pair with $t = -1$ (if the next agreed pair is *ending_inversion*) or $t = t'$ (if the next agreed pair is *continuing_inversion*). Otherwise, if the input pair (t', r') is of type *ending_inversion*, it returns the pointer to the $ISA\_table\_x[][i + 1]$ as a new inversion is supposed to start from $i + 1$. Similar actions are performed for y if T_y is the input.

Procedure 2. *PairUp_xColl_yColl(collection_x, collection_y, i):* This step is called at iteration i, with the matched pairs for index $i + 1$ as input. It sets the $\langle collection\_y \rangle \equiv \langle y\_cont\_inv, y\_end\_inv \rangle$ as $Y sibling$ of $\langle collection\_x \rangle \equiv \langle x\_cont\_inv, x\_ending\_inv \rangle$. Thus it lets the alignment $\theta_x(x)$ and $\theta_y(y)$ up to i, proceed one step forward, i.e., from i to $i + 1$. It executes following steps.

step a: If x_end_inv and y_end_inv both exist, then pair them up and insert into $ICA\_table[i + 1]$.
step b: Insert a pointer to the y_cont_inv into the $Y sibling$ list of $collection\_x$.
step c: Insert a pointer to the y_end_inv into the $Y sibling$ list of $collection\_x$.

Procedure 3. *PairUp_xColl_ySingle(collection_x, single_y, i):* It works as above but here the *single_y* is a single pair (t, r). If both *collection_x* and *single_y* are nonempty (*Compatibility Check*), it performs the following steps.

step a: If *single_y* is an *ending_inversion* and *collection_x* has x_end_inv pair, then pair them up and insert into $ICA\_table[i + 1]$
step b: Insert a pointer to *single_y* into the $Y sibling$ list of *collection_x*.

Procedure 4. *next_calculation_collection(x_cont_inv, x_next_atcg[], i):* Find the next agreed pairs of x_cont_inv and keep those in a child table x_next $\_atcg[]$ such that $x\_next\_atcg[\alpha]$ holds the agreed pairs yielding α. For example, lets say, $x\_cont\_inv = \langle (t', r_1), (t', r_2), \dots, (t', r_p) \rangle$. For each of these pairs we call $next\_calculation((t', r'), i, T\_x)$, $r' = 1, 2, \dots, p$. Each time as soon as one unique next agreed pair is returned, we add that to $x\_next\_atcg[]$ as follows.

case 1: If the next agreed pair is a *continuing_inversion* pair, yielding α, then insert into the x_cont_inv of $next\_atcg[\alpha]$.
case 2: If the next agreed pair ends at $i + 1$ (has $t = -1$) yielding α, then set it as the x_end_inv of $x\_next\_atcg[\alpha]$.

Now we explain the algorithm using the procedures stated above. The main algorithm iterates over $i = 1$ to $n - 1$. The column i of each of the tables described above actually represent the alignment of $\theta_x(x)$ and $\theta_y(y)$ up to index i for some θ_x and θ_y. So at each iteration i, it processes the rows in three tables: $ICA\_table[i]$, $ISA\_table[][i]$, and $ISB\_table[i]$ to calculate the next agreed pairs, pair up the matched pairs and insert those into the column $i+1$ of the appropriate table. If for any row $\langle Xsibling\rangle$-$\langle Ysibling\rangle$, next agreed pairs of $Xsibling$ does not get matched pair from next agreed pairs of $Ysibling$, then it means no alignment with the inverted sequence of x presented by that $Xsibling$ exists in y. Thus this alignment $\langle Xsibling\rangle$-$\langle Ysibling\rangle$ is not passed forward anymore and rather dropped here. We will explain the algorithm using an illustrative example. Consider, $x = AGCCAGCT$ and $y = TCGGGCTT$ given in Figure 1.

3.1   Initialization

$ISA\_table\_x[1]$ and $ISA\_table\_y[1]$ are shown in the Figure 2. It executes the following loop to start alignments by pairing up these two tables. Input parameters are set as: $table\_x = ISA\_table\_x[1]$, $table\_y = ISA\_table\_y[1]$, and $i = 1$.

Procedure 5. *Four Iteration Loop(table_x, table_y, i):*
For each base letters $\alpha \in \{A, T, C, G\}$, if $table\_x[\alpha]$ has non empty $Xsibling$ and $table\_y[\alpha]$ has non empty $Ysibling$ (*Compatibility Check*), then it calls $PairUp\_xColl\_yColl(collection\_x, collection\_y, i)$ with $collection\_x=table\_x[\alpha]$, and $collection\_y=table\_y[\alpha]$.

(i) Before Initialization

j\i	1
1	(1,1).A
2	(1,1).T
3	(1,2).C
4	(1,3).C
5	(1,4).G
6	(1,5).T
7	(1,6).C
8	(1,7).G
9	(1,8).A
T_x[1]	

j\i	1
1	(1,1).T
2	(1,1).A
3	(1,2).G
4	(1,3).C
5	(1,4).C
6	(1,5).C
7	(1,6).G
8	(1,7).A
9	(1,8).A
T_y[1]	

α	x_end_inv	x_cont_inv	Ysibling
A	(-1,1)	(1,9)	-
T	(-1,2)	(1,6)	-
C	-	(1,3), (1,7)	-
G	-	(1,4),(1,5),(1,8)	-

ISA_table_x[1]

α	y_end_inv	y_cont_inv
A	(-1,2)	(1,8), (1,9)
T	(-1,1)	
C		(1,4),(1,5),(1,6)
G		(1,3),(1,7)

ISA_table_y[1]

(ii) After Initialization

Xsibling	Ysibling
(-1,1)	(-1,2)
(-1,2)	(-1,1)

ICA_table [1]

x_end_inv	x_cont_inv	Ysibling
-	-	-

ISB_table [1]

α	x_end_inv	x_cont_inv	Ysibling
A	(-1,1)	(1,9)	⟨(1,8), (1,9)⟩, ⟨(-1,2)⟩
T	(-1,2)	(1,6)	⟨(-1,1)⟩
C	-	(1,3), (1,7)	⟨(1,4), (1,5), (1,6)⟩
G	-	(1,4),(1,5),(1,8)	⟨(1,3), (1,7)⟩

ISA_table_x[1]

i   ii

Fig. 2. (i)Before Initialization; (ii)After Initialization

3.2   Iteration

For each iteration $i = 1, 2, \ldots, n - 1$, following steps are performed.

Step 1   Process ICA_table[i]: For the first row $\langle Xsibling\rangle$-$\langle Ysibling\rangle=\langle(-1,r')\rangle$-$\langle(-1,r'')\rangle$, we call Procedure 1, i.e., $next\_calculation((-1,r'), T\_x, i)$ and

$next\_calculation((-1, r''), T\_y, i)$. They return pointers to $ISA\_table\_x[i+1]$ and $ISA\_table\_y[i+1]$ respectively. After that, we call the *Four Iteration Loop()* with these two tables as input. Other rows of $ICA\_table[i]$ are not processed as they involve doing the same assignments. See Figure 3 an illustration.

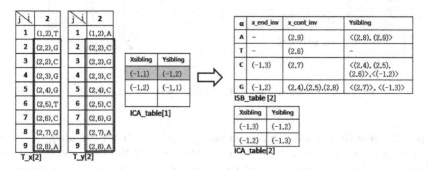

Fig. 3. Demonstration of Step 1 for iteration 1

Step 2   Process ISA_table_x[][i]: For each $\alpha \in \{A, T, C, G\}$ we perform Step 2.1, Step 2.2 and Step 2.3.

Step 2.1: It calls Procedure 4, with x_cont_inv of $ISA\_table\_x[\alpha][i]$, which finds its next agreed pairs and keeps those in a child table $x\_next\_atcg[]$ (see Figure 4)

Step 2.2: For each list item $Ysibling[p]$, in this step we find the alignment of the pairs in $x\_next\_actg[]$ (calculated in the previous step) with the next agreed pairs found from $Ysibling[p]$. We need to deal with one of the following cases.

Step 2.2 Case 1: The $Ysibling[p]$ is of type y_cont_inv having $size > 1$:
  Step 2.2.1: If $y\_next\_atcg[]$ of $Ysibling[p]$ is not calculated yet, then call Procedure 4, i.e., $next\_calculation\_collection(Ysibling[p], y\_next\_actg[], i)$.
  Step 2.2.2: Now both the $x\_next\_atcg[]$ and $y\_next\_actg[]$ are ready to be paired up. So we call the *Four iteration loop()* with these two sets as input.
  Step 2.2.3: If $Xsibling$ has x_end_inv pair, and $y\_next\_actg[]$ has not been paired with $ISA\_table\_x[][i+1]$ yet, then pair them up by calling the same *Four Iteration Loop()* with input tables: $ISA\_table\_x[][i+1]$ and $y\_next\_actg[]$. Please refer to Figure 5 for an illustration.

Step 2.2 Case 2: The $Ysibling[p]$ is of type y_cont_inv having $size = 1$:
  Step 2.2.4: Call $next\_calculation((t', r'), i, T_y)$, where $(t', r')=y\_cont\_inv$. Let the returned unique next agreed pair yield α and name it *pair_y*.
  Step 2.2.5: We call $PairUp\_xColl\_ySingle(x\_next\_actg[\alpha], single\_y, i)$.
  Step 2.2.6: If $Xsibling$ has x_end_inv pair, and the *pair_y* has not been paired with $ISA\_table\_x[\alpha][i+1]$ yet, then we call
$PairUp\_xColl\_ySingle(ISA\_table\_x[\alpha][i+1], pair\_y, i)$

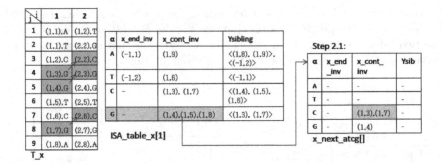

Fig. 4. Demonstration of Step 2.1 for $\alpha = G$ in iteration 1

Step 2.2 Case 3: If $Ysibling[p]$ is of type y_end_inv: If $x\_next\_atcg[]$ has not been paired up with $ISA\_table\_y[][i+1]$ yet then we call the *Four Iteration Loop* with input tables: $x\_next\_atcg[]$ and $ISA\_table\_y[i+1]$. Please refer to Figure 5.

Step 2.3 Update the ISB_table[i+1]: For each new $x\_next\_atcg[\alpha]$ created in Step 2.1, if it has non empty $Ysibling$ list, then we insert it into $ISB\_table[i+1]$ as new rows, where $\alpha \in \{A, T, C, G\}$. Please see the Figure 5 for an illustration.

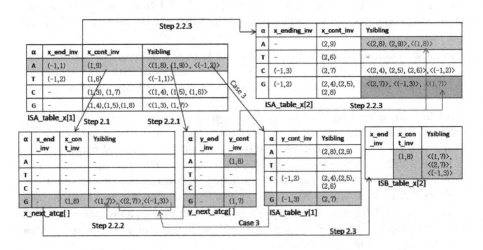

Fig. 5. Demonstration of Steps 2.2 and 2.3 for $\alpha = A$ in iteration 1

Step 3 Process ISB_table[i]: For each row p of $ISB\_table[i]$: $\langle Xsibling \rangle$ - $\langle Ysibling \rangle$, we execute the Steps 3.1, 3.2, and 3.3. They are identical to Step 2.1, 2.2, 2.3 except the fact that the row items $ISB\_table[p][i]$ are used instead of $ISA\_table\_x[\alpha][i]$.

3.3 Termination

After the iterations complete, if the $ICA\_table[n]$ contains no row, we return NO indicating the absence of the consensus sequence between x and y. Otherwise return YES, indicating existence of the consensus sequence between x and y.

Note here that, algorithm presented by Cho et al. [6] does not return correct results because of not tracking the prefix of common ancestors properly which is kept in our algorithm using the ISA_table, ISB_table and ICA_table.

4 Analysis and Experimental Result

The proofs of correctness, polynomiality, and complexity analysis of the algorithm are not presented here because of the space constraints, but will be available soon in the journal version. Theoretical worst case and average case time complexity of the algorithm are $O(n^4)$ and $O(n^3)$ respectively. However, practical run time of the algorithm in the worst and average case are $O(n^3)$ and $O(n^2)$ respectively. This is apparent from the experimental results reported in Figure 6. In our experiments, x and y are selected such that both contain the same number of bases from the complement category, $A-T$ and $G-C$. We define *performance factor* (equivalent to the runtime), a counter that keeps track of the total number of statements executed for finding next agreed *Pair (t, r)* and pairing the matched agreed pairs. *So runtime of a test case is found by adding the performance factor of the dominating steps, i.e., Step 2 and Step 3 in the algorithm.* For each length n (ranging from 30 to 120), we run the experiment under six categories (column 3 to 8 of the table in Figure 6) based on the percentage of similarity between the input sequences x and y. Under each category, we generate twenty sets of test cases by randomly choosing x and y. Then we calculate the average runtime of the test cases. No comparisons with previous works is provided as there exists no other works on the same problem that we are working on (algorithm by Cho et al. [6] is inaccurate thus not considered). We plan for experiments with real datasets in future.

n^3	Similarity Length, n	Runtime					
		Result: YES					Result: NO
		100% $(O(n^3))$	90% $(O(n^3))$	80% $(O(n^2))$	50% $(O(n^2))$	Arbitrary $(O(n^2))$	Arbitrary $(O(n^2))$
27000	30	15244.92	8134.25	6350.92	4876.08	4436.30	3126.20
125000	50	70473.00	33074.50	21750.70	15574.60	15353.90	7299.40
343000	70	197616.20	72018.70	47676.20	33668.80	27614.20	20505.75
729000	90	417153.00	117320.80	83799.90	62009.80	52154.40	23087.36
1728000	120	987613.10	223927.00	162498.80	113256.50	99694.60	58938.85

Fig. 6. Runtime of the algorithm for $n = 30, 50, 70, 90, 120$

5   Steps for Detecting Allelic Heterogeneity

Now, lets see how this algorithm can help in detecting alellic heterogeneity. As we proceed from index i to $i+1$, we maintain successor links among the agreed pairs for either x or y. After the algorithm terminates, these connections resemblance a tree type structure (from left to right). So applying DFS on the structure gives us all those paths starting at index 1 and ending at index n. These actually represents the consensus strings. Now lets think we have an unknown disease χ and we want to see if it is allelic heterogeneous with the disease Cystic Fibrosis, i.e., if both of these are mutated from the same gene CFTR[1]. For this purpose we input gene sequence of χ and Cystic Fibrosis as x and y respectively. Then if the algorithm terminates with 'NO', it means no common gene sequence exists from which both χ and Cystic Fibrosis can be derived. So they are definitely not allelic heterogeneous. On the other hand, if algorithm returns 'YES', it means there is a possibility of allelic heterogeneity among those two diseases. So now we have to perform DFS and compare the retrieved consensus strings with CFTR. If match is found then we can perform additional clinical diagnostic approaches to validate the output. However, if no match is found, there is no need of performing expensive clinical diagnostic tests, which saves huge energy and costs.

6   Conclusion

This paper maps the consensus string problem to the biomedical problem of detecting allelic heterogeneity. The proposed algorithm finds the common ancestor sequence given two mutated sequences where mutation involves only non overlapping inversions. Future research endeavor could be directed towards other mutation operations such as insertion, deletion and translocation. Finding minimum consensus string distance for two input sequences and improving time complexity remain as future works as well. Besides these, we introduce an open problem: determine the complexity class of the consensus string problem for N input sequences under non overlapping inversion metric.

Acknowledgment. This research work has been conducted at CSE, BUET as part of the M.Sc. Engg. Thesis of Zohora under the supervision of Rahman.

References

1. Prasun, P., Pradhan, M., Agarwal, S.: One gene, many phenotypes. Journal of Postgraduate Medicine 53(4) (2007)
2. Meisenberg, G., Simmons, W.H.: Allelic heterogeneity is the greatest challenge for molecular genetic diagnosis. Elsevier Health Sciences (2011)
3. Sutcliffe, J.S., Delahanty, R.J., Prasad, H.C., McCauley, J.L., Han, Q., Jiang, L., Li, C., Folstein, S.E., Blakely, R.D.: Allelic heterogeneity at the serotonin transporter locus (slc6a4) confers susceptibility to autism and rigid-compulsive behaviors. The American Journal of Human Genetics 77(2), 265–279 (2005)

---

[1] [http://www.jpgmonline.com/viewimage.asp?img=jpgm_2007_53_4_257_33968_1 .jpg]

4. Castellani, C.: Cftr2: How will it help care? Paediatric Respiratory Reviews 14, 2–5 (2013)
5. Chen, Z.Z., Ma, B., Wang, L.: A three-string approach to the closest string problem. Journal of Computer and System Sciences 78(1), 164–178 (2012)
6. Cho, D.-J., Han, Y.-S., Kim, H.: Alignment with non-overlapping inversions on two strings. In: Pal, S.P., Sadakane, K. (eds.) WALCOM 2014. LNCS, vol. 8344, pp. 261–272. Springer, Heidelberg (2014)
7. Schöniger, M., Waterman, M.S.: A local algorithm for DNA sequence alignment with inversions. Bulletin of Mathematical Biology 54(4), 521–536 (1992)
8. Vellozo, A.F., Alves, C.E.R., do Lago, A.P.: Alignment with non-overlapping inversions in $o(n^3)$-time. In: Bücher, P., Moret, B.M.E. (eds.) WABI 2006. LNCS (LNBI), vol. 4175, pp. 186–196. Springer, Heidelberg (2006)
9. Charnoz, A., Agnus, V., Malandain, G., Soler, L., Tajine, M.: Tree matching applied to vascular system. In: Brun, L., Vento, M. (eds.) GbRPR 2005. LNCS, vol. 3434, pp. 183–192. Springer, Heidelberg (2005)

Continuous Time Bayesian Networks for Gene Network Reconstruction: A Comparative Study on Time Course Data

Enzo Acerbi[1,2] and Fabio Stella[3]

[1] Singapore Immunology Network, A*STAR, Singapore
[2] DIMET, University of Milano-Bicocca, Italy
[3] DISCo, University of Milano-Bicocca, Italy
`enzo_acerbi@immunol.a-star.edu.sg`

Abstract. Dynamic aspects of regulatory networks are typically investigated by measuring relevant variables at multiple points in time. Current state-of-the-art approaches for gene network reconstruction directly build on such data, making the strong assumption that the system evolves in a synchronous fashion and in discrete time. However, omics data generated with increasing time-course granularity allow to model gene networks as systems whose state evolves in continuous time, thus improving the model's expressiveness. In this work continuous time Bayesian networks are proposed as a new approach for regulatory network reconstruction from time-course expression data. Their performance is compared to that of two state-of-the-art methods: dynamic Bayesian networks and Granger causality. The comparison is accomplished using both simulated and experimental data. Continuous time Bayesian networks achieve the highest F-measure on both datasets. Furthermore, precision, recall and F-measure degrade in a smoother way than those of dynamic Bayesian networks and Granger causality, when the complexity of the gene regulatory network increases.

Keywords: gene network reconstruction, time-course data, continuous time Bayesian networks.

1 Introduction

Biological networks are processes involving many heterogeneous and interacting molecular species. Typically, biological networks are conceptually categorized into a few main types of sub-networks such as signalling, regulatory, metabolic and protein-protein which, despite being highly interconnected and overlapping, are at the moment still treated independently. Our experiments are concerned with gene regulatory networks (GRNs). The relevance of studying GRNs comes from the fact that through regulatory interactions, the cell can modify its genetic transcriptional state in response to internal and external stimuli. Therefore, from the understanding of GRNs complex biological mechanisms can be deciphered.

Network reconstruction, also known as network *"reverse-engineering"* or network inference, is a challenging research area in systems biology, which aims

M. Basu, Y. Pan, and J. Wang (Eds.): ISBRA 2014, LNBI 8492, pp. 176–187, 2014.
© Springer International Publishing Switzerland 2014

to uncover the underlying structure of such mechanisms. Despite having led to significant discoveries, current reconstruction methods still suffer from several limitations, such as having to deal with a large search space: this makes the task of finding the optimal model extremely hard, thus often leading to models with poor accuracy. Moreover, the availability of experimental datasets has historically been limited either in terms of measured variables or time-course length. In recent years, in molecular biology there has been an increase in the amount of time-course omics data, that is, genome-wide measurements with a significantly rich time dimension, made available by high throughput techniques. Measuring thousands of variables simultaneously and over time offers the opportunity for in depth study of the dynamic evolution of a system. Most computational methodologies for network reconstruction have been conceived before the advent of omic technologies and may not always be suitable for these new types and magnitudes of data. So far a number of approaches have been applied, each one making use of different strategies and assumptions. Detailed reviews of the existing approaches can be found in [1,2]. A common feature of current state-of-the-art approaches is that they directly build on expression data which is obtained by measuring the system variables at multiple points in time (e.g. through expression microarray). Thus, they make the strong assumption that the system under study evolves in a synchronous fashion and in discrete time. Nowadays, omics data are being generated with increasing time-course granularity and length. Consequently, modellers are allowed to increase models' expressiveness and to represent the system's state as evolving in continuous time.

In this paper continuous time Bayesian networks (CTBNs) [8] are proposed as a new approach for solving the GRN reconstruction problem from time-course data. In a CTBN the state of a variable evolves in continuous time as a function of a conditional Markov process. This setting brings many advantages to the description of the temporal aspect of a system, some of them directly relevant to the GRN reconstruction task. Firstly, the structural learning problem for CTBNs can be solved locally in polynomial time once the maximum number of regulators for each gene is set. Secondly, CTBNs can handle variables evolving at different time granularities (i.e. fast evolving variables and slow evolving variables). Thirdly, through inference CTBNs allow us to answer queries directly involving the quantification of temporal aspects such as *"How long does gene X have to remain up-regulated to have an effect on the regulation of gene Y?"*. Finally, continuous time Bayesian networks share all of the advantages which are characteristic of probabilistic graphical models and which make them suitable for the analysis of biological networks. This paper is focused on a comparison of the performance achieved by the structural learning of CTBNs to that achieved by two state-of-the-art models, namely dynamic Bayesian networks (DBNs) [3] and Granger causality analysis (GC) [4,5]. The comparison with DBNs originates from the fact that CTBNs model time explicitly and thus theoretically overcome the limitation associated with the observational model assumption on which DBNs are built. GC has been selected because of its historical and current relevance when faced with the task of inferring causal relations from time series data. Since its early introduction, GC has been successfully applied to a multitude of domains such as economics, neuroscience and biology. Since DBNs and GC are well documented in the literature, their formal definition is omitted. DBNs and GC were also directly compared for GRN reconstruction in [6]: the authors showed that when the length of the time-course is smaller than a given threshold, DBNs tend to outperform GC; while vice-versa when the length of the time-course is greater than the given threshold. The analysis and comparisons performed here are based on an extensive and robust set of numerical experiments run on simulated data and include tests on an experimental dataset as well. The study with simulated data has been conducted on networks of increasing dimension (10, 20, 50 and 100 genes) in order to investigate how the approaches perform on small and big networks. The networks were extracted from the known transcriptional networks of two different organisms: *E. coli* and *S. cerevisiae*. To ensure robustness, the performance is not calculated on a single network instance but it is estimated by the average value computed over a set of 10 randomly sampled network instances of the same dimension. The time-course data generated from the networks includes a rich set of interventional data. The study on experimental data, based on a five genes regulatory network synthetically constructed in the yeast *S. cerevisiae* [7], provides a gold standard for precise evaluation.

2   Continuous Time Bayesian Networks

DBNs model dynamical systems without representing time explicitly. They discretize time to represent a dynamical system through several time slices. In [8] the authors pointed out that *"since DBNs slice time into fixed increments, one must always propagate the joint distribution over the variables at the same rate"*. Therefore, if the system consists of processes which evolve at different time granularities and/or the obtained observations are irregularly spaced in time, the inference process may become computationally intractable. CTBNs overcome the limitations of DBNs by explicitly representing temporal dynamics and thus allow us to recover the probability distribution over time when specific events occur. CTBNs have been used to discover intrusion in computers [9], to analyse the reliability of dynamical systems [10], for learning social network dynamics [11] and to model cardiogenic heart failure [12]. However, CTBNs have never been applied to the analysis of biological data.

A continuous time Bayesian network (CTBN) is a probabilistic graphical model, whose nodes are associated with random variables and whose state evolves in continuous time. The evolution of each variable depends on the state of its parents in the graph associated with the CTBN model. A CTBN consists of two main components: *i)* an initial probability distribution and *ii)* the dynamics which rule the evolution over time of the probability distribution associated with the CTBN.

Definition 1. *(Continuous time Bayesian network). [8]. Let* **X** *be a set of random variables* $X_1, X_2, ..., X_N$. *Each* X_n *has a finite domain of values* $Val(X_n) =$

$\{x_1, x_2, ..., x_{I(n)}\}$. *A continuous time Bayesian network B over $\mathbf{X}$ consists of two components: the first is an initial distribution $P_{\mathbf{X}}^0$, specified as a Bayesian network $\mathcal{B}$ over $\mathbf{X}$. The second is a continuous transition model, specified as:*

- *a directed (possibly cyclic) graph $\mathcal{G}$ whose nodes are $X_1, X_2, ..., X_N$; $Par_{\mathcal{G}}(X_n)$, often abbreviated $\mathbf{U}_n$, denotes the parent set of X_n in $\mathcal{G}$.*
- *a conditional intensity matrix, $\mathbf{Q}_{X_n}^{Par_{\mathcal{G}}(X_n)}$, for each variable $X_n \in \mathbf{X}$.*

Given the random variable X_n, the *conditional intensity matrix* (CIM) $\mathbf{Q}_{X_n}^{Par(X_n)}$ $= \mathbf{Q}_{X_n|\mathbf{U}_n}$ consists of a set of intensity matrices, one intensity matrix

$$\mathbf{Q}_{X_n|\mathbf{u}} = \begin{bmatrix} -q_{x_1|\mathbf{u}} & q_{x_1 x_2|\mathbf{u}} \cdot & q_{x_1 x_{I(n)}|\mathbf{u}} \\ q_{x_2 x_1|\mathbf{u}} & -q_{x_2|\mathbf{u}} \cdot & q_{x_2 x_{I(n)}|\mathbf{u}} \\ & \cdots & \\ q_{x_{I(n)} x_1|\mathbf{u}} & q_{x_{I(n)} x_2|\mathbf{u}} \cdot & -q_{x_{I(n)}|\mathbf{u}} \end{bmatrix},$$

for each instantiation $\mathbf{u}$ of the parents $\mathbf{U}_n$ of node X_n, where $q_{x_i|\mathbf{u}} = \sum_{x_j \neq x_i} q_{x_i x_j|\mathbf{u}}$ is the rate of leaving state x_i for a specific instantiation $\mathbf{u}$ of $\mathbf{U}_n$, while $q_{x_i x_j|\mathbf{u}}$ is the rate of arriving to state x_j from state x_i for a specific instantiation $\mathbf{u}$ of $\mathbf{U}_n$. Matrix $\mathbf{Q}_{X_n|\mathbf{U}_n}$ can equivalently be summarized by using two types of parameters, $q_{x_i|\mathbf{u}}$ which is associated with each state x_i of the variable X_n when its parents are set to $\mathbf{u}$, and $\theta_{x_i x_j|\mathbf{u}} = \frac{q_{x_i x_j|\mathbf{u}}}{q_{x_i|\mathbf{u}}}$ which represents the probability of transitioning from state x_i to state x_j, when it is known that the transition occurs at a given instant in time.

Learning the structure of a CTBN from a data set $\mathcal{D}$ consists of finding the structure $\mathcal{G}$ which maximizes the *Bayesian score* [13]:

$$\mathbf{score}_B(\mathcal{G} : \mathcal{D}) = \ln P(\mathcal{D}|\mathcal{G}) + \ln P(\mathcal{G}). \tag{1}$$

Efficiency of the search algorithm for finding the optimal structure $\mathcal{G}^*$ is significantly increased if we assume *structure modularity* and *parameter modularity*. The prior over the network structure $P(\mathcal{G})$ satisfies the structure modularity property if $P(\mathcal{G}) = \prod_{n=1}^{N} P(Par(X_n) = Par_{\mathcal{G}}(X_n))$, while the prior over parameters satisfies the parameter modularity property, if for any pair of structures $\mathcal{G}$ and $\mathcal{G}'$ such that $Par_{\mathcal{G}}(X) = Par_{\mathcal{G}'}(X)$ we have that $P(\mathbf{q}_{\mathbf{X}}, \theta_{\mathbf{X}}|\mathcal{G}) = P(\mathbf{q}_{\mathbf{X}}, \theta_{\mathbf{X}}|\mathcal{G}')$. In [13] the authors combined parameter modularity, parameter independence, local parameter independence and assumed a Dirichlet prior over θ parameters and a beta prior over q parameters to obtain the following expression of the Bayesian score for a CTBN B:

$$\mathbf{score}_B(\mathcal{G} : \mathcal{D}) = \sum_{n=1}^{N} FamScore(X_n, Par_{\mathcal{G}}(X_n) : \mathcal{D}) \tag{2}$$

where

$$FamScore(X_n, Par_{\mathcal{G}}(X_n) : \mathcal{D}) = \ln P(Par(X_n) = Par_{\mathcal{G}}(X_n)) + \tag{3}$$
$$\ln MargL^q(X_n, \mathbf{U}_n : \mathcal{D}) + \tag{4}$$
$$\ln MargL^\theta(X_n, \mathbf{U}_n : \mathcal{D}). \tag{5}$$

According to [13] $MargL^q(X_n, \mathbf{U}_n : \mathcal{D})$ can be written as follows:

$$\prod_{\mathbf{u}} \prod_{x} \frac{\Gamma\left(\alpha_{x|\mathbf{u}} + M[x|\mathbf{u}] + 1\right) \tau_{x|\mathbf{u}}^{\alpha_{x|\mathbf{u}}+1}}{\Gamma\left(\alpha_{x|\mathbf{u}} + 1\right) \left(\tau_{x|\mathbf{u}} + T[x|\mathbf{u}]\right)^{\alpha_{x|\mathbf{u}}+M[x|\mathbf{u}]+1}} \tag{6}$$

while $MargL^\theta(X_n, \mathbf{U}_n : \mathcal{D})$ can be written as follows:

$$\prod_{\mathbf{u}} \prod_{x} \frac{\Gamma\left(\alpha_{x|\mathbf{u}}\right)}{\Gamma\left(\alpha_{x|\mathbf{u}} + M[x|\mathbf{u}]\right)} \times \prod_{x' \neq x} \frac{\Gamma\left(\alpha_{xx'|\mathbf{u}} + M[x, x'|\mathbf{u}]\right)}{\Gamma\left(\alpha_{xx'|\mathbf{u}}\right)}. \tag{7}$$

where Γ is the Gamma function, $M[x, x'|\mathbf{u}]$ represents the count of transitions from state x to state x' for the node X_n when the state of its parents $\mathbf{U}_n$ is set to $\mathbf{u}$, while $T[x|\mathbf{u}]$ is the time spent in state x by the variable X_n when the state of its parents $\mathbf{U}_n$ is set to $\mathbf{u}$. Furthermore, $M[x|\mathbf{u}] = \sum_{x' \neq x} M[x, x'|\mathbf{u}]$, $\alpha_{x|\mathbf{u}}$ and $\tau_{x|\mathbf{u}}$ are hyperparameters over the CTBN's q parameters while $\alpha_{xx'|\mathbf{u}}$ are hyperparameters over the CTBN's θ parameters. However, $Par(\mathcal{G})$ does not grow with the amount of data. Therefore, the significant terms of $FamScore(X_n, Par_{\mathcal{G}}(X_n) : \mathcal{D})$ are $MargL^q(X_n, \mathbf{U}_n : \mathcal{D})$ and $MargL^\theta(X_n, \mathbf{U}_n : \mathcal{D})$. Thus, given a dataset $\mathcal{D}$, the optimal CTBN's structure is selected by solving the following problem:

$$\max_{\mathcal{G} \in \mathbf{G}} \sum_{n=1}^{N} \ln MargL^q(X_n, \mathbf{U}_n : \mathcal{D}) + \ln MargL^\theta(X_n, \mathbf{U}_n : \mathcal{D}), \tag{8}$$

where $\mathbf{G} = \{\mathbf{U}_n \in \mathbf{X} : n = 1, ..., N\}$ represents all possible choices of parent set $\mathbf{U}_n$ for each node X_n, $n = 1, ..., N$. Optimization problem (8) is over the space $\mathbf{G}$ of possible CTBN structures, which is significantly simpler than that of BNs and DBNs. Indeed, learning optimal BN's structure is NP-hard, while the same does not hold true in the context of CTBN in the case where the maximum number of parents for each node is limited. In fact, in CTBN all edges are across time and represent the effect of the current value of one variable to the next value of other variables. Therefore, no acyclicity constraints arise, and it is possible to optimize the parent set $\mathbf{U}_n$ for each variable X_n, $n = 1, ..., N$, independently. In [13] the authors proved that, if the maximum number of parents is restricted to k, then learning the optimal CTBN's structure is polynomial in the number of nodes N. However, we usually do not want to exhaustively enumerate all possible parent sets $\mathbf{U}_n$ for each variable X_n, $n = 1, ..., N$. In this case we resort to *greedy hill-climbing* search by using operators that add/delete/reverse edges to the CTBN structure $\mathcal{G}$. It is worthwhile to mention that family scores of different variables do not interact. Therefore, the *greedy hill-climbing* search on CTBNs can be performed separately on each variable X_n, thus making the overall search process much more efficient than on BNs and DBNs.

3   Data

The simulated dataset was generated with the GeneNetWeaver tool [14] which has previously been used to generate datasets for the network inference challenge of the international Dialogue for Reverse Engineering Assessments and Methods (DREAM) competition [15]. The tool allows to extract sub-networks from the known *in vivo* regulatory network structures of *E. coli* and *S. cerevisiae*, endowing them with dynamic models of regulation. In order to generate the simulated datasets, sub-networks consisting of 10, 20, 50 and 100 genes were randomly extracted from the full networks of both organisms. To ensure robustness, 10 different network instances were sampled for each gene set dimension. This resulted in 10 networks of 10 genes, 10 networks of 20 genes, etc. For the sake of brevity, the sets of 10 networks consisting of 10, 20, 50 and 100 genes will be referred to as 10-NETs, 20-NETs, 50-NETs and 100-NETs respectively. When extracting the 10-NETs and 20-NETs, no constraint on the minimum number of regulators to include (i.e. nodes that have at least one outgoing link in the full network) was added, while for the 50-NETs and 100-NETs the minimum number of regulators to include was set to 20. This choice was made to avoid the generation of networks characterized by a large number of leaf nodes and thus with a too simple structure. Extraction of nodes was made randomly and with preference for the addition of neighbor nodes which maximize the modularity of the network structure. This procedure is referred to as *greedy neighbor selection*. We have noticed that with this procedure the network's modularity tends to proportionally increase with the number of nodes. A network with a high degree of modularity is a network with less complex regulation interactions, and therefore it is relatively easy to infer. This is the reason why with 50-NETs and 100-NETs we forced the procedure to add a minimum number of regulator nodes. No constraint was set on the maximum number of parents allowed per node. Given each extracted network structure, GeneNetWeaver combined ordinary and stochastic differential equations to generate the corresponding dataset. Generated datasets consist of 21 time points and 10 replicates for each experiment. Apart from the wild type time-course data (unperturbed network), interventional time-course data including knockouts, knockdowns, multifactorial perturbations and (when applicable) dual knockouts were generated extensively for every single gene in the network. The multiplicative constant of the white noise term in the stochastic differential equations was set to 0.05 as in DREAM4. Finally, all expression values were normalized between 0 and 1. This dataset aims to simulate the amount of data that high-throughput techniques will soon allow to generate while maintaining a realistic time-course magnitude: expression microarray experiments repeated with this many time points are uncommon today, but possible. On the other hand, the dataset is still unrealistically rich in terms of number of perturbations and replicates. Such a comprehensive dataset is necessary to fairly compare the analyzed methods.

Due to the current lack of reliable large scale gold standards, *in vivo* evaluation is a critical point for GRN reconstruction methods, which often rely on less quantifiable evaluations such as comparison with the existing literature and information available in public databases. The *in vivo* benchmarking was performed on a small but *certified network*: a five-gene network synthetically constructed in the yeast *S. cerevisiae* [7] and shown in Figure 2. This network, despite its small size, contains a representative set of interconnections, including regulator chains and feedback loops. The dynamic behaviour of the network was studied by shifting cells from glucose to galactose and vice-versa and by collecting samples every 20 min up to 280 min in the first case and every 10 min up to 190 min in the second case. 4 to 5 biological replicates were analyzed respectively, expression levels were measured through RT-PCR. No interventional data was used.

4   Results

4.1   Simulated Data

Prior to learning CTBNs, an extensive parameter optimization was run on the 10-NETs and 20-NETs, where the required learning time was still feasible. Optimal parameters values were subsequently applied to the 50-NETs and 100-NETs. Because the Bayesian models cannot handle continuous data, a discretization was applied. Discretization of continuous data is known to be a critical task: too few bins (levels) of discretization lead to a loss of important information. However, when increasing the number of bins it is known that the required amount of data and computational resources increase as well. To find the optimal number of bins, tests were performed with data discretized into 3, 4, 5, 6 and 7 bins equally sized and equidistant. Best performances were obtained when using 5 bins. It is worthwhile to note that discretization intervals were chosen individually for each variable (gene), based on the *maximum* and *minimum* value of expression levels of each variable among the whole set of data generated. In order to preserve the significance and comparability of the results, one needs to keep track of the discretization intervals applied to each variable. An analysis on the importance of the discretization strategy can be found in [16]. Regarding the hyperparameters α and τ, introduced in section 2, best values were found to be 0.01 and 5 respectively. These values were identified to be *optimal* by running the learning algorithm several times with different values for α and τ. The computational nature of the exact structural learning problem lent itself to greedy learning. However, preliminary tests on the 10-NETs returned the same results for both exhaustive and greedy learning, although it cannot be established whether the exhaustive learning on the larger networks would have returned better results. The last parameter investigated was the maximum number of parents allowed for each node: since the greater this value is, the longer is the computational time; sequential tests with an increasing value of this parameter were run. It was observed that CTBNs were never able to detect more than 3 parents per node, even when the true network contains nodes with a number of parents greater than 3. Indeed, it is known that score based structural learnings can fail to detect causal links when the dataset size is small compared to the number of parameters to be estimated [17].

For DBNs [1] parameter optimization on the number of discretization bins was re-run, and results confirmed that what is optimal for CTBNs may not be the

---

[1] Run with the Bayesian Net toolbox [18] version 1.07.

Table 1. Performance comparison on simulated data for organism *E.coli*

Method	NETs size	Mean Precision	Mean Recall	**Mean F1**
CG	10	0.46	0.68	**0.54**
	20	0.40	0.70	**0.49**
	50	0.24	0.82	**0.37**
	100	0.16	0.82	**0.27**
DBNs	10	0.90	0.29	**0.41**
	20	0.55	0.42	**0.47**
CTBNs	10	0.66	0.58	**0.61**
	20	0.72	0.48	**0.57**
	50	0.53	0.57	**0.54**
	100	0.45	0.51	**0.48**
Random	10	0.16	0.55	**0.24**
	20	0.11	0.51	**0.18**
	50	0.03	0.49	**0.06**
	100	0.02	0.50	**0.04**

best option for DBNs. Indeed, results indicated 3 as the optimal number of discretization bins for DBNs. Discretization intervals were selected individually for each variable, as was done for CTBNs. Model selection has been performed by using the BIC criterion, which is known to trade off data fitting with model complexity, thus controlling overfitting [17]. Analogously to what was observed with CTBNs, DBNs were never able to detect more than 3 parents per node. Experiments with 50-NETs and 100-NETs are not shown because DBNs learning became intractable.

For GC [2] analysis no discretization was required since the approach can handle continuous data. Best value for the model order parameter, i.e. the number of past observations to incorporate into the regression model, was discovered to be equal to 1. Covariance stationary (CS) is a requirement for the GC to be applied. Data resulted to be CS according to the ADF criterium, but not according to KPSS. However, when differencing was applied to correct this condition, data interpretation may have become more difficult, and in fact performances significantly worsen; consequently no differencing has been applied. Pre-processing steps of de-trending and de-meaning have been applied. After performing the analysis and obtaining the matrix of magnitudes of GC interactions, the statistically significant set of interactions has been individuated. The optimal significance cut-off resulted to be 0.01, with a Bonferroni multiple test correction.

Results on *E. coli* dataset are summarized in Table 1 and Fig. 1A; performance is evaluated in terms of precision and recall, maximizing their harmonic mean, i.e. the F-measure (F1). Aggregate values are calculated as the arithmetic average over the sets of 10 network instances. As reported previously, for 50-NETs and 100-NETs learning optimal DBNs became computationally intractable; thus corresponding results are not available. CTBNs outperformed

---

[2] Run using the toolbox for Granger causal connectivity analysis (GCCA) [19] v2.9.

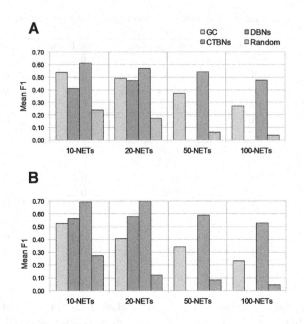

Fig. 1. Performance comparison on *E. coli* (A) and *S. cerevisiae* (B) simulated data. Precision, recall and F1 values are averaged over sets of 10 networks.

DBNs and GC for 10-NETs, 20-NETs, 50-NETs and 100-NETs in terms of F1 values. Moreover, CTBNs also showed to be robust with respect to increasing network dimension. Indeed, their performance smoothly degrades as the number of nodes of the network increases. Because of this, the initial F1 gap between CTBNs and GC which was limited to 0.07 for 10-NETs, progressively increases to 0.21 for 100-NETs. The table corresponding to *S. cerevisiae* is omitted, while the corresponding results are shown in Fig. 1B. CTBNs outperformed DBNs and GC for all the network dimensions. Here, the F1 gap between CTBNs and GC is greater that that of *E. coli* and ranges from 0.17 for 10-NETs up to 0.29 for 100-NETs. Analogously to what was observed on *E. coli*, also on *S. cerevisiae* the performance achieved by CTBNs is slightly affected by the dimension of the regulatory network to be learned: this can be explained by the local nature of the CTBNs learning algorithm, although this can not be definitely stated. The *Random* method refers to the performance of an algorithm that, starting from an empty graph, tries to infer the structure of the network by randomly adding edges. As expected, the precision of the random algorithm is always poor while its recall stabilizes around 0.50. As shown in Table 1 and Figs. 1A, 1B the performances of the three methods are always better than performance values achieved by the random algorithm, thus confirming their effectiveness.

4.2 Experimental Data

Parameter optimization was run for GC, DBNs and CTBNs. Optimal number of bins resulted to be 3 for DBNs and CTBNs, while the maximum number of parents was set to 5. Optimal prior values for CTBNs were equal to those on simulated data. Learning criteria for DBNs was set to BIC. For GC all the pre-processing steps listed for the simulated data were applied, finding a p-value cutoff of 0.05 with an approximation of the False Discovery Rate (FDR) correction being the optimal ones. On the *S. cerevisiae* experimental dataset the results were coherent with those obtained on simulated datasets: CTBNs outperformed DBNs and GC. A graphical representation of the true network compared to the ones inferred by DBNs, GC and CTBNs is provided in Fig. 2. CTBNs achieved both the maximum value of true positives (5) and the minimum value of false negatives (3), while all methods generated exactly one (1) false positive prediction. As reported before, no time-course perturbational data was available for this experiment.

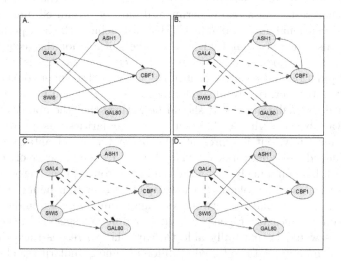

Fig. 2. Performance comparison on *S. cerevisiae* experimental data. True network (A), network inferred by GC (B), DBNs (C), CTBNs (D). Green arcs represent true positives, red arcs false positives and dotted lines false negatives.

5 Discussion

For the first time CTBNs were applied to the gene regulatory network reconstruction task from gene expression time-course data. Results from simulated and experimental data show that CTBNs outperform state-of-the-art methods, i.e. dynamic Bayesian networks and Granger causality analysis. CTBNs achieved the greatest value of the F1 measure. Performance loss with respect to network's dimension is limited and smooth. This suggests that CTBNs, when applied to regulatory networks larger than those analyzed in this paper, can still be effective to discover causal structures. Better accuracy values achieved by CTBNs versus DBNs could be explained by the local nature of the structural learning for the former as opposed to the global nature of the learning for the latter. This feature also offers an explanation for the limited loss of performance of CTBNs when the network dimension, i.e. the number of nodes, increases. Moreover, CTBNs remained computationally tractable even when learning with DBNs became infeasible. These aspects make CTBNs a good candidate for solving the problem of reconstructing gene regulatory networks, which are systems characterized by a large number of variables. In accordance with what was shown in [6], DBNs and GC were found to perform similarly. In particular, it was not possible to determine if one of these methods was definitively better than the other: for simulated data, GC performed better than DBNs on *E. coli* (Fig. 1A) while on *S. cerevisiae* DBNs performed better than GC (Fig. 1B). The simulated time-course dataset used for the analysis is at present unrealistically rich in terms of the number of perturbations and replicates. However, continuous improvement in experimental technologies will allow researchers to reach this level in the near future. In addition, when tested on a real experimental dataset of limited dimension and with no interventional data available, CTBNs still achieve the best performance. This result also suggests that CTBNs could be suitable for the reconstruction of other types of biological networks as well, such as signaling cascades, where direct manipulation and measurement of the individual members of the cascade are difficult. Moreover, this result underlines how CTBNs can perform well also when applied to small datasets. It should be noticed that the ability of CTBNs to handle variables evolving at different time granularities was not directly exploited in this study, but it represents another potential advantage over other methods. Indeed, gene networks are characterized by the presence of gene interactions which happen quickly, e.g. within minutes from a given triggering event, and others that can take place at a lower pace, e.g. within hours or days. To reconstruct such a regulatory network, one may want to integrate data coming from experiments measuring genes evolving at different speeds. In such a context, CTBNs would be naturally able to learn, i.e. to reconstruct, the causal network by combining data collected at different time granularities. The encouraging results achieved in this investigation suggest that the structural learning of CTBNs should be considered as a new gene network reconstruction method when time-course expression data is available. The results obtained call for more studies aimed at verifying the effectiveness of CTBNs in more complex settings. In particular, it would be interesting to investigate the performance of CTBNs starting from a genome wide experimental dataset and on more sophisticated organisms such as humans. Another relevant aspect which remains to be explored is the inference task, which would allow a deeper analysis of the dynamic aspects of the reconstructed gene network.

Acknowledgments. The authors would like to thank M. Poidinger, V. Narang, A. La Torraca, T. Zelante, F. Cordero, F. Zolezzi, the anonymous reviewers for their valuable help and Y. Amer for the CTBN Matlab Toolbox.

References

1. Acerbi, E., et al.: Computational reconstruction of biochemical networks. In: 2012 15th Int. Conf. Information Fusion (FUSION), pp. 1134–1141. IEEE (2012)
2. Karlebach, G., et al.: Modelling and analysis of gene regulatory networks. Nature Reviews Molecular Cell Biology 9(10), 770–780 (2008)
3. Dean, T., Kanazawa, K.: A model for reasoning about persistence and causation. Computational Intelligence 5(2), 142–150 (1989)
4. Granger, C.W.: Investigating causal relations by econometric models and cross-spectral methods. In: Econometrica, pp. 424–438 (1969)
5. Ding, M., et al.: Granger causality: Basic theory and application to neuroscience. In: Handbook of Time Series Analysis, p. 437 (2006)
6. Zou, C., et al.: Granger causality vs. dynamic bayesian network inference: a comparative study. BMC Bioinformatics 10(1), 122 (2009)
7. Cantone, I., et al.: A yeast synthetic network for in vivo assessment of reverse-engineering and modeling approaches. Cell 137(1), 172–181 (2009)
8. Nodelman, U., et al.: Continuous time bayesian networks. In: Proc. of the 18th Conf. on Uncertainty in AI, pp. 378–387. Morgan Kaufmann Publishers Inc. (2002)
9. Xu, J., Shelton, C.R.: Continuous time bayesian networks for host level network intrusion detection. In: Daelemans, W., Goethals, B., Morik, K. (eds.) ECML PKDD 2008, Part II. LNCS (LNAI), vol. 5212, pp. 613–627. Springer, Heidelberg (2008)
10. Boudali, H., et al.: A continuous-time bayesian network reliability modeling, and analysis framework. IEEE Trans. on Reliability 55(1), 86–97 (2006)
11. Fan, Y., et al.: Learning continuous-time social network dynamics. In: Proc. of the 25th Conf. on UAI, pp. 161–168. AUAI Press (2009)
12. Gatti, E., et al.: A continuous time bayesian network model for cardiogenic heart failure. Flexible Services and Manufacturing J., 1–20 (2011)
13. Nodelman, U., et al.: Learning continuous time bayesian networks. In: Proc. of the 19th Conf. on UAI, pp. 451–458. Morgan Kaufmann Publishers Inc. (2002)
14. Marbach, D., et al.: Generating realistic in silico gene networks for performance assessment of reverse engineering methods. J. of Comp. Biology 16(2), 229–239 (2009)
15. Stolovitzky, G., et al.: Dialogue on reverse-engineering assessment and methods. Annals of the New York Academy of Sciences 1115(1), 1–22 (2007)
16. Friedman, N., et al.: Using bayesian networks to analyze expression data. J. of Comp. Biology 7(3-4), 601–620 (2000)
17. Koller, D., et al.: Probabilistic graphical models: principles and techniques. MIT Press (2009)
18. Murphy, K., et al.: The bayes net toolbox for matlab. Computing Science and Statistics 33(2), 1024–1034 (2001)
19. Seth, A.K., et al.: A matlab toolbox for granger causal connectivity analysis. J. of Neuroscience Methods 186(2), 262 (2010)

Drug Target Identification Based on Structural Output Controllability of Complex Networks

Lin Wu[1], Yichao Shen[1], Min Li[2], and Fang-Xiang Wu[1,3,*]

[1] Division of Biomedical Engineering, University of Saskatchewan, Saskatoon, SK,
S7N 5A9, Canada
[2] School of Information Science and Engineering, Central South University,
Changsha, Hunan, 410083, China
[3] Department of Mechnical Engineering, University of Saskatchewan, Saskatoon, SK,
S7N 5A9, Canada
faw341@mail.usask.ca

Abstract. Identifying drug target is one of the most important tasks in systems biology. In this paper, we develop a method to identify drug targets in biomolecular networks based on the structural output controllability of complex networks. The drug target identification has been formulated as a problem of finding steering nodes in networks. By applying control signals to these nodes, the biomolecular networks can be transited from one state to another. According to the control theory, a graph-theoretic algorithm has been proposed to find a minimum set of steering nodes in biomolecular networks which can be a potential set of drug targets. An illustrative example shows how the proposed method works. Application results of the method to real metabolic networks are supported by existing research results.

1 Introduction

The one-target one-drug approach in drug discovery was dominant for a long time[1]. However, there are many limitations for drug design against single target in the aspects of drug efficiency and safety. Drugs in clinical treatment may not be as efficient as predicted in the experiment because of the interactions between pathways in biomolecular networks. Many biomolecular networks are robust so that the changes of a single target would be offset by the interactions in the networks, which makes the phenotypes of the biological systems unchanged[2]. Side effects often come from the undesired effects of drugs to the biomolecular networks. For example, a single-target drug may affect the states of non-target biomolecules and causes unexpected effects that can not be eliminated[3]. To overcome the weaknesses of single-target drugs, multi-target drugs design has attracted growing attention in recent years. Systems biology, which uses network-based approaches to study the properties and functions of biomolecular networks as a whole system, plays a vital role in multi-target drug design [4–6].

---

* Corresponding author.

M. Basu, Y. Pan, and J. Wang (Eds.): ISBRA 2014, LNBI 8492, pp. 188–199, 2014.

The last decade has witnessed an exceptional development in biomolecular interaction data and most attention has been paid to the biomolecular networks [6]. Cellular systems can be represented as biomolecular networks which are graphs with nodes and edges. Protein and protein interaction (PPI) networks, metabolic networks and gene regulatory networks (GRN) represent the undirected or directed interactions between biomolecules at different levels within cells [7]. Based on the biomolecular network analyses, past studies have made great progresses in discovering diseases related information, such as disease biomolecules and drug target sets. Disease biomolecules are the biomolecules in the networks whose abnormalities would lead to the changes of phenotypes and cause diseases. Drug target sets are combinations of biomolecules in the networks and by changing their states, the biomolecular networks in disease states can be driven to healty ones. To identify disease biomolecules based on networks, three general methods have been developed, which are linkage-based, module-based and diffusion-based methods [5, 8–10]. These approaches are based on different assumptions of the properties the disease biomolecules may have in biomolecular networks. For example, linkage-based method tends to associate a protein with a disease if its directly interacted proteins relate to the disease [8]. By this method, Janus kinase 3 (JAK3) has been correctly predicted to involve in combined immunodeficiency syndrome.

Abounds of approaches have been proposed to discover potential drug target sets in different biomolecular networks. Some approaches focus on the topological properties of biomolecular networks. Csermely et al. [2] firstly introduce network based analysis in the performance of multi-target drugs. Hwang et al. [11] proposed a concept named bridging centrality for each node and suggested that the biomolecules with high bridging centralities are potential drug targets. Some approaches are based on the dynamics of biomolecular networks. The analyses of dynamics consider the state changes of disease or non-disease biomolecules as efficiency or side effects. Yang et al. [3] construct an arachidonic acid metabolic network (AAnetwork) in which all the interactions between biomolecules are expressed by ordinary differential equations. Based on the model, not only optimal target sets have been identified, but also mechanisms for the side effects of existing drugs NSAIDs and Vioxx have been found. Li et al. [12, 13] use a flux balance analysis based linear program model to deal with the same problem, which does not require much accurate knowledge of networks dynamics compared to Yang's method. Wu et al. [14] develop a network dynamical model to identify effective drugs combinations and successfully detected the combination of Metformin and Rosiglitazone, which is actually Avandamet, an effective drug to cure Type 2 Diabetes.

Recently, many studies have applied the network control strategies to the identification of drug target sets in biomolecular networks. In the view of control theory, drug targets in the networks can be interpreted as steering nodes. By applying signals to the set of steering nodes the networks can be steered to desired states. Based on the Boolean networks model, Kim et al. [15] apply the optimization algorithm to search for a minimum steering node set so that the networks can be steered from any state to the desired states. Based on nonlinear dynamics network model, Cornelius *et al.* [16] study the T-cell survival signalling network that governs the development of T-LGL leukemia and ranks the nodes according to their potential to be steering nodes. In this paper, we applied the concept of structural output controllability [17–20] to the problem of drug target identification and developed an efficient algorithm to find out the minimum set of steering nodes. Moreover, modifications to our methods make it available to verify the performance of existing drugs as well as identify effective drug combinations.

2   Problem Formulation

If we are able to control a system, we can design appropriate inputs in order to steer the system from any state to another state that we desired. For a biological system at an abnormal state, if state changes of some biomolecules can affect other biomolecules and steer the system to a healthy state, the biomolecules be perturbed can be considered as drug targets. Thus the problem of identifying drug targets can be formulated as the problem of finding sets of nodes in systems. By applying the control signals to these nodes, the systems can be steered from certain states to others.

2.1   Dynamical Model of Biomolecular Networks

Even though most of complex dynamical processes are nonlinear, the controllability of nonlinear systems is structurally similar to that of linear systems in many aspects [18, 21]. In this paper, we use the linear time-invariant dynamical model, which is the most popular and canonical model to represent the dynamics of a biological system:

$$\begin{cases} \dot{\mathbf{x}}(t) = A\mathbf{x}(t) + B\mathbf{u}(t) \\ \mathbf{y}(t) = C\mathbf{x}(t) \end{cases} \tag{1}$$

where $\mathbf{x}(t) = (x_1(t), ..., x_n(t))^T$ is a state vector that describes the state of each node in the network. A is an $n \times n$ matrix which represents the interactions between nodes in the network. The $n \times m$ matrix B is an input matrix that identifies the nodes controlled directly by the external input signals represented by $\mathbf{u}$. $\mathbf{u}(t) = (u_1(t), ..., u_m(t))^T$ is an input vector of m control signals. $\mathbf{y}(t) = (y_1(t), ..., y_p(t))^T$ is an output vector of the network and each entry of it represents an output. The outputs of the system are linear combinations of node states in the networks which are indicated by the $p \times n$ matrix C.

From the perspective of the biomolecular networks to interpret system dynamical model, states of nodes can be concentrations of metabolites or enzymes in a metabolic network or can be expression levels of genes in a gene regulatory network. Each entry a_{ij} in matrix A indicates the strength of influence from biomolecule j to biomolecule i in the biological system. Then the structure of a biomolecular network can be represented by matrix A.

Drugs can be one type of external inputs to the biological systems and represented by $\mathbf{u}$. Input matrix B indicates biomolecules to which external inputs are directly applied and b_{ik} indicates the strength of external input k to node i in the network. In biomolecular networks, the nodes that external inputs directly applied to are considered to be drug targets. For the output matrix C, in our research, we consider there is one and only one entry which has nonzero value in each row and each column of C while other entries are zero. With this assumption, each entry in $\mathbf{y}(t)$ represents the state of one node which is indicated by a corresponding row of C. The outputs $\mathbf{y}(t)$ are the states of a set of biomolecules in the biological system. In the problem of drug target identification, we concern about the states of the disease biomolecules as well as the biomolecules whose state changes would lead to side effects. The objective of drug design is to steer the states of disease biomolecules healthy while keeping the state changes of side effect causing biomolecules minimal. So we consider the outputs of the biological systems are the states of these two types of biomolecules. The ability of external inputs to control the outputs of the systems means the efficiency of the drugs to cure the diseases while minimizing the side effects.

Suppose the external inputs $\mathbf{u}$ can be chosen arbitrarily, input matrix B should be well designed in order to control the outputs of the system. Finding drug targets becomes the problem of finding an input matrix B. For a biological system that has been modeled by matrix A and the output matrix C can be identified by the specific nodes, we try to find a matrix B and for such the matrix B, there exist appropriate inputs $\mathbf{u}$ which can make outputs of the system as we desired. In this case, the system is said to be output controllable. Here we develop an algorithm to find the matrix B which ensures the output controllability of the system while having a small number of nodes that need to be directly controlled as well as a small state subspace which will be affected [19].

2.2   Controllability of Networks

According to control theory, a system is completely controllable if for any initial state $\mathbf{x}(t_0) = \mathbf{x_0}$ and any other final state $\mathbf{x_1}$, there exists a finite time t_f and inputs $\mathbf{u}(t)$, such that $\mathbf{x}(t_f) = \mathbf{x_1}$ [21, 22]. According to the Kalman's controllability rank condition, a system is completely controllable if and only if the controllability matrix

$$\mathfrak{C} = \begin{bmatrix} B & AB & A^2B & ... & A^{n-1}B \end{bmatrix} \tag{2}$$

of the system has full row rank n [23]. To completely control a system, the matrix B should be properly designed to satisfy the Kalman's condition, which is equivalent to find an appropriate set of nodes in the network to which control signals should be applied directly. However, there are too many nodes that control signals should be applied to in order to make a biomolecular network completely controllable. Specifically, in gene regulatory networks, independent control signals should directly apply to at least about 80% of nodes to completely control the networks [18]. For the drug target identification problem, it is difficult and unnecessary to completely control the networks. In reality, we are more concerned about the states of the nodes that will cause the disease or bring side effects. So we focus on output controllability of biomolecular systems in this paper.

Output controllability is related to the outputs of the system, which describes the ability of external inputs to steer the output from any initial condition to any other final condition in finite time. A system is output controllable if for any initial output $\mathbf{y_0} = \mathbf{y}(t_0)$ and any other final output $\mathbf{y_1}$, there exists a finite time t_f and inputs $\mathbf{u}(t)$, such that $\mathbf{y}(t_f) = \mathbf{y_1}$. For a dynamic system model described by equation (1), the $p \times mn$ output controllability matrix has been defined as:

$$o\mathfrak{C} = \begin{bmatrix} CB & CAB & CA^2B & ... & CA^{n-1}B \end{bmatrix}. \tag{3}$$

Based on the control theory, a system is output controllable if and only if rank $(o\mathfrak{C}) = p$ [24].

For most biological systems, we often only know whether there is an interaction between two biomolecules, but don't know the strength of the connection between them. So we can only judge whether an entry in matrix A is zero or not. A matrix M is said to be a structural matrix if its entries are either fixed zeros or independent free parameters. And a system described by (1) is called a structural system (A, B, C) if A, B and C are structural matrices. "Structural controllability" is a concept which measures the controllability of structural systems [17]. For a structural system (A, B, C), it is structural controllable if the generic rank of $\mathfrak{C}$ is n, where the generic rank of the structural matrix $\mathfrak{C}$ is defined to be the maximal rank that the matrix may achieve by freely choosing values of free parameters [19]. The generic rank of $\mathfrak{C}$ is denoted by $GDCS(A, B)$, it measures the dimension of controllable subspace of the structural system. The smaller value of $GDCS(A, B)$, the lower dimension the states of nodes in the system will be changed in, which means the fewer side effects[25]. So we try to find the matrix B that has small $GDCS(A, B)$ on the basis that the system is structural output controllable, which means the generic rank of $o\mathfrak{C}$ is p.

Calculating the generic rank of matrix $\mathfrak{C}$ and $o\mathfrak{C}$ can be achieved by graph-theoretic approaches. Before we state the method, firstly we introduce some concepts. $G(A, B)$ is a digraph related to system (A, B, C), which contains sets of nodes $V_A \bigcup V_U$, where $V_A = \{v_1, ..., v_n\}$ and $V_U = \{u_1, ..., u_m\}$. The edge set is the pairs

$$v_j v_i \qquad \text{for } a_{ij} \neq 0$$

and

$$u_j v_i \qquad \text{for } b_{ij} \neq 0.$$

The nodes in V_A and the edges between them make up the biomolecular network. The nodes in V_U represent input nodes. Each input node corresponds to an input signal u and edges from a node in V_U to V_A correspond to a column in B which indicate nodes in the biomolecular network to which the control signal directly applied. For a graph, a sequence of edges $\{(v_1 \to v_2), (v_2 \to v_3), ..., (v_{k-1} \to v_k)\}$ is called a path. A path is called a simple path when all the nodes $\{v_1, v_2, ..., v_k\}$

are distinct. When $v_1 = v_k$ and other nodes are distinct, the path is called a simple cycle. A node v_i in V_A is called inaccessible from input nodes if and only if there are no paths that can reach it from nodes in V_U. For a set of simple paths and simple cycles have no nodes in common, if all the paths start from distinct input nodes in V_U and cycles are accessible from input nodes, this set is defined to be a path and cycle covering of $G(A, B)$ by us.

In graph-theoretic representation, the generic rank of $\mathfrak{C}$ is the maximum number of nodes in V_A that can be covered by a path and cycle covering [17, 19], which can be calculated by a maximum matching algorithm in corresponding bipartite graph representation of $G(A, B)$. Nodes $r_i \in V_R$ and $c_j \in V_C$ in the bipartite graph correspond to the row i and column j of matrix A, respectively. If $a_{ij} \neq 0$, there is an undirected edge between r_i and c_j in the bipartite graph. Each input node $u_i \in V_U$ corresponds to one node u_i in bipartite graph and if $b_{ij} \neq 0$, there is an undirected edge between r_i and u_j. The maximum matching algorithm finds out the maximum number of nodes in V_R that can be matched to the nodes in V_C or V_U. The nodes labeled R which have been matched in the matching correspond to the nodes in V_A of $G(A, B)$, each of which is covered by a simple path or cycle.

However, the generic rank of $o\mathfrak{C}$ can not be calculated accurately by any graph-theoretic method yet. Murota and Poljak [20] have developed a method to calculate the upper and lower bounds of generic rank($o\mathfrak{C}$). In this paper, we need to make sure the designed matrix B can make the system structural output controllable, so we use the lower bound. Based on our former assumption, each row in C corresponds to one node in the biomolecular network which is a node in V_A. p rows of C correspond to p different nodes which is a subset of V_A and denoted as S. According to [20], the lower bound of generic rank($o\mathfrak{C}$) is the maximum number of nodes in S that can be covered by a path and cycle covering in $G(A, B)$. The paper [18] uses maximum matching algorithm to find out the minimum number of nodes in V_U needed to control the system based on the method of calculating the generic rank of $\mathfrak{C}$. Similarly, we develop a new algorithm to identify a minimum set of drug targets needed to control the outputs of the system based on the calculation of lower bound of generic rank($o\mathfrak{C}$).

2.3 Algorithm Description

Given a biomolecular network whose structural matrices A and C have been determined, we would like to develop an algorithm to find a matrix B which makes the generic rank of $o\mathfrak{C}$ equal to p. At the same time, the algorithm tends to have fewer steering nodes and smaller value of $GDCS(A, B)$. Since there is no algorithm to calculate the generic rank of matrix $o\mathfrak{C}$ accurately, the developed algorithm makes sure the lower bound of the generic rank of $o\mathfrak{C}$ equals to p, which guarantees that the system is structural output controllable.

Firstly we construct a weighted bipartite graph that corresponds one on one to the network A and the output C. Nodes $r_i \in V_R$ and $c_j \in V_C$ in the bipartite graph correspond to row i and column j of matrix A, respectively. If $a_{ij} \neq 0$, there is an undirected edge between r_i and c_j in the bipartite graph. There are p input nodes u_i we called potential input nodes because we only need to directly control at most p nodes in the network to control the output of the system. For each potential input nodes u_i, there are edges between it and all the nodes in V_R which are eligible potential drug targets. Eligible potential drug targets are the nodes in the network within which we want to find drug targets. The edges and the weights of them have been assigned as follows:

$$w_{r_i c_j} = \begin{cases} \lambda & \text{if } a_{ij} \neq 0 \text{ and } v_i \in S \\ 1 & \text{if } a_{ij} = 0, \ i = j \text{ and } v_i \notin S \\ 0 & \text{if } a_{ij} \neq 0 \text{ and } v_i \notin S \end{cases}$$

and

$$w_{r_i u_j} = \begin{cases} 0 & \text{if } v_i \text{ is an eligible target, } v_i \in S \text{ and } u_j \in V_U \\ -\lambda & \text{if } v_i \text{ is an eligible target, } v_i \notin S \text{ and } u_j \in V_U \end{cases}$$

where $S \subseteq V_A$, which is the set of output nodes corresponding to C, λ is an integer constant larger than n.

Then we use Kuhn-Munkres (KM) algorithm to find out the maximum weight complete matching of the constructed bipartite graph [26]. The maximum weight complete matching, which means every node in V_R has been matched to a distinct node in V_C or V_U while maximizing the sum of the weight of edges in the matching. A complete matching always exists except one case that there exists a node $v_i \in S$, which is not an eligible target itself and inaccessible from any other eligible targets. For this case, the KM algorithm does not work because it requires the existing of at least one complete matching. Then the system is not able to be output controllable with these eligible targets. A complete matching in the bipartite graph corresponds to a covering consisting of nodes distinguished simple paths starting from input nodes and cycles that covers all the nodes in S. To understand this, we can see that if r_i matches $c_j (i \neq j)$ or u_j, which means there is an edge from v_j or u_j to v_i in $G(A, B)$. Also, there is no edge between r_i and c_i for $v_i \in S$ in the corresponding bipartite graph, which makes sure that r_i can not match c_i and thus v_i must be covered by a path or cycle in $G(A, B)$. After the maximum weight matching, we search for all the nodes in S covered by cycles and then check whether if all these cycles are accessible from input nodes. If so, the covering is a path and cycle covering. Otherwise, we can add more input nodes to make the inaccessible cycles which contain nodes in S accessible from input nodes. This can be accomplished by a graph searching algorithm based on strong connected components (SCCs) identification. We choose some nodes from higher hierarchy SCCs as drug targets to make all cycles which contain nodes in S accessible from input nodes [27, 28], which guarantees that the system is structural output controllable.

Here we prove that the number of input nodes q found in the maximum matching is minimum. There are q input nodes in the matching, which means there are q nodes in V_R that match nodes in V_U in the bipartite graph. From the weight of each edge, we can see that the score of the maximum matching is in the form of $(p - q)\lambda + k$, while k is an integer smaller than λ. If there is a matching with $q - q'$ input nodes, where q' is a positive integer number, the score of the matching will be in form of $(p - q + q')\lambda + k'$. That score is obviously larger than $(p - q)\lambda + k$, which contradicts with the fact that $(p - q)\lambda + k$ is the score of a maximum matching. Since the term $(p - q)\lambda$ in score of the maximum matching is determined, the final value of the maximum matching depends on k, which is the number of nodes for each $v_i \notin S$ having $a_{ii} = 0$ and r_i matches c_i. Actually these v_i are the uncovered nodes in the original graph, the maximum matching tends to make as large k as possible, which reduces the $GDCS(A, B)$.

3 Examples

3.1 A Numerical Example

Fig. 1 is an illustrative example of a system with 6 kinds of biomolecules and 6 interactions between them, where we assume that biomolecules 2 and 5 are related to a certain disease and the abnormality of biomolecule 4 will cause serious

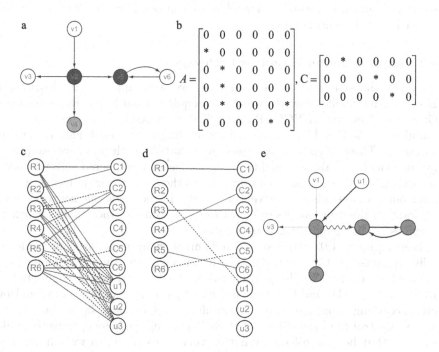

Fig. 1. A numerical example. a) A biomolecular network with 6 biomolecules and 6 reactions. The outputs of the system are the states of the nodes filled in red and orange. b)Corresponding structural matrix. "∗" in matrices A and C represent the free parameters while 0 represents the fixed parameters. c) The constructed weighted bipartite graph. The weights of red lines, black lines, dash lines and blue lines are λ, 1, 0 and $-\lambda$, respectively. d)One of the maximum matching results. e)The corresponding simple path and cycle in original graph.

side effects. In Fig. 1a, the nodes filled with red represent disease biomolecules and others filled with orange represent the biomolecules whose state changes would cause side effects. Fig. 1b shows the structural matrices A and C correspond to the system and outputs, respectively. Symbols "$*$" in matrices A and C represent free parameters.

Based on the algorithm described in last section and an assumption that all the nodes in the networks are eligible drug targets, we can construct a corresponding bipartite graph shown in Fig. 1c. The edges with different weights are color-coded. Fig. 1d shows a matching of Fig. 1c that all nodes in V_R are matched while having the maximum value of edges in the matching. From the matching in bipartite graph we can get a set of paths and cycles in the original network, which are indicated by the bold edges in Fig. 1e. The wave edge indicates that node 5 is accessible from input node u_1, which means the simple path $\{(u_1 \to v_2), (v_2 \to v_4)\}$ and cycle $\{(v_5 \to v_6), (v_6 \to v_5)\}$ is already a path and cycle covering and there is no need to process the following steps.

From the above analysis we can see the node 2 is connected by input node directly, which means by directly applying control signal to node 2, the system is structural output controllable. Therefore we can consider node 2 to be the drug target in the biological system.

3.2   Experiment on H.sapiens Pathways

To verify the significance of our method, we have applied our algorithm to *H.sapiens* metabolic networks which are acquired from Kyoto Encyclopedia of Genes and Genomes (KEGG) Pathway [29]. In KEGG, unique identifiers with initial letters D, C and EC represent different drugs, compounds and enzymes, respectively. Three drugs and corresponding metabolic pathways have been tested by our model (See Table 1) and the results have been compared with the former works [13, 30]. When applying our method to these metabolic pathways, we consider both compounds and enzymes are nodes in the metabolic networks. Since we want all the drug targets to be enzymes, we consider those nodes which are enzymes as eligible potential drug targets.

Benoxaprofen (D03080) is an anti-inflammatory drug inhibiting arachidonate 5-lipoxygenase (EC:1.13.11.34), which involves in the arachidonic acid metabolism (hsa00590). In pharmacology, the biosynthesis of compounds of LTB4, LTC4, LTD4 and LTE4 will decrease by the using of benoxaprofen. However, according to our model, only controlling arachidonate 5-lipoxygenase is not able to control the biosynthesis of these four compounds independently, which suggests that benoxaprofen may not be very efficient. When we run our algorithm on the metabolism network and define the outputs of the system to be the states of LTB4, LTC4, LTD4 and LTE4, we have found that the optimal drug target sets are LTA4H (EC:3.3.2.6) and LTC4 synthase (EC:4.4.1.20). In the view of our model, this set of drug targets is able to control the states of four output compounds independently as well as have a small value of $GDCS$, which means higher efficiency and fewer side effects comparing to the single target of arachidonate 5-lipoxygenase. This result is identical to the works of Li *et al.*

[13] and Sridhar *et al.* [30] by using their own models. In addition, several other researches have suggested that the anti-inflammatory activity can be affected by the levels of LTA4H and LTC4 synthase [31, 32].

Rasagiline mesylate (D02562) is a monoamine oxidase inhibitor which is used as an anti-parkinsonian drug. Severity of Parkinson disease is affected by the level of pros-methylimidazoleacetic acid in the histidine metabolism network (hsa00340). Then we choose methylimidazoleacetic acid and methylimidazole acetaldehyde as the output of the system, which is the same to the experiment in [13, 30], we have found that the optimal drug target is monoamine oxidase. This result verifies the validity of Rasagiline mesylate and it is the same as that in [13, 30].

Febuxostat (D01206) is a drug used to treat hyperuricemia by inhibiting xanthine dehydrogenase (EC:1.17.1.4) and xanthine oxidase (EC:1.17.3.2). Excess of urate will lead to hyperuricemia. We have applied our algorithm to the related metabolic pathway purine metabolism network (hsa00230) and defined the state of urate as the system output. We find that enzyme EC:2.4.2.16 is the optimal drug target based on our model, which is different from the existing drug of febuxostat and the result from Li *et al.* [13]. Li *et al.* suggest that the drug target set consists of three enzymes which are EC:1.17.1.4, EC:1.17.3.2 and EC:2.4.2.16 and urate can be affected by all these enzymes. However, according to our model, we consider that by controlling only one of these three enzymes we can control the state of urate. Choosing EC:2.4.2.16 can achieve the smallest $GDCS$, which means its side effect is minimum. In fact, choosing EC:1.17.1.4 and EC:1.17.3.2 as targets will not only lead to a large controllable subspace, but also affect many other unrelated compounds in the network comparing to choosing EC:2.4.2.16, which means much more side effects.

Table 1. Experiments on real metabolic networks

Pathways	Number of nodes	Number of edges	Output nodes	Drug targets
hsa00590	135	144	LTB4, LTC4, LTD4, LTE4	LTA4H, LTC4 synthase
hsa00340	73	97	methylimidazoleacetic acid, methylimidazole acetaldehyde	monoamine oxidase
hsa00230	265	355	urate	EC:2.4.2.16

4   Conclusion

Identifying drug targets is a major challenge in new drug development where both drug efficiency and side effects should be taken into consideration. Previous network based approaches have made some achievements in researches of diseases. In this paper, we apply the concepts in complex network control to the drug identification problem. Based on these concepts, we are able to define the measurements of drug efficiency and its side effects. On the basis of the measurements, we design an algorithm to identify the drug target sets as well as verify the performance of existing drugs and to find multiple drugs combinations for a certain disease. Compared with other network based methods, our approach does not need the exact values of parameters in a network which are always unknown or hardly to be estimated accurately. In fact, our method only requires the structure of the biomolecular networks and its results are based on the general property which is the controllability of complex networks. Indeed, by applying our method to the real metabolic networks, we can see our results are supported by existing research results.

Acknowledgement. This research is supported by Natural Sciences and Engineering Research Council of Canada (NSERC) and Chinese Scholarship Council (CSC).

References

1. Lindsay, M.A.: Target discovery. Nat. Rev. Drug. Discov. 2, 831–838 (2003)
2. Csermely, P., Agoston, V., Pongor, S.: The efficiency of multi-target drugs: the network approach might help drug design. Trends Pharmacol. Sci. 26, 178–182 (2005)
3. Yang, K., Bai, H., Ouyang, Q., Lai, L., Tang, C.: Finding multiple target optimal intervention in diseaserelated molecular network. Molecular Systems Biology 4, 228 (2008)
4. Azmi, A.S., Wang, Z., Philip, P.A., Mohammad, R.M., Sarkar, F.H.: Proof of concept: Network and systems biology approaches aid in the discovery of potent anticancer drug combinations. Mol. Cancer Ther. 9(12), 3137–3144 (2010)
5. Kotlyar, M., Fortney, K., Jurisica, I.: Network-based characterization of drug-regulated genes, drug targets, and toxicity. Methods 57, 499–507 (2012)
6. Barabási, A., Gulbahce, N., Loscalzo, J.: Network medicine: a network-based approach to human disease. Nature Reviews Genetics 12, 55–68 (2011)
7. Chen, L., Wang, R., Zhang, X.: Biomolecular Networks: Methods and Applications in Systems Biology. Wiley (2009)
8. Oti, M., Snel, B., Huynen, M.A., Brunner, H.G.: Predicting disease genes using protein-protein interactions. J. Med. Genet. 43(8), 691–698 (2006)
9. van den Akker, E.B., Verbruggen, B., Heijmans, B., Beekman, M., Kok, J., Slagboom, P., Reinders, M.J.: Integrating protein-protein interaction networks with gene-gene co-expression networks improves gene signatures for classifying breast cancer metastasis. J. Integr. Bioinform. 8(2), 188 (2011)
10. Vanunu, O., Magger, O., Ruppin, E., Shlomi, T., Sharan, R.: Associating genes and protein complexes with disease via network propagation. PLoS Comput. Biol. 6(1) (2010)
11. Hwang, W.C., Zhang, A., Ramanathan, M.: Identification of information flow-modulating drug targets: A novel bridging paradigm for drug discovery. Clinical Pharmacology & Therapeutics 84, 563–572 (2008)
12. Li, Z., Wang, R.-S., Zhang, X.-S.: Two-stage flux balance analysis of metabolic networks for drug target identification. BMC Systems Biology 5(suppl. 1), 11 (2011)

13. Li, Z., Wang, R.S., Zhang, X.S., Chen, L.: Detecting drug targets with minimum side effects in metabolic networks. IET Syst. Biol. 3(6), 523–533 (2009)
14. Wu, Z., Zhao, X.M., Chen, L.: A systems biology approach to identify effective cocktail drugs. BMC Systems Biology 4(suppl. 2), 57 (2010)
15. Kim, J., Park, S.M., Cho, K.H.: Discovery of a kernel for controlling biomolecular regulatory networks. Sci. Rep. 3, 2223 (2013)
16. Cornelius, S.P., Kath, W.L., Motter, A.E.: Realistic control of network dynamics. Nat. Commun. 4, 1942 (2013)
17. Lin, C.: Structural controllability. IEEE Trans. Auto. Contr. AC-19, 201–208 (1974)
18. Liu, Y., Slotine, J., Barabási, A.: Controllability of complex networks. Nature 473, 167–173 (2011)
19. Hosoe, S.: Determination of generic dimensions of controllable subspaces and its application. IEEE Trans. Auto. Contr. AC-25, 1192–1196 (1980)
20. Murota, K., Poljak, S.: Note on a graph-theoretic criterion for structural output controllability. IEEE Trans. Auto. Contr. AC-35, 939–942 (1990)
21. Slotine, J., Li, W.: Applied Nonlinear Control. Prentice-Hall (1991)
22. Nise, N.: Control System Engineering, 6th edn. Wiley (2011)
23. Kalman, R.: Mathematical description of linear dynamical systems. J.S.I.A.M Control Ser. A 1, 152–192 (1962)
24. Ogata, K.: Modern Control Engineering, 3rd edn. Prentice-Hall (1997)
25. Wu, F.X., Wu, L., Wang, J., Liu, J., Chen, L.: Transittability of complex networks and its applications to regulatory biomolecular networks. Sci. Rep. (accepted)
26. Jungnickel, D.: Graphs, Networks and Algorithms, 3rd edn. Springer (2005)
27. Liu, Y.Y., Slotine, J.J., Barabási, A.L.: Control centrality and hierarchical structure in complex networks. PLoS ONE 7(9), e44459 (2012)
28. Cowan, N.J., Chastain, E.J., Vilhena, D.A., Freudenberg, J.S., Bergstrom, C.T.: Nodal dynamics, not degree distributions, determine the structural controllability of complex networks. PLoS ONE 7(6), e38398 (2012)
29. Kanehisa, M., Goto, S.: Kegg: Kyoto encyclopedia of genes and genomes. Nucleic Acids Res. 28(1), 27–30 (2000)
30. Sridhar, P., Song, B., Kahveci, T., Ranka, S.: Mining metabolic network for optimal drug targets. In: Pac. Symp. Biocomput., vol. 13, pp. 291–302 (2008)
31. Rao, N.L., Dunford, P.J., Xue, X., Jiang, X., Lundeen, K.A., Coles, F., Riley, J.P., Williams, K.N., Grice, C.A., Edwards, J.P., Karlsson, L., Fourie, A.M.: Anti-inflammatory activity of a potent, selective leukotriene a4 hydrolase inhibitor in comparison with the 5-lipoxygenase inhibitor zileuton. J. Pharmacol. Exp. Ther. 321(3), 1154–1160 (2007)
32. Torres-Galván, M.J., Ortega, N., Sánchez-García, F., Blanco, C., Carrillo, T., Quiralte, J.: Ltc4-synthase a-444c polymorphism: lack of association with nsaid-induced isolated periorbital angioedema in a Spanish population. Ann. Allergy Asthma Immunol. 87(6), 506–510 (2001)

NovoGMET: *De Novo* Peptide Sequencing Using Graphs with Multiple Edge Types (GMET) for ETD/ECD Spectra

Yan Yan[1], Anthony J. Kusalik[1,2], and Fang-Xiang Wu[1,3,*]

[1] Division of Biomedical Engineering, University of Saskatchewan, Canada
[2] Department of Computer Science, University of Saskatchewan, Canada
[3] Department of Mechanical Engineering, University of Saskatchewan, Canada
faw341@mail.usask.ca

Abstract. *De novo* peptide sequencing using tandem mass spectrometry (MS/MS) data has become a major computational method for sequence identification in recent years. With the development of new instruments and technology, novel computational methods have emerged with enhanced performance. However, there are only a few methods focusing on ECD/ETD spectra, which mainly contain variants of c-ions and z-ions. A *de novo* sequencing method for ECD/ETD spectra, NovoGMET, is presented here and compared with another successful *de novo* sequencing method, pNovo+, which has an option for ECD/ETD spectra. The proposed method applies a new spectrum graph with multiple edge types (GMET), considers multiple peptide tags, and integrates amino acid combination (AAC) and fragment ion charge information. Experiments conducted on three different datasets show that the average full length peptide identification accuracy of NovoGMET is as high as 88.70%, and that NovoGMET's average accuracy is more than 20% greater on all datasets as compared to pNovo+.

Keywords: *De novo* peptide sequencing, ECD/ETD spectra, graph with multiple edge types, peptide tags, fragment ion charge determination.

1 Introduction

Tandem mass spectrometry (MS/MS) has emerged as a major technology for peptide identification. In a typical MS/MS experiment, protein mixtures are first digested into suitably sized peptides, and then the peptides are ionized via an ionization process. After that, selected peptides are further broken into fragment ions, and their tandem mass spectra (MS/MS spectra) are collected [1]. MS/MS spectra usually contain two kinds of information, the mass-to-charge (m/z) value and intensity of each ion detected.

In MS/MS, peptide ions are broken into various kinds of fragment ions. There are six kinds of commonly observed ions, namely a-, b-, c-, x-, y-, and z-ions. Different fragmentation techniques used in MS/MS yield differing dominant types

---

* Corresponding author.

M. Basu, Y. Pan, and J. Wang (Eds.): ISBRA 2014, LNBI 8492, pp. 200–211, 2014.
© Springer International Publishing Switzerland 2014

of fragment ions. Collision-induced dissociation (CID) and higher-energy collisional dissociation (HCD) yield b-ions and y-ions as dominating ions. Electron capture dissociation (ECD) and electron transfer dissociation (ETD) preferentially produce variants of c-ions and z-ions, and occasionally a-ions [2–4]. ETD [5] is a modification of ECD technique [6] which was designed for dissociation of multiply protonated peptide ions in MS/MS. It produces high quality MS/MS spectra for multi-charged peptides, has no strong cleavage preferences based on peptide sequences, and preserves labile post-translational modifications (PTMs) [7–9]. These features yield spectra containing more useful information, which has the potential to give better performance when used in peptide sequencing.

Currently, there are three main kinds of methods used for peptide sequencing from MS/MS data: database searching, peptide tagging and *de novo* sequencing. In database searching, theoretical spectra are computed from an existing protein database and peptides are identified by matching the theoretical spectra to experimental spectra. The major disadvantage of database searching is that it cannot identify new or unknown peptides. Peptide tagging [10, 11] usually produces partial sequences, often called tags, from an MS/MS spectrum, and then uses these tags to search against a protein database or to help with *de novo* sequencing. The use of tags can dramatically reduce the search space and time needed, and has the potential to improve *de novo* peptide sequencing. *De novo* sequencing automatically interprets spectra using the masses of amino acids. It can identify new proteins, proteins resulting from mutations, proteins with unexpected modifications and so on. With the recent development of high mass-accuracy MS/MS and alternative fragmentation techniques, *de novo* sequencing has shown promising developments[12]. Therefore, this study focuses on *de novo* peptide sequencing.

ECD/ETD is a new technology and has different properties from CID/HCD. The studies of MS/MS spectra produced by ECD/ETD have recently focused on the characteristics of the spectra[3, 4, 7], various PTMs in spectra[8], and the performance of database search techniques for such spectra [13]. On the other hand, less attention has been paid to *de novo* sequencing methods for ECD/ETD spectra although many *de novo* sequencing methods has been developed for CID and HCD spectra. Considering the unique features of these spectra and the shortage of corresponding algorithms, a suitable model designed for ECD/ETD spectra would be very useful and noteworthy.

In this paper, we present a modified spectrum graph with multiple edge types (GMET) to model ECD/ETD spectra. Apart from amino acid combinations (AACs), the order-independent amino acid composition information of a peptide, and the peptide tags which have already been integrated in our previous method [14], this new model considers unique features in ECD/ETD spectra. Since it is quite common to acquire ECD/ETD spectra from multi-charged peptide samples, multi-charged fragment ions are frequently observed. Wisely considering and utilizing these ions can be a great help in peptide sequencing.

2   Methods

In this section, a *de novo* peptide sequencing algorithm for ECD/ETD spectra is proposed. The proposed algorithm considers unique features of these spectra including alternate dominant ions and multi-charged fragment ions, and borrows a previously presented graph model [14]. The model incorporates a new kind of graph, a graph with multiple edge types (GMET), by considering multiple peptide tags and fragment ion charge information.

The whole method is summarized in Figure 1. To start, the algorithm first searches the low mass region of a spectrum, typically less than 200 Da, to find any consecutive, same-type ions of the first two cleavage points; to be specific, ion pairs $\{c_1, c_2\}$, $\{z_{\dot{1}}, z_{\dot{1}}\}$, $\{c-1_{\dot{1}}, c-1_{\dot{1}}\}$, and $\{(z+1)_1, (z+1)_1\}$. In the previous notion for ions, "+1" and "-1" represent addition or loss of 1Da in mass, respectively; subscript numbers indicate the cleavage positions on a peptide backbone from either N-terminal or C-terminal; an ion with "." means a radical fragment ion, which is a free radical species carrying a charge. If any pair exists, the two amino acids at the ends of a peptide sequence can be inferred and the sequencing will be conducted on the rest of the peptide sequence. The motivation of this preprocessing step is to reduce the potential sequencing length and make the problem easier to solve. This step is shown in Figure 1 in a dashed box because one may not find any ion pairs for some spectra to limit the sequencing length.

After this preprocessing step, a tag finding-and-ranking algorithm is applied to find length-3 tags and their associated scores. The tags are sorted in decreasing order of score, and placed in a set T. A higher score corresponds to greater confidence that a tag belongs to the peptide that generated the specific spectrum. The algorithm then uses the top peptide tag in T to separate prospective whole sequence into smaller pieces and stores them in *Part*. A predefined threshold *thres* determines whether further separation of a mass region in *Part* is needed. After that, the algorithm calculates amino acid compositions (AACs) of each region in *Part* and applies GMET on them to find partial peptide sequences. Finally, all suitable parts are assembled together to form the final peptide candidates, and a ranking scheme is applied to select and output the best ones. These steps in the algorithm and the various variables appearing in Figure 1 are explained in details in the following subsections.

2.1   Basic ion Types Considered in ECD/ETD Spectra

Since ECD/ETD spectra have dominant ion types different from CID spectra, we first investigate the ion types considered suitable for the proposed method. Through study of the literature about the frequency of different fragment ions [3–5, 7, 9, 15], the ions listed in Table 1 were selected based on their frequency observed in MS/MS spectra. Here, m_{H_2O}, m_{NH_3}, m_H, m_{CO} denote the mass of H_2O, NH_3, H, and CO- group, respectively. $\sum(residue\ mass)$ is the mass sum of all amino acids from the end amino acid of a peptide sequence to the amino acid of the current cleavage point. Similarly, $\sum(previous\ residue\ mass)$ is the mass sum of all amino acids from the end amino acid of a peptide sequence to

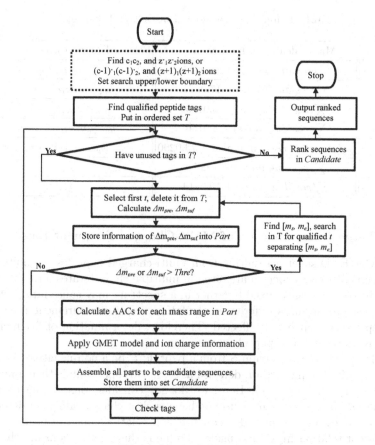

Fig. 1. Method flow chart

amino acid one previous to the current cleavage point. b_m and y_m are the masses of the b-ion and y-ion, respectively, at the current cleavage point. $x_{previous}$ is the mass of the x-ion at the cleavage point one position previous of the current one. An ion with "." means that it is a radical fragment ion. In addition, the complementary ion relationships in Equations (1)-(3) hold for the ions in Table 1.

$$b_i + y_{N-i} = m_p + 2m_H, \tag{1}$$

$$c_i + z^\bullet_{N-i} = m_p + 3m_H, \tag{2}$$

$$c - 1^\bullet_i + (z+1)_{N-i} = m_p + 3m_H, \tag{3}$$

where m_p is the mass of parent peptide P, N is peptide length, and $i \in \{1, 2, \ldots, N\}$. Δ_i is the ion mass of the ith Δ-ion, where $\Delta = \{\Delta_i | \Delta_i \in \{c, c-1^\bullet, z^\bullet, z+1, b, y\}\}$.

Table 1. Ion types considered in ECD/ETD spectra

Ion Type	Mass calculation from residues	Mass calculation from other ions
c	$\sum(residue\ mass) + 18.0344$	$b_m + m_{NH3}$
$c-1^{\bullet}$	$\sum(residue\ mass) + 17.0265$	$b_m + m_{NH3} - m_H$
$z^{\bullet}$	$\sum(residue\ mass) + 3.0156$	$y_m - m_{NH3} + m_H$
$z+1$	$\sum(residue\ mass) + 4.0156$	$y_m - m_{NH3} + 2m_H$
w	$\sum(previous\ residue\ mass) + 73.0290$	$x_{previous} + m_{CO}$
b	$\sum(residue\ mass) + 1.0078$	b_m
y	$\sum(residue\ mass) + 19.0814$	y_m

2.2 Charge Determination of Multi-Charged Fragment Ions

Since ECD/ETD spectra resulting from multi-charged peptides are quite common, fragment ions in spectra are often in multiple charge states. Fragment ions having different charges mixed together in a spectrum may cause interpretation problems since the mass difference calculation, a key measurement used in *de novo* peptide sequencing, is effected. Therefore, determination of fragment ion charges is quite useful in peptide sequencing.

Let the charge of a spectrum S from a peptide P be a non-negative integer n. The charge of a fragment ion i, denoted as ξ_i, in this spectrum is then a member of $\Xi = \{+1, +2, \ldots, n-1\}$. Here, we do not consider ions having charge n since the majority of ions in S do not have such a charge value. To determine ξ_i without knowing the peptide sequence, ion relationships are needed. Here, two different relationships are considered. One is that the two ions are the same fragment ion with different charges, e.g. $+1$ and $+2$ of the same c-ion. The other is that the two ions are complementary but with different charges, e.g. a $+1$ c-ion and complementary $z^{\bullet}$-ion having charge $+2$. These relationships are used to determine (part of) the ion charges in a spectrum.

In a spectrum S with charge n, suppose fragment i has charge $\xi_i \in \Xi$, $\Xi = \{+1, +2, \ldots, n-1\}$. Therefore, we have $n-1$ possibilities for the charge of an ion $i \in S$. Then for S, we build $n-1$ scenarios assuming all ions in S are in charge state $\xi_i \in \Xi$, and calculate the associated $n-1$ spectra having all ions with charge state $+1$, denoted as $S_{\xi_i to1}$. The set of such spectra is denoted $SA = \{S_{\xi_i to1} | \forall j \in S_{\xi_i to1}, \xi_j = 1; \xi_i \in \Xi\}$. The ion charges can be inferred by considering the above two relationships. If two ions $p \in S_{\xi_p to1}$ and $q \in S_{\xi_q to1}$ satisfy either of the relationships, where $\xi_p, \xi_q \in \Xi$, then the associated ions of p and q in the original spectrum S, denoted as i_p and i_q, are in charge state ξ_p and ξ_q, respectively. The detailed steps of charge determination of multi-charged fragment ions are shown in the following. Here, the m/z value of an ion $i \in S$ is denoted as $(m/z)_i$.

1. For a spectrum S with charge n, calculate spectrum $S_{\xi_i to1}$ consisting of only charge $+1$ ions j, where $(m/z)_j = (m/z)_i \cdot \xi_i - (\xi_i - 1)$, $\forall i \in S$, and $\forall \xi_i \in \Xi$.

2. For ions $p \in S_{\xi_p tol}$ and $q \in S_{\xi_q tol}$, calculate $M_{diff1} = |(m/z)_p - (m/z)_q|$, where $\xi_p \neq \xi_q$. If $M_{diff1} < \delta_1$, then the associated ions of p and q in S, denoted as i_p and i_q, are in charge state ξ_p and ξ_q, respectively. δ_1 is a small valued threshold. This calculation is between two of the $n-1$ spectra in SA indicating the same ion with different charges.

3. For ions $p \in S_{\xi_p tol}$ and $q \in S_{\xi_q tol}$, calculate $M_{diff2} = |(m/z)_p - (m/z)_q - m_p - 3mH|$, where $\xi_p + \xi_q = n$. If $M_{diff2} < \delta_2$, then the associated ions of p and q in S, denoted as i_p and i_q, are in charge state ξ_p and ξ_q, respectively. δ_2 is a small valued threshold. This calculation can be on the same spectrum in SA or two spectra in SA forming a complementary ion pair.

4. Label all identified ξ_i on i, and set $\xi_j = 1$ for the rest of the ions $j \in S$; output results.

2.3 Model of GMET

This model of GMET is derived from [14]. Here, we give a brief introduction to the former model and focus on the improvements of GMET.

The new graph type is of the form $GMET = (V, E, \Xi)$, each peak (corresponding to a fragment ion) in the experimental spectrum is represented as a vertex $v \in V$ and its m/z value is denoted as $(m/z)_v$; each v has a charge value $\xi \in \Xi$, which is determined by the previous charge determination process introduced in subsection 2.2. A denotes the set of 20 amino acids, and $a_i \in A$ is a certain amino acid. a_i is also used to represent its residue mass. m_{loss} is defined to be the mass of some small molecules or groups lost from fragment ions, which typically include H_2O, NH_3, CO- and NH-. $\forall u, v \in V$, if m_u and m_v denote their mass values, then we have $m_u = ((m/z)_v \cdot \xi_u) - (\xi_u - 1)$ and $m_v = ((m/z)_v \cdot \xi_v) - (\xi_v - 1)$. The following types of edges in E are considered in $GMET$.

1. Type I edge, e_{uv}^I: A directed edge e_{uv}^I is drawn from u to v when $|(m_v - m_u) - a_i| \leq \theta$, where θ is a given threshold. e_{uv}^I is labeled with a_i. This type of edge is similar to that defined in a traditional spectrum graph but with ion charges considered here.

2. Type II edge, e_{uv}^{II}: An undirected edge e_{uv}^{II} is drawn between u and v when $|m_p + 3m_H - (m_v + m_u)| \leq \theta$. Here u and v are viewed as complementary ions.

3. Type III edge, e_{uv}^{III}: An undirected edge e_{uv}^{III} is drawn between u and v when $|(m_v + m_u) - (m_p + 3m_H) - a_i| \leq \theta$. e_{uv}^{III} is labeled with a_i.

4. Type IV edge, e_{uv}^{IV}: An undirected edge e_{uv}^{IV} is drawn between u and v when $|m_p + 3m_H - (m_v + m_u) - a_i| \leq \theta$. e_{uv}^{IV} is labeled with a_i.

5. Type V edge, e_{uv}^V: A directed edge e_{uv}^V is drawn from u to v when $|(m_v - m_u) - m_{loss}| \leq \theta$.

In the experiments conducted in this paper, $\theta = 0.01 Da$. However, it can also determined by users according to their needs. From a GMET, peptide sequences can be inferred by building an induced graph $GMET_{AA}$ just as in [14]. The procedure for building the induced graph and inferring the sequence are not described here; details can be found in [14].

2.4   Integration of Peptide Tags and Amino Acid Composition (AAC)

Peptide tags and AAC information are incorporated in this model as they were in previous work [14], but with adjustment for the ECD/ETD spectra. AAC, which consists of order-independent amino acid composition information of a peptide, is helpful [16, 17] in limiting the edges in a spectrum graph. Multiple length-3 peptide tags are considered here to separate the whole peptide into suitable sized parts for sequencing based on GMET. The approach also yields a solution to the problem of increased numbers of possible AACs.

The results in [14] showed that the single tag strategy is no longer effective when the peptide length is over 15. Based on the frequency of each amino acid occurring in a peptide [18], the average mass of a peptide of length 15 can be determined to be approximately 1600Da. Therefore, the threshold to determine whether more tags are needed to separate a peptide is set to be $Thre = 1600$. With this value of $Thre$, the AAC database can be limited to masses of no more than 1600Da, which reduces the number of possible AACs considered in the GMET case dramatically.

An effective method named DirecTag [11] and the ranking criterion proposed in [14] are used here to output and rank tags. In the following, detailed steps of integrating peptide tags and AACs are shown.

1. Generate 3-tags using DirecTag [11] and put them into a set T according to their ranking scores in decreasing order.
2. Select the first tag, calculate the mass of its prefix and suffix separated by the tag, denoted as Δm_{pre} and Δm_{suf}. Store the two mass intervals related to Δm_{pre} and Δm_{suf} into $Part$ and delete t from T.
3. If Δm_{pre} and $\Delta m_{suf} \leq Thre$, go to Step 5. Otherwise, find the mass interval(s) larger than $Thre$, denoted as $[m_s, m_e]$; go to Step 4.
4. Search in T for the tag whose first and last ion in the interval of $[m_s, m_e]$. If there is such a tag, denoted as t', go back to Step 2 using t' to separate $[m_s, m_e]$; else, go to Step 5.
5. Find out all possible AACs of mass intervals in $Part$ using the in-house AAC database.
6. Construct a GMET from both sides of each 3-tag. AAC information is applied to restrict the choices of edges in the GMET. Specifically, only amino acids included in the AAC are considered when forming edges.
7. Combine all partial sequences to be whole candidate peptides, and put them into set $Candidate$.
8. If $T \neq \emptyset$, go back to Step 2 to use other tags in T; else, rank sequences in $Candidate$ and output results.

2.5   Candidate Peptide Ranking Scheme

When all candidate peptides are put in set $Candidate$ in the proposed algorithm, peptide ranking is the last step. Apart from the mass difference between parent

Table 2. Number of spectra and charges in each dataset used in the experiments

Dataset	Number of spectra	Charge of spectra
SwedECD	1414	+2
SCX_ETDFT_no_decon	1298	+2, +3, +4, +5
SCX_ETDFT_decon	612	+2, +3, +4, +5, +6

peptide mass m_p and the candidate peptide mass (denoted as Δm_p), unique features in ECD/ETD spectra are considered. According to literature [4, 15], x-ions (y-ion+CO) and z-ions (y-ion-NH_3, not the $z^{\cdot}$-ion) are absent from ECD/ETD spectra. Therefore, if a candidate peptide P_c is the correct peptide that produced a specific spectrum S, x-ions and z-ions calculated from P_c sequence should not be observed from S. That is to say, the fewer the number of x-ions and z-ions that are calculated from a candidate peptide sequence observed in S, the more likely it is the correct peptide.

Therefore, in the proposed ranking scheme, each candidate peptide P_c can be represented as a vector $CP(\Delta m_p, CZ_{match})$, where CZ_{match} is the total number of x-ions and z-ions calculated from P_c observed in S. The ranking scheme first orders CP according to Δm_p in a decreasing order, and then secondarily orders CP according to CZ_{match} in a decreasing order. After this process, the candidate peptides are output with their final ranking.

3 Experiments and Results

3.1 Datasets

Three ECD/ETD spectral datasets were used to investigate the performance of NovoGMET. Another newly developed *de novo* peptide sequencing algorithm, pNovo+ [19], was used for comparison. It has been shown that pNovo+ achieved superior sequencing results on various testing data compared to another successful *de novo* sequencing software PEAKS [20]. pNovo+ [19] can be used for HCD and (or) ETD/ECD spectra either alone or together. Here, the ETD/ECD option in pNovo+ is used for peptide sequencing.

The first dataset SwedECD contains ECD MS/MS spectra of doubly charged tryptic peptides [3]. The other two datasets, SCX_ETDFT_no_decon and SCX_ETDFT_decon, are from the same research paper [21]. The original datasets contain various fragmentation MS/MS spectra including CID, HCD, and ETD spectra. The ETD spectra were selected for the experiments here. For all spectra in these three datasets, a sequence is associated with each spectrum which is viewed as the correct sequence of the peptide producing this spectrum. The number and charges of spectra in each dataset are summarized in Table 2.

3.2 *De novo* Peptide Sequencing Performance

All three datasets were used to investigate the performance of NovoGMET and pNovo+. For each spectrum in each dataset, the top three candidates output by

Table 3. Average full length peptide sequencing accuracy comparison

Dataset	pNovo+	GMET without multi-tags and ion charge information	Proposed NovoGMET
SwedECD	52.86%	**72.25%**	N/A
SCX_ETDFT_no_decon	63.02%	64.10%	**85.83%**
SCX_ETDFT_decon	64.62%	59.15%	**88.70%**

the two methods were output. If any one of the three candidates from a spectrum was correct (the same as the sequence associated with this spectrum), it was deemed that full length accuracy was achieved for this spectrum and for the given method. Threshold value Δ in the experiments is 0.01Da. In order to analyze the contribution of multi-tags and ion charges in NovoGMET, sequencing results of using the GMET model without multi-tags and ion charge information are also shown. Detailed comparison is presented in Table 3.

From Table 3 one can see that for all three datasets the proposed method achieves higher full length accuracy than pNovo+, and the improvements are over 20%. Since the SwedECD dataset consists of +2 spectra with average peptide length of 10.2 [3], most of the fragment ions have charge +1, and a single tag is sufficient for peptide sequencing. Therefore, we did not use multi-tags and ion charges in the comparison; doing so is expected to yield results to those shown. From Table 3 it is evident that the GMET model achieves better performance than pNovo+ on the SwedECD dataset.

In order to see the contribution of multi-tags and ion charges, further experiments were conducted using the two datasets consisting of highly charged ($\geq$ +3) spectra, SCX_ETDFT_no_decon and SCX_ETDFT_decon. The results in Table 3 again illustrate that NovoGMET obtains the highest full length accuracy among all three methods, and the GMET without multi-tags and ion charges has performance similar to that of pNovo+. Therefore, using multi-tags and ion charges dramatically improves the identification accuracy of highly charged spectra ($\geq$ +3). Specifically, accuracy is improved 22.81% and 24.08% compared to pNovo+, and 21.73% and 32.55% compared to the GMET model without multi-tags and ion charges.

Furthermore, we considered the relationship between the number of correctly identified peptides and peptide length. Figures 2 summarizes the results of comparing the output of NovoGMET and pNovo+ on SCX_ETDFT_no_decon. Figures 2 shows that NovoGMET outperforms pNovo+ on almost every peptide length except lengths of 11. For length greater than 16, NovoGMET achieves almost perfect identification on this dataset, while pNovo+ has lower identification accuracy. The lower accuracy of NovoGMET at length 11 may be due to the threshold controlling whether multi-tags are used in the algorithm. Some peptides may require multi-tags for better performance but do not achieve the pre-defined threshold. In future study, we will explore suitable thresholds for the use of multi-tags.

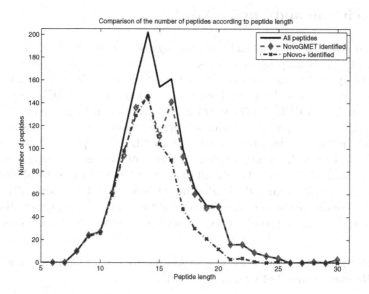

Fig. 2. Comparison of the number of correctly identified peptides and peptide length between NovoGMET and pNovo+ using the SCX_ETDFT_no_decon dataset

Table 4. Comparison between the number of correctly identified peptides and spectrum charge using the SCX_ETDFT_no_decon and SCX_ETDFT_decon datasets

Spectra charge	Dataset	pNovo+ identified	NovoGMET identified	Total spectra
+3	SCX_ETDFT_no_decon	488	583	682
+3	SCX_ETDFT_decon	181	204	247
+4	SCX_ETDFT_no_decon	311	457	532
+4	SCX_ETDFT_decon	199	295	320
+5	SCX_ETDFT_no_decon	1	9	9
+5	SCX_ETDFT_decon	5	28	28
+6	SCX_ETDFT_decon	0	2	2

Finally, the relationship between the number of correctly identified peptides and spectrum charge is examined. Table 4 summarizes the results of the comparison between NovoGMET and pNovo+ on the SCX_ETDFT_no_decon and SCX_ETDFT_decon datasets. Table 4 shows that NovoGMET outperforms pNovo+ under every spectrum charge. Thus for the two datasets consisting primarily of +3 and +4 charged spectra, NovoGMET identifies more peptides, which indicates the importance of using ion charge information for MS/MS peptide sequencing, especially with higher charged spectra.

4   Conclusion and Future Work

In this paper, a *de novo* peptide sequencing method for ECD/ETD spectra, NovoGMET, has been proposed. It uses a new spectrum graph model, considers multiple peptide tags to separate a peptide into small mass regions, and integrates fragment ion charge and amino acid composition (AAC) information.

Three datasets of ECD/ETD spectra were used to investigate the performance of NovoGMET by comparing it with another successful *de novo* peptide sequencing method, pNovo+, which has an option for ECD/ETD spectra. Experimental results showed that the improvements in terms of average full length peptide sequencing accuracy is over 20% on all datasets when compared to pNovo+.

In future, we will expand the evaluation of NovoGMET to more ECD/ETD datasets and optimize the parameters in the method. In addition, we are planning to modify the model and apply it to the multiple spectra sequencing problem.

Acknowledgments. This work was supported by Natural Sciences and Engineering Research council of Canada (NSERC).

References

1. Wysocki, V.H., Resing, K.A., Zhang, Q., Cheng, G.: Mass spectrometry of peptides and proteins. Methods 35(3), 211–222 (2005)
2. He, L., Ma, B.: ADEPTS: advance peptide *de novo* sequencing with a pair of tandem mass spectra. Journal of Bioinformatics and Computational Biology 8, 981–994 (2010)
3. Fälth, M., Savitski, M.M., Nielsen, M.L., Kjeldsen, F., Andren, P.E., Zubarev, R.A.: Analytical utility of small neutral losses from reduced species in electron capture dissociation studied using SwedECD database. Analytical Chemistry 80(21), 8089–8094 (2008)
4. Chalkley, R.J., Medzihradszky, K.F., Lynn, A.J., Baker, P.R., Burlingame, A.L.: Statistical analysis of Peptide electron transfer dissociation fragmentation mass spectrometry. Analytical Chemistry 82(2), 579–584 (2010)
5. Syka, J., Coon, J.: Peptide and protein sequence analysis by electron transfer dissociation mass spectrometry. Proceedings of the National Academy of Sciences 101(26), 9528–9533 (2004)
6. Zubarev, R., Kelleher, N.L., McLafferty, F.W.: Electron capture dissociation of multiply charged protein cations. A nonergodic process. Journal of the American Chemical Society 120(16), 3265–3266 (1998)
7. Mikesh, L.M., Ueberheide, B., Chi, A., Coon, J.J., Syka, J.E., Shabanowitz, J., Hunt, D.F.: The utility of ETD mass spectrometry in proteomic analysis. Biochimica et Biophysica Acta (BBA) - Proteins and Proteomics 1764(12), 1811–1822 (2006)
8. Wiesner, J., Premsler, T., Sickmann, A.: Application of electron transfer dissociation (ETD) for the analysis of posttranslational modifications. Proteomics 8(21), 4466–4483 (2008)
9. Kim, M.S., Pandey, A.: Electron transfer dissociation mass spectrometry in proteomics. Proteomics 12(4-5), 530–542 (2012)

10. Pan, C., Park, B., McDonald, W., Carey, P., Banfield, J., VerBerkmoes, N., Hettich, R., Samatova, N.: A high-throughput *de novo* sequencing approach for shotgun proteomics using high-resolution tandem mass spectrometry. BMC Bioinformatics 11(1), 118 (2010)

11. Tabb, D.L., Ma, Z.Q., Martin, D.B., Ham, A.J.L., Chambers, M.C.: DirecTag: Accurate sequence tags from peptide MS/MS through statistical scoring. Journal of Proteome Research 7(9), 3838–3846 (2008)

12. Lu, B., Chen, T.: Algorithms for de novo peptide sequencing using tandem mass spectrometry. Drug Discovery Today: BIOSILICO 2(2), 85–90 (2004)

13. Shen, Y., Tolić, N., Xie, F., Zhao, R., Purvine, S.O., Schepmoes, A.A., Moore, R.J., Anderson, G.A., Smith, R.D.: Effectiveness of CID, HCD, and ETD with FT MS/MS for degradomic-peptidomic analysis: Comparison of peptide identification methods. Journal of Proteome Research 10(9), 3929–3943 (2011)

14. Yan, Y., Kusalik, A.J., Wu, F.X.: A multi-edge graph based *de novo* peptide sequencing method for HCD spectra. In: IEEE International Conference on Bioinformatics and Biomedicine (BIBM), pp. 176–181 (December 2013)

15. Medzihradszky, K.F., Chalkley, R.J.: Lessons in *de novo* peptide sequencing by tandem mass spectrometry. Mass Spectrometry Reviews (2013)

16. Nefedov, A.V., Mitra, I., Brasier, A.R., Sadygov, R.G.: Examining troughs in the mass distribution of all theoretically possible tryptic peptides. Journal of Proteome Research 10(9), 4150–4157 (2011)

17. Hansen, T.A., Kryuchkov, F., Kjeldsen, F.: Reduction in database search space by utilization of amino acid composition information from electron transfer dissociation and higher-energy collisional dissociation mass spectra. Analytical Chemistry 84(15), 6638–6645 (2012)

18. Beals, M., Gross, L., Harrell, S.: Amino acid frequency (1999), http://www.tiem.utk.edu/~gross/bioed/webmodules/aminoacid.htm

19. Chi, H., Chen, H., He, K., Wu, L., Yang, B., Sun, R.X., Liu, J., Zeng, W.F., Song, C.Q., He, S.M., Dong, M.Q.: pNovo+: *De Novo* peptide sequencing using complementary HCD and ETD tandem mass spectra. Journal of Proteome Research 12(2), 615–625 (2013)

20. Ma, B., Zhang, K., Hendrie, C., Liang, C., Li, M., Doherty-Kirby, A., Lajoie, G.: PEAKS: powerful software for peptide *de novo* sequencing by tandem mass spectrometry. Rapid Communications in Mass Spectrometry 17(20)

21. Frese, C.K., Altelaar, A.F.M., Hennrich, M.L., Nolting, D., Zeller, M., Griep-Raming, J., Heck, A.J.R., Mohammed, S.: Improved peptide identification by targeted fragmentation using CID, HCD and ETD on an LTQ-Orbitrap velos. Journal of Proteome Research 10(5), 2377–2388 (2011)

Duplication Cost Diameters

Paweł Górecki[1], Jarosław Paszek[1], and Oliver Eulenstein[2]

[1] Department of Mathematics, Informatics and Mechanics, University of Warsaw, Poland
{gorecki,j.paszek}@mimuw.edu.pl
[2] Department of Computer Science, Iowa State University, USA
oeulenst@cs.iastate.edu

Abstract. The gene duplication problem seeks a species tree that reconciles given gene trees with the minimum number of gene duplication events, called gene duplication cost. To better assess species trees inferred by the gene duplication problem we study diameters of the gene duplication cost, which describe fundamental mathematical properties of this cost. The gene duplication cost is defined for a gene tree, a species tree, and a leaf labeling function that maps the leaf-genes of the gene tree to the leaf-species. The diameters of this cost are its maximal values when one topology or both topologies of the trees involved are fixed under all possible leaf labelings, and are fundamental in understanding how gene trees and species trees relate. We describe the properties and formulas for these diameters for bijective and general leaf labelings, and present efficient algorithms to compute the diameters and their corresponding leaf labelings. Moreover, we provide experimental evaluations demonstrating applications of diameters for the gene duplication problem.

1 Introduction

A basic tenet of all biological disciplines is the common evolutionary history of all life forms, including all extant species. The evolutionary relationships among such entities are usually represented as species trees and are a key tool in understanding evolution and its complex events that have engineered the enormous species and phenotypic diversity to date. Species trees are fundamental to evolutionary biology, but are also essential tools for an array of other disciplines such as agronomy, biochemistry, conservation biology, epidemiology, and medical sciences. For example, evolutionary trees are increasingly used to study the dynamic range of patients' cancer progressions, and to tailor corresponding treatments [22]. Species trees were also used to develop pesticides [17], to control invasive species [16], and to predict outbreaks of infectious diseases [14]. Invariably common to such studies is that large-sale species trees need to be accurately inferred.

One approach to construct large-scale trees is to utilize the rapidly growing availability of *gene trees*, i.e. the evolutionary history of genes. Gene trees describe how parts of the species genomes have evolved, and thus can be assembled into larger evolutionary trees of species. Unfortunately, evolutionary mechanism can cause conflicting evolutionary relationships between gene trees and the topology of the species tree along whose branches they have evolved [26]. Resolving such conflicts has become a grand

M. Basu, Y. Pan, and J. Wang (Eds.): ISBRA 2014, LNBI 8492, pp. 212–223, 2014.

challenge in the field of phylogenetic tree inference. There has been considerable interest in inference approaches that account for conflict involving gene duplication [5], which is a major and frequently occurring evolutionary mechanism [25].

One such approach is solving the gene duplication problem, which is well studied [5]. Given a collection of gene trees, this problem seeks a species tree that is a median tree of the given trees under the gene duplication cost, i.e. the fewest number of duplication events that can explain the conflict between a gene tree and a species tree [12,23]. Despite the NP-hardness of the gene duplication problem [19], effective local search heuristics [1,30] have produced credible estimates for this problem [4,20,21,24]. Recently, exact dynamic programming solutions have been developed for the gene duplication problem that were able to solve instances of up to 22 taxa within a few hours on a standard workstation [3]. Furthermore, there are modifications of the gene duplication problem that handle various types of input trees, such as erroneous trees [8], unrooted trees [11], and non-binary trees [18].

Unfortunately, despite ongoing work on the gene duplication problem, little is known about the mathematical properties of the gene duplication cost that is at the heart of the gene duplication problem. Here we investigate into diameters of this cost, that is, the maximal values of this cost when one shape or both shapes of the input trees are fixed. Diameters of the duplication cost are fundamental in understanding how gene trees and species trees relate under this cost [12].

Related Work. Goodman et al.'s pioneering work [7] introduced the gene duplication cost between a gene tree and a species tree that are both rooted and full binary. It is also assumed that the leaves of the gene tree are related to the leaves of the species tree by a function that is called *leaf labeling*. The leaf labeling is thought to relate the leaves of the species tree by a leaf of the gene tree from which it was sampled. An extension of the leaf labeling, called *least common ancestor (LCA) mapping*, relates every node of the gene tree to the most recent species in the species tree that could have contained this gene. A node in the gene tree is a *gene duplication* if it has the same lca mapping as one of its children. The *duplication cost* between a gene tree and a species tree under a given leaf labeling is the number of gene duplications. While diameter for other cost functions used for tree inference are well understood [10,28], diameters for the gene duplication cost have not been investigated yet.

Our Contribution. Under the gene duplication cost we study diameters of trees when leaf labelings are constrained to be bijective and when they are unconstrained. In particular, we study under the assumption that only the topologies of the trees involved are given (i) the diameter of a gene tree and a species tree, (ii) the diameter of a gene tree, and (iii) the diameter of a species tree. For example, the bijective diameter of a gene tree is the maximal duplication cost among all gene and species tree pairs such that the gene tree is fixed and its leaf labeling is bijective.

For bijective leaf labelings we show that the diameter of a gene tree and a species tree is equal to the number of non-cherry nodes (a node that is not the parent of two leaves) in the gene tree. While it follows that this diameter is linear time computable, we show that a leaf labeling that induces this diameter is also linear time computable. We also provide the exact conditions for a gene tree and a species tree to establish this diameter. For other types of diameters we describe their formulas. Moreover, we study the properties of the expected values of the bijective diameters and provide formulas for them. Based on this theoretical results, we evaluate computationally the expected values of bijective diameters. We also provide two experiments showing applications of unconstrained diameters in the gene duplication problem for two empirical datasets [13,27].

2   Preliminaries

2.1   Basic Definitions

We begin by recalling some basic definitions from phylogenetic theory. A *(phylogenetic) tree* is a connected acyclic graph such that exactly one of its nodes has a degree of two (root), and all its remaining nodes have a degree one (leaves) and three. The nodes with degree at least two are called internal. By $\leq$ we denote the partial order in a tree T such that $a \leq b$ if b is a vertex on the shortest path between a and the root of T. Note that $a \prec b$ is equivalent to $a \leq b$ and $a \neq b$. The least common ancestor of a and b in T is denoted by $a \oplus b$. We write that a and b are *comparable* if $a \leq b$ or $b \leq a$. If a is not the root of T, then the *parent* of a, denoted by πa, is the least node v such that $a \prec v$. If two nodes a and b have the same parent, then a is called *a sibling* of b. A sibling of a will be denoted by σa. $T(a)$ is the maximal subtree of T rooted at a. $|T|$ denotes the number of leaves in T. A *cherry* in T is a subtree of T that has exactly two leaves. A leaf that is not an element of a cherry is called *free leaf*. By L_T we denote the set of all leaves in T.

A *species tree* is a tree whose leaves are called *species*. A *gene tree* over a set of species X, is a triple $\langle V_G, E_G, \Lambda_G \rangle$ such that $\langle V_G, E_G \rangle$ is a tree and Λ_G is the leaf labeling function $\Lambda_G \colon L_G \to X$, called *labeling*. For simplicity, if the species tree S is known, we write that the gene tree is over S instead of L_S. Traditionally, gene and species trees are defined by nested parenthesis notation. For a species tree S and a gene tree G over S, let $M \colon V_G \to V_S$ be the *least common ancestor (LCA) mapping* between G and S that preserves the labeling of the leaves. In other words, $M|_{L_G} = \Lambda_G$, and for any non-root node a we have $M(\pi a) = M(a) \oplus M(\sigma a)$.

Parenthesis may be omitted in formulas for more clarity. For instance, instead of writing $M(\pi x)$ to denote the LCA mapping of πx, we write $M\pi x$. If $P = \langle p_1, p_2, \ldots, p_k \rangle$ is a sequence of nodes of G, then by MP, we denote the sequence $\langle Mp_1, Mp_2, \ldots, Mp_k \rangle$. If Q is a set of nodes, we define MQ to be $\{Mq \colon q \in Q\}$.

A *path* P in a tree T is a non-empty sequence of nodes without repetitions such that for every adjacent v and w in P, $\{v, w\} \in E_T$. Note that every path in a tree T has a unique $\leq$-maximal element. We denote it by $\max P$. A path P is called *simple* (in T) if its vertexes are comparable and its unique $\leq$-minimal element will be denoted by $\min P$. A *path partition* Π of T is a set of pairwise disjoint paths in T such that $\bigcup \Pi = V_T$.

A tree is called *caterpillar* if it contains exactly one cherry. A tree T is called *trivial* if T is a single noded tree, i.e., $|V_T| = 1$. By $\mathcal{C}(T)$ we denote the set of all cherry roots from a tree T. By χ_T we denote the number of all cherries in a tree T.

We call an internal node g from G a *S-duplication (node)*, or *duplication*, if $Mg = Mc$ for some child c of g. The *duplication cost* (D) from G to S, denoted by $\mathrm{D}(G, S)$, is defined as the total number of duplication nodes present in G.

2.2 Duplication Diameters

For a species tree S, we denote by $\mathbb{G}(S)$ the set of all gene trees over S. By $\mathbb{B}(S)$ we denote the set of all gene trees over S with bijective labeling.

Let G be a gene tree. By $\hat{G}$ we denote the unlabeled tree obtained from G by forgetting the labeling. We define several types of *bijective diameters*. For trees T and S with the same number of leaves we define *the bijective duplication diameter for fixed shapes* as:

$$b^D(T, S) = \max\{D(G, S) \colon G \in \mathbb{B}(S), \hat{G} = T\}$$

We define *the bijective duplication diameter for a fixed species tree*:

$$b^D(\star, S) = \max_{G \in \mathbb{B}(S)} b^D(G, S).$$

Next, we define *the bijective duplication diameter for a fixed shape of a gene tree*:

$$b^D(T, \star) = \max_S b^D(T, S).$$

Similarly to $b^D(T, S)$, $b^D(T, \star)$ and $b^D(\star, S)$ we introduce $u^D(T, S)$, $u^D(T, \star)$ and $u^D(\star, S)$, respectively, to denote the *unconstrained duplication diameters* by replacing $\mathbb{B}(S)$ in the above definitions with $\mathbb{G}(S)$, b^D with u^D and relaxing the assumption that T and S have the same size. We omit straightforward definitions.

3 Results

3.1 Bijective Duplication Diameters

In this section all labelings are bijections. First we define problems related to bijective diameters.

Problem 1. Given a tree T and a species tree S with the same number of leaves. Find a gene tree G such that $\hat{G} = T$ and $D(G, S) = b^D(T, S)$.

Problem 2. Given a tree T. Find a species tree S and a gene tree G such that $\hat{G} = T$ and $D(G, S) = b^D(T, \star)$.

Problem 3. Given a species tree S. Find a gene tree $G \in \mathbb{B}(S)$ such that $D(G, S) = b^D(\star, S)$.

First, we show the upper bound for the diameter:

Lemma 1. *For every species tree S and every tree G both with n leaves:* $b^D(G, S) \leqslant n - 1 - \chi_G$.

Proof. It is obvious that a gene tree can have at most $n - 1$ duplication nodes. Additionally, a cherry root in G cannot be a duplication node. Thus, the upper bound for $b^D(\hat{G}, S)$ is $n - 1 - \chi_G$.

We now show that the upper bound is reached by showing a simple procedure that induces the maximal number of gene duplications. First we introduce a notion of a *trunk* [10], that will be used to assign mappings to a cherry leaves.

A *trunk* of a species tree S is a non-empty sequence $\Upsilon = \langle \Upsilon_1, \Upsilon_2, \ldots, \Upsilon_k \rangle$, of internal nodes from S that starts in the root of S and satisfies (I) both children of Υ_k have the same number of descendant leaves, and (II) for $i > 1$, Υ_i is the child of Υ_{i-1} having more descendant leaves than its sibling. A subtree T of S is called a *limb*, if the root of T is not a trunk node, while its parent is a trunk node. S can be represented in *limb format* by using the standard nested parenthesis notation: $S = (S_1, (S_2, \ldots, (S_{|\Upsilon|}, S_{|\Upsilon|+1})))$, where S_i is a limb of S, for each i. For a leaf a in S let Υa be the lowest trunk node, whose subtree contains a.

Algorithm 1. Inference of a gene tree that induces the diameter b^D

---

1. **Input:** A species tree S and a tree T both with $n > 1$ leaves. **Output:** A gene tree G such that $T = \hat{G}$ and $D(G, S) = b^D(T, S)$.
2. **Comment** We define the labeling function Λ for T.
3. **Let** $\langle \xi_1, \xi_2, \ldots, \xi_{\chi_G} \rangle$ be a sequence of all cherry roots from T. Let $(S_1, (S_2, \ldots, (S_k, S_{k+1})))$ be a limb representation of S. Let $v_1, v_2, \ldots, v_n$ be the sequence of all leaves of S, such that, for every $i < j$, if $v_i \in S_k$ and $v_j \in S_l$, then $k \leqslant l$.
4. **For** $i = 1, 2, \ldots, \chi_G$ **Do** if x and y are the leaves of $G(\xi_i)$, set $\Lambda(x) := v_i$ and $\Lambda(y) := v_{n-\chi_G+i}$.
5. **Let** $v_{\chi_G+1}, v_{\chi_G+2}, \ldots, v_{n-\chi_G}$ be the labels of the remaining $n - 2 * \chi_G$ free leaves of G.
6. **Return** $\langle V_T, E_T, \Lambda \rangle$.

Theorem 1. *For every species tree S and every tree T both with n leaves:* $b^D(T, S)$ *equals the number of non-cherry nodes from T, that is,* $n - 1 - \chi_T$.

Proof. We show that Algorithm 1 infers a labeling that induces the maximal number of gene duplications. Let Λ be the function inferred by the algorithm. It is easy to see that Λ is a well defined labeling for T. Then, for every $1 \leqslant i < j \leqslant |L_S|$, we have $M\xi_i \geqslant M\xi_j$, thus all cherry mappings are comparable. We show that every non-cherry node is a S-duplication in a gene tree $\langle V_T, E_T, \Lambda \rangle$. Let g be a non-cherry node and let i be the minimal index of a cherry root from ξ that is present in $G(g)$. All mappings of internal nodes are comparable, thus $M\xi_i = Mg$. Moreover, at least one child of g is mapped to $M\xi$, thus g is a duplication. This completes the proof.

Now, we introduce a notion of a cherry partition, that is crucial for the bijective diameters of the duplication cost. A path partition Π of a tree G is called *cherry partition* *(of G)* if every path that contains an internal node is a simple path whose minimal element is a cherry root. We say that the LCA mapping M between a species tree S and a gene tree G over S *induces a cherry partition* if there is a cherry partition Π of G such that for every path P in Π, MP consists of a single element. Let $A \subseteq V_G$, then by $G \triangle A$ we denote the set of all maximal subtrees of G that do not contain any nodes from A.

First we classify cherry partitions.

Lemma 2. *Let S be a species tree, G be a gene tree over S and M be the LCA mapping between G and S. Then, M induces a cherry partition of G if and only if for every internal node g of G there is a cherry K in $G(g)$ such that $MK = Mg$.*

Proof. ($\Rightarrow$). Assume that M induces a cherry partition Π. Let g be an internal node of G. The proof is by induction on the structure of G. We consider two cases. (1) If g is a cherry root, then the property is trivially satisfied. (2) Otherwise, g has a child c such that g and c belongs the the same path from Π. Thus, $Mg = Mc \preceq Mo c$. By the inductive hypothesis there is a cherry K in $G(c)$ such that $Mc = MK$. By the previous observation, we have $MK = Mg$. This completes the first part of the proof. ($\Leftarrow$). We show that there exists a cherry partition of G induced by M. Consider the following procedure, where for each i, X_i is a set of subtrees of G: (I) Let $X_0 = \{G\}$. (II) For $0 < i \leqslant \chi_G$, let $X_i = (T \triangle r) \cup (X_{i-1} \backslash \{T\})$ and t_i is the root of T, where T is some non-trivial subtree of G from X_{i-1} and r is a cherry root from T such that $Mr = Mt_i$ (in G). It is easy to proof that for each $i \geqslant 0$, X_i is a set of disjoint subtrees of G satisfying $\bigcup_{T \in X_i} C(T) = C(G) \backslash \{r_1, r_2, \ldots, r_i\}$. Thus, $X_{\chi_G} = \bigcup_{g \in L_G} G(g)$, i.e., it is composed of all trivial subtrees of G. For $i = \{1, 2, \ldots, \chi_G\}$, let P_i be the simple and the shortest path connecting r_i and t_i. Note, that MP_i is a one-element set. Now, it should be clear that $\bigcup P_i$ is the set of all internal nodes of G, and the following family of simple paths $\bigcup_{g \in L_G} \langle g \rangle \cup \{P_1, P_2, \ldots, P_{\chi_G}\}$ is a cherry partition of G induced by M.

Theorem 2. *Let S be a species tree, G be a gene tree over S and M be the LCA mapping between G and S. Then, $b^D(\hat{G}, S) = D(G, S)$ if and only if M induces a cherry partition of G.*

Proof. ($\Rightarrow$). Assume that $b^D(\hat{G}, S) = D(G, S)$. Then by Theorem 1 every non-cherry node is a duplication. By Lemma 2 it is sufficient to show that for every internal g, there is a cherry K in $G(g)$ such that $Mg = MK$. The proof is by induction on the structure of G. If g is a cherry root the condition is trivially satisfied. Otherwise, g is a duplication and has a non-leaf child c such that $Mg = Mc$. Then by the inductive hypothesis, there is a cherry K in $G(c)$ such that $MK = Mc$. Thus, $MK = Mg$ which completes the inductive proof. ($\Leftarrow$). Assume that there is a cherry partition induced by M. Let g be a non-cherry node of G. Then, by Lemma 2 there is a cherry root r in $G(g)$ such that $Mg = Mr$. Thus, for some child c of g, $Mc = Mg$. We conclude that g is a duplication. We proved that every non-cherry node is a duplication. Thus, by Theorem 1, $D(G, S) = b^D(\hat{G}, S)$.

Theorem 3. *Given a species tree S and a tree T both with n leaves, Algorithm 1 infers a gene tree G such that $\hat{G} = T$ and $D(G, S) = b^D(T, S)$. The time complexity of Algorithm 1 is $O(n)$.*

Proof. Algorithm 1 is adopted from [10]. Correctness of Algorithm 1 follows from Theorem 1. It should be clear that every step of the algorithm can be completed in $O(n)$ number of steps. For more details please refer to [10].

In the remaining part of this section, we study other bijective diameters.

Theorem 4. *For every tree G and every species tree S with the same number of leaves, $b^D(G, S) = b^D(G, \star)$.*

Proof. The proof follows immediately from Theorem 1.

We conclude that Problem 2 can be solved by choosing any species tree with $|G|$ leaves and applying Algorithm 1.

Theorem 5. *For a species tree S with $n > 1$ leaves $b^D(\star, S) = n - 2$.*

Proof. It follows from Theorem 1, that the gene tree G that maximizes duplication cost should have the minimal number of cherries (which is 1 in this case). Thus, such a tree is a caterpillar tree. Moreover, by Lemma 2 and Theorem 2 the root of the only cherry present in G has to be mapped to the root of S.

We conclude that Problem 3 has a simple solution.

Theorem 6. *For a species tree S and a gene tree $G \in \mathbb{B}(S)$, $D(G, S) = b^D(\star, S)$ if and only if G is a caterpillar tree such that the only cherry of G is mapped to the root of S.*

Proof. ($\Leftarrow$). It easy to see that $D(G, S) = n - 2$. Thus by Theorem 5 $D(G, S) = b^D(\star, S)$. ($\Rightarrow$). See the proof of Theorem 5.

3.2   Unconstrained Duplication Diameters

In this section, we show similar results for the unconstrained diameters. To avoid repetitions we skip formal definitions of unconstrained problems.

Theorem 7. *For trees T and S, we have $u^D(T, S) = u^D(T, \star) = |T| - 1$.*

Proof. For the first diameter, choose the labeling for T that is a constant function. Then, every internal node is a duplication. Thus, we have $|T| - 1$ gene duplications, which is the maximal possible duplication cost. The same holds true for $u^D(T, \star)$.

Theorem 8. *For a species tree S, $u^D(\star, S) = +\infty$.*

Proof. Assume that we have a sequence of trees $T_1, T_2, \ldots$ over S such that T_n has n leaves. Then, by Theorem 7 $\lim_{n \to +\infty} u^D(T_n, S) = \lim_{n \to +\infty} n - 1 = +\infty$. We conclude, $u^D(\star, S) = +\infty$.

3.3   Expected Number of Gene Duplications

In this section we show the formulas for the expected values of diameters considered in this paper.

In this section we assume $n!! = 1$ for $n \leqslant 2$. For a set X consisting of $n > 0$ species, by $\mathbb{S}(X, c)$, we denote the set of all pairwise non-isomorphic bijectively labeled gene trees over X (i.e., having n leaves uniquely labeled by elements from X) and c cherries. By $t_{n,c}$ we denote the size of $\mathbb{S}(X, c)$.

Lemma 3. *For $n > 1$ and $c \geqslant 0$, we have*

$$t_{n,c} = \begin{cases} (2c-1)!!(2c-3)!! & \text{if } n = 2c \geqslant 2, \\ t_{n-1,c-1}(n - 2c + 1) + t_{n-1,c}(n + 2c - 2) & \text{if } n > 2c > 0, \\ 1 & \text{if } n = 1, c = 0, \\ 0 & \text{otherwise.} \end{cases}$$

Proof. We omit the proof for brevity.

Based on the recurrence for the number of rooted binary leaf-labeled trees with n leaves [6], we can derive the following recurrence for $t_{n,c}$

Lemma 4. *If $n \geqslant 3$ and $c \in \{1, 2, \ldots, \lfloor n/2 \rfloor\}$, then*

$$2t_{n,c} = \sum_{k=1}^{n-1} \binom{n}{k} \sum_{i=0}^{c} t_{k,i} t_{n-k,c-i}.$$

Proof. Every species tree $S \in \mathbb{S}(X, c)$ has the following form $S = (S', S'')$, where $S' \in \mathbb{S}(A, i)$ and $S'' \in \mathbb{S}(X \backslash A, c - i)$ for some $\varnothing \neq A \subsetneq X$, $k = |A|$ and $i \in \{0, 1, \ldots, c\}$. Thus, for every $k \in \{1, 2, \ldots, n - 1\}$ we distribute k species from X in the left subtree of S. This can be done in $\binom{n}{k}$ ways. Then, for the left subtree we choose shapes with i cherries, while for the right subtree we choose shapes with $c - i$ cherries. Additional, every possibility is counted twice due to the symmetry.

Lemma 5. *For every $n \geqslant 1$, $\sum_c t_{n,c} = (2n - 3)!!$.*

Proof. It follows easily from the fact that the number of rooted binary leaf-labeled trees with n leaves equals $(2n - 3)!!$ [6].

Theorem 9. *Under the assumption of a uniform distribution of gene trees with n leaves, the expected value of the bijective diameter for fixed shape of gene tree equals*

$$\frac{1}{(2n - 3)!!} \sum_c t_{n,c}(n - c - 1), \tag{1}$$

where n is the number of species.

Proof. It follows easily from Theorem 1 and Lemma 5.

The same result can be obtained for b^D diameter (under a uniform distribution of gene and species tree pairs). Finally, it is straightforward to proof that under the assumption of a uniform distribution of species trees, the expected value of the bijective diameter of a species tree equals $n - 1$. We omit details.

4   Experimental Evaluation and Discussion

We performed two types of experimental evaluations related to the diameters studied in this article. The first experiment analyzes the expected value of the bijective diameter of the gene tree, while the second experiment studies the effect of inferring species trees by solving the GTP problem when the cost involved is normalized by the diameter.

4.1   Expected Value of the Bijective Diameter of a Gene Tree

We computed values of the expected value of $b^D(T, \star)$ diameter according the the formula 1 from Theorem 9. By using a dynamic programming for efficient computation of the recurrence from Lemma 4, implemented in a python script, we computed the expected value of this diameter for $n = 10, 20, \ldots, 990$ (99 values in total). The overall time of computation was approximately 3 hours of a standard PC workstation (a single core, AMD processor, 1400MHz). The result is depicted in Figure 1. Based on Theorem 1 and Theorem 4, we present minimal (i.e., $\lceil n/2 \rceil - 1$) and maximal (i.e., $n - 1$) values of this diameter for fixed n. The main conclusion from this experiment is that the expected value can be well approximated by the average between these two values.

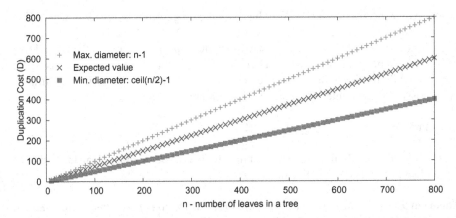

Fig. 1. The expected value of the bijective diameter for a fixed shape of a gene tree (middle line) with values of maximal and minimal values of this diameter. Here ceil(x) denote the ceil function, i.e., the smallest following integer function.

4.2   GTP Evaluation

We studied the gene tree parsimony problems (GTP) [3,9,19,29] under duplication cost and its normalized variant. The problems are defined as follows:

Problem 4 (GTP-DUP). Given a collection of gene trees Q. Find a species tree S that minimizes the total cost $\sum_{G \in Q} D(G, S)$.

Problem 5 (GTP-DUP-NORM). Given a collection of gene trees Q. Find a species tree S that minimizes the total normalized cost $\sum_{G \in Q} D(G, S) / b^D(\hat{G}, \star)$.

Since Problem 4 is known to be NP-hard, we implemented in Java a classical hill climbing heuristic based on the nearest neighborhood interchange (NNI) local search algorithm [2,9]. We used our computer program to perform four computational experiments on two publicly available datasets under the standard duplication cost (GTP-DUP) and the normalized cost (GTP-DUP-NORM).

Guigó dataset. The first dataset consists of 53 gene trees from 16 eukaryotes from [13] (median size of a gene tree is 4.66). It is known from [3], that there are exactly 71 optimal species trees having the minimal total duplication cost equal to 36. Our heuristic was able to infer all these optimal trees. The heuristic run for the normalized variant yielded the same set of species trees and the normalized cost 8.9809.

TreeFam dataset. The second dataset consists of 1274 curated gene family trees from TreeFam v7.0 [27] spanning 25 mostly animal species. The gene trees were rooted by using FastUrec [9] with the species tree based on the NCBI taxonomy (see Figure 2b). Median size of a gene tree in this dataset is 31.80. Multiple runs of our program inferred a single optimal species tree with 7451 gene duplications in total (see Figure 2a). The same tree was inferred for the normalized variant of the duplication cost with 177.3693.

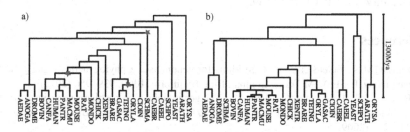

Fig. 2. Species trees for TreeFam dataset. Left: an optimal tree inferred by our heuristic. Right: Species tree based on NCBI taxonomy with branch lengths obtained from diversification dates of the TreeTime database [15]. Stars in the left tree denote the differences between both trees. Note that lengths of the branches of the left tree are not proportional to the time.

The total runtime for these computational experiments was approximately 3 hours on a server with 64 cores (8 Opteron AMD processors, 1400 MHz). The prototype computer programs are available on request.

5   Conclusions and Future Outlook

In this article we investigated into diameters of the duplication cost under several variants and two types of leaf labelings. We proved mathematical properties describing these diameters. Based on these properties we proposed simple formulas for the diameters and efficient algorithms to compute the diameters and their corresponding leaf labelings. In particular we presented an optimal, linear time, algorithm for the bijective case when the shapes of both input trees are fixed.

This is a continuation of our previous research on the deep coalescence diameters [10], that lays foundations for further study on the diameters of other reconciliation based cost functions, e.g., duplication-loss or loss costs [11,23,28].

Our GTP experiments for the duplication cost show no difference between optimal species trees inferred for the standard and normalized variants of the cost. This is likely due to the low resolution of the duplication cost function. However, our experiments are only based on two examples, and future studies will include more complex analyses. Furthermore, we will also investigate in expected values of diameters under various phylogenetic models.

Acknowledgements. The authors wish to thank the three anonymous referees whose constructive suggestions and criticisms have helped considerably to improve the quality of this paper. We also would like to thank Prof. J. Tiuryn for helpful comments. This work was conducted as a part of the Gene Tree Reconciliation Working Group at the National Institute for Mathematical and Biological Synthesis, sponsored by the U.S. National Science Foundation, the U.S. Department of Homeland Security, and the U.S. Department of Agriculture through NSF Award #EF-0832858, with additional support from The University of Tennessee, Knoxville. Support was provided by the grant of NCN (2011/01/B/ST6/02777), and to O. Eulenstein by the U.S. National Science Foundation Award #CCF-1017189.

References

1. Bansal, M.S., Eulenstein, O.: Algorithms for genome-scale phylogenetics using gene tree parsimony. IEEE/ACM Trans. Comput. Biol. Bioinform. 10(4), 939–956 (2013)
2. Bansal, M.S., Eulenstein, O., Wehe, A.: The gene-duplication problem: near-linear time algorithms for nni-based local searches. IEEE/ACM Trans. Comput. Biol. Bioinform. 6(2), 221–231 (2009)
3. Chang, W.-C., Górecki, P., Eulenstein, O.: Exact solutions for species tree inference from discordant gene trees. J. Bioinform. Comput. Biol. 11(05), 1342005 (2013)
4. Cotton, J.A., Page, R.D.M.: Going nuclear: gene family evolution and vertebrate phylogeny reconciled. Proc. Biol. Sci. 269(1500), 1555–1561 (2002)
5. Eulenstein, O., Huzurbazar, S., Liberles, D.A.: Reconciling Phylogenetic Trees. In: Evolution after Gene Duplication. John Wiley, Hoboken (2010)
6. Felsenstein, J.: The number of evolutionary trees. Syst. Zool. 27, 27–33 (1978)
7. Goodman, M., Czelusniak, J., Moore, G.W., Romero-Herrera, A.E., Matsuda, G.: Fitting the gene lineage into its species lineage. A parsimony strategy illustrated by cladograms constructed from globin sequences. Systematic Zoology 28, 132–163 (1979)
8. Górecki, P., Eulenstein, O.: Algorithms: simultaneous error-correction and rooting for gene tree reconciliation and the gene duplication problem. BMC Bioinformatics 13(suppl. 10), S14 (2012)
9. Górecki, P., Burleigh, J.G., Eulenstein, O.: GTP supertrees from unrooted gene trees: Linear time algorithms for NNI based local searches. In: Bleris, L., Măndoiu, I., Schwartz, R., Wang, J. (eds.) ISBRA 2012. LNCS, vol. 7292, pp. 102–114. Springer, Heidelberg (2012)
10. Górecki, P., Eulenstein, O.: Maximizing deep coalescence cost. In: Accepted to IEEE/ACM Trans. Comput. Biol. Bioinform. (2014), preprint is available at http://dx.doi.org/10.1109/TCBB.2013.144
11. Górecki, P., Eulenstein, O., Tiuryn, J.: Unrooted tree reconciliation: A unified approach. IEEE/ACM Trans. Comput. Biol. Bioinform. 10(2), 522–536 (2013)
12. Górecki, P., Tiuryn, J.: DLS-trees: A model of evolutionary scenarios. Theor. Comput. Sci. 359(1-3), 378–399 (2006)
13. Guigó, R., Muchnik, I.B., Smith, T.F.: Reconstruction of ancient molecular phylogeny. Mol. Phylogenet. Evol. 6(2), 189–213 (1996)
14. Harris, S.R., Cartwright, E.J.P., Török, M.E., Holden, M.T.G., Brown, N.M., Ogilvy-Stuart, A.L., Ellington, M.J., Quail, M.A., Bentley, S.D., Parkhill, J., Peacock, S.J.: Whole-genome sequencing for analysis of an outbreak of meticillin-resistant staphylococcus aureus: a descriptive study. Lancet Infect. Dis. 13(2), 130–136 (2013)
15. Hedges, S.B., Dudley, J., Kumar, S.: Timetree: a public knowledge-base of divergence times among organisms. Bioinformatics 22(23), 2971–2972 (2006)

16. Hufbauer, R.A., et al.: Population structure, ploidy levels and allelopathy of Centaurea maculosa (spotted knapweed) and C. diffusa (diffuse knapweed) in North America and Eurasia. In: Proceedings of the International Symposium on Biological Control of Weeds, pp. 121–126 (2003)

17. Jackson, A.P.: A reconciliation analysis of host switching in plant-fungal symbioses. Evolution 58(9), 1909–1923 (2004)

18. Lafond, M., Swenson, K.M., El-Mabrouk, N.: An optimal reconciliation algorithm for gene trees with polytomies. In: Raphael, B., Tang, J. (eds.) WABI 2012. LNCS (LNBI), vol. 7534, pp. 106–122. Springer, Heidelberg (2012)

19. Ma, B., Li, M., Zhang, L.: From gene trees to species trees. SIAM Journal on Computing 30(3), 729–752 (2000)

20. Martin, A.P., Burg, T.M.: Perils of paralogy: using hsp70 genes for inferring organismal phylogenies. Syst. Biol. 51(4), 570–587 (2002)

21. McGowen, M.R., Clark, C., Gatesy, J.: The vestigial olfactory receptor subgenome of odontocete whales: phylogenetic congruence between gene-tree reconciliation and supermatrix methods. Syst. Biol. 57(4), 574–590 (2008)

22. Nik-Zainal, S., et al.: The life history of 21 breast cancers. Cell 149(5), 994–1007 (2012)

23. Page, R.D.M.: Maps between trees and cladistic analysis of historical associations among genes, organisms, and areas. Systematic Biology 43(1), 58–77 (1994)

24. Page, R.D.M.: Extracting species trees from complex gene trees: reconciled trees and vertebrate phylogeny. Mol. Phylogenet. Evol. 14, 89–106 (2000)

25. Page, R.D.M., Holmes, E.C.: Molecular evolution: a phylogenetic approach. Blackwell Science (1998)

26. Rokas, A., Williams, B.L., King, N., Carroll, S.B.: Genome-scale approaches to resolving incongruence in molecular phylogenies. Nature 425, 798–804 (2003)

27. Ruan, J., et al.: TreeFam: 2008 Update. Nucleic Acids Res. 36, D735–D740 (2008)

28. Than, C.V., Rosenberg, N.A.: Mathematical properties of the deep coalescence cost. IEEE/ACM Trans. Comput. Biol. Bioinform. 10(1), 61–72 (2013)

29. Wehe, A., Bansal, M.S., Burleigh, G.J., Eulenstein, O.: DupTree: a program for large-scale phylogenetic analyses using gene tree parsimony. Bioinformatics 24(13), 1540–1541 (2008)

30. Wehe, A., Burleigh, J.G., Eulenstein, O.: Efficient algorithms for knowledge-enhanced supertree and supermatrix phylogenetic problems. IEEE/ACM Trans. Comput. Biol. Bioinform. 10(6), 1432–1441 (2013)

Computational Identification of De-Centric Genetic Regulatory Relationships from Functional Genomic Data

Zongliang Yue[2,4], Ping Wan[4], Zhan Xie[4], Jake Y. Chen[1,2,3,*]

[1] Institute of Biopharmaceutical Informatics and Technology, Wenzhou Medical University,
Wenzhou, Zhejiang Province, China
[2] School of Informatics and Computing, Indiana University, Indianapolis, IN 46202, USA
[3] Indiana Center for Systems Biology and Personalized Medicine, Indiana University,
Indianapolis, IN 46202, USA
[4] Bioinformatics Lab, Capital Normal University, Beijing, China
jakechen@iupui.edu

Abstract. We developed a new computational technique to identify de-centric genetic regulatory relationship candidates. Our technique takes advantages of functional genomics data for the same species under different perturbation conditions, therefore making it complementary to current computational techniques including database search, clustering of gene expression profiles, motif matching, structural modeling, and network effect simulation methods. It is fast and addressed the need of biologists to determine activation/inhibition relationship details often missing in synthetic lethality or chip-seq experiments. We used GEO microarray data set GSE25644 with 158 different mutant genes in *S. cerevisiae*. We screened out 83 targets with 610 activation pairs and 93 targets with 494 inhibition pairs. In the Yeast Fitness database, 33 targets (40%) with 126 activation pairs and 31 targets (33%) with 97 inhibition pairs were identified. To be identified further are 50 targets with 484 activation pairs and 62 targets with 397 inhibition pairs. The aggregation test confirmed that all discovered de-centric regulatory relationships are significant from random discovery at a p-value=0.002; therefore, this method is highly complementary to others that tend to discover hub-related regulatory relationships. We also developed criteria for rejecting genetic regulator candidates x as a candidate regulator and assessing the ranking of the regulator-target relationship identified. The top 10 high suspected regulators determined by our criteria were found to be significant, pending future experimental verifications.

Keywords: Genetic Regulation Relationship, Functional Genomics, Yeast, Computational Prediction.

1 Introduction

Identifying novel gene regulatory relationship from large-scale functional genomics data has been a major theme for the characterization of complex biomolecular systems. Gene

---

[*] Corresponding author.

M. Basu, Y. Pan, and J. Wang (Eds.): ISBRA 2014, LNBI 8492, pp. 224–235, 2014.
© Springer International Publishing Switzerland 2014

regulators identification is confirmed in several organisms by monitoring of gene expression using DNA microarrays. Since the expression profiles have now been performed for a large number of organisms and thousands of microarray experiments have been performed for yeast, C. elegans, Drosophila, Arabidopsis, mice, and humans, one may reconstruct molecular interaction or regulation relationships from mining the data without conducting specific experiments to test whether a candidate regulator-target relationship exists. The DNA expression changing level in microarray can be used to identify novel regulators. James et al [1] examined that temporal gene expression patterns during chondrogenic differentiation in a mouse micromass culture system using Affymetrix microarrays. They showed transcriptional regulation can be certified by changing the expression of molecules indicated in gene ontology annotations for molecular function. The microarray analysis was also used in identifying regulators in the stress response pathway. Lorenz et al [2] showed microarray analysis and scale-free gene networks can be used to identify candidate regulators in drought-stressed roots of loblolly pine. Although molecular evidence of gene regulation is surging since the introduction of high throughput technologies, the full identification of regulators of genes, even in lower eukaryotes, remains largely incomplete.

Traditional experimental gene regulator finding methods using gene knockouts, synthetic lethality, or chip-seq in eukaryotes are often incomplete due to the large combinatorial space to explore [3-5]. To implement the experimental method, identifying gene regulators from large databases of functional genomics data sets has been a main research area in bioinformatics and computer-based methods sprang up. Some popular methods include the following: homologous gene regulator database search, clustering of gene expression profiles and transcription factor binding site pattern matching, physics-based modeling of candidate transcription factors and target binding relationships, and network based methods [6, 7]. Ru-Fang Yeh et al [8] introduced an accurate and efficient automated approach about the homologous gene regulator database search for identifying genes in higher eukaryotic genomes and provided a first-level annotation of the draft human genome. Stephane et al [9] established a rigorous significance test and demonstrated its use on publicly available transcript profiles that implement clustering of gene expression profiles. Their theory links the threshold of selection of putatively regulated genes to the fraction of false positive clones one is willing to risk. Gerhard et al [10] developed a system named ANREP for finding matches to patterns composed of spacing constraints called "spacers" and approximate matches to "motifs" that are, recursively, patterns composed of "atomic" symbols. Niu et al [11] reviewed several physics-based methods, like MM-based methods, for studying the protein-ligand. Albert et al [6] mentioned that network medicine offers a platform to identify drug targets and biomarkers for complex diseases. Systematic approaches were especially popular to reconstruct transcription modules and identify conditions of perturbations under which a particular transcription module is activated/deactivated [12]. However, few computer-based methods focus on non-hub genes' prediction through expression profiles. hence, our aim is developing a new model to identify the regulators of non-hub genes as well as screening out highly suspected target candidates from expression profiles in Yeast.

In this study, our aim is to developed a new computational technique to identify de-centric genetic regulatory relationship candidates. Here, a genetic regulatory relationship is more complicated than the relationship between gene regulators and targets. Gene regulator is the gene coding for transcription factors which are activator/repressor proteins. An activator binds to a site on the DNA molecule and causes an increase in transcription of a nearby gene. A repressor binds to operators or promoters, preventing RNA polymerase from transcribing RNA. In comparison, the genetic regulatory relationship includes the gene control transcription factors like enhancer or silencer that indirectly affect target gene's expression additionally. Since the network with genetic regulatory relationship was generated, the high-connectivity nodes called hub nodes dominate the topology of scale-free networks. And those hub genes are popular and several methods using different models were applied to them since they are usually of great biological significance[13]. However, the method to detect the low-connectivity nodes called peripheral genes which are validated in experiments within biological significance remains blank. Our method is able to find the genetic regulatory relationship about peripheral genes which we defined as the de-centric genetic regulatory relationship, therefore it is complementary to other methods that tend to discover the hub-related regulatory relationship. In addition, the input data is only microarray which are simple and widely used in various research areas like discovering the de-centric regulatory relationship under perturbation and miRNA analysis. The computational efficiency of our method is shown in that we have a faster order of convergence than Monte Carlo simulations using random or pseudorandom sequences. When applying our method into N genes' samples, the computational complexity is $o(N^2)$. When using the traditional Monte Carlo method which randomly selects nodes, the method separates them into hub nodes and de-centric nodes. Adding one node each time will call for updating of method results of complexity of $o(N!)$

2   Method

We first normalized the input data. First, the input data was normalized. Biological independent replicates on microarrays were averaged and groups without mutation used as control were filtered off. Second, the de-centric targets were selected by clustering of gene expression profiles and transcription factor binding site pattern matching. Third, a Pearson Correlation model was applied to extract the de-centric activation/inhibition pairs and a criteria was used to reject low confidence de-centric genetic regulatory pairs. Fourth, the de-centric genetic regulatory network was generated using de-centric genetic regulatory pairs which were significant as confirmed by an aggregation test. Fifth, the genetic regulatory pairs were validated in several published databases. The validation also involved a hypergeometric test which confirmed that the top 10 suspected de-centric targets were significant.

2.1   Preparation of Functional Genomics Data

We used raw data GSE25644 required from the Gene Expression Omnibus (GEO) database (http://www.ncbi.nlm.nih.gov/geo/). GSE25644 is a DNA microarray gene expression profile with all 158 viable protein kinase/phosphatase deletions in S. cerevisiae under a single growth condition [14]. Each mutant was profiled four times, from two independent cultures on dual-channel microarrays using a batch of wild-type (WT) RNA as a common reference. The GSE25644 was normalized by averaging each of the two independent cultures' results on microarrays, and the wild-type groups were filtered off.

2.2   Selection of De-Centric Targets

The selection of de-centric targets is based on clustering of gene expression profiles and transcription factor binding site pattern matching. We modeled each candidate target gene as being controlled by N binary-state regulators that lead to $k'+1 \leq 2^N$ observable states ("step-levels") for the target. Here, k' is the number of steps generated from each gene RNA expression. If there are sufficiently large collections of functional genomics experiments, each being performed under a heterogeneous perturbation condition, the method will search each target's genetic regulator candidates to test if a significant switch-responder pattern existed before ranking candidate genetic regulators for final result reporting.

First, the average increasing value Δ of each gene was calculated and averaged values were sorted. The average increasing value Δ was calculated by the maximum value of the gene expression (we assume the target gene t) minus the minimum value and then divided by the number of samples n.

Second, the average increasing value Δt of a target was regarded as a standard to seek steps which means the number of dj larger than Δt, we defined as k value. For each target, gene expression was sorted from low to high and then we calculated the difference between adjacent samples, dj. If the difference dj is larger than corresponding Δt, k value plus one. If the difference dj is smaller than corresponding Δt, k will be retained.

Third, iteration was performed for every target to find out each k of targets. k values were sorted from low to high and the largest kmax corresponds to target t'.

Fourth, every target was normalized by the new average increasing value Δmax as a standard to seek for new step levels (k') of each target.

Fifth, the binary state N was calculated, and N means the number of genetic regulators of each target. For instance, assuming that a target's binary state N is 2, this target would have less than 3 steps within 4 step-levels theoretically. The formula of N binary-state is shown below:

$$2^N = k'+1$$

We can use the cluster k' to calculate the theoretical number of genetic regulators of the de-centric targets.

2.3   Identification of Genetic Regulatory Relationship among Genes

The genetic regulatory relationship of regulators to targets was predicted based on the step-levels associated with Pearson Correlation. First, the model of activation/inhibition is determined by the low gene expression step-level and the high gene expression step-level. If there existed a positive genetic regulator activating its target, the correlation in low gene expression's step-level would be low to 0 and the correlation in high gene expression's step-level would be high to 1. Second, to explore the potential activation/inhibition genetic regulatory relationship between regulator x and target t, each gene x regarded as a genetic regulator candidate was aligned to each step level of the target t. The low step levels of t_0 and the high step levels of t_h were used to compute Pearson correlation R_0 and R_h respectively.

The pearson correlation is a measure of the strength of the linear relationship between two variables. Since two independent genes expression relationship is linear, the correlation coefficient adequately represent the strength of the relationship between these two genes.

The cutoff values of the R_0 and the R_h are 0.2 and 0.8 respectively. For instance, the R_0 located in (-0.2,0.2) and the R_h located in the (0.8,1) indicated the x is a positive genetic regulator of t. For illustrative purposes, we show a simple form to explicitly explain the activation/inhibition models of the genetic regulator x to target t.

Form 1. Activation/Inhibition pairs predicted by model PC

	R_0	R_h
x→(+)t	→0(-0.2,0.2)	→1(0.8,1)
x→(-)t	→0(-0.2,0.2)	→-1(-0.8,-1)

Third, to reject the incorrect genetic regulator x, a criteria we developed is that the middle gene expression step-levels of genetic regulator x shouldn't have a significant larger change value than the average change value of gene expression. Assuming that one genetic regulator x has one middle gene expression step-level from step i to step j and the lowest gene expression value among xi to xj is xlow, the change of gene expression from xn(n in i to j) to xlow shouldn't be more than the genetic regulator's average change level of the value Δx.

Fourth, to examine the step-level correlation between low step levels of the t (R_0) and high step levels of the t (R_h) between the target t and x, all genetic regulator candidates were ordered to each target using max $f(t, x)$: $|R_o - R_h|$. Then we choose top 10 high $|R_o - R_h|$ genetic regulator pairs, since the number of nodes generating the genetic regulatory network should be balance and moderate to present the de-centric targets.

2.4   Generation of De-Centric Genetic Regulatory Network and Testing of Network Significance

The de-centric genetic regulatory networks were generated by the top 10 high $|R_o - R_h|$ genetic regulator pairs and the statistical data analysis tests were applied to detect the significance of the connected network. Our hypothesis of this statistical evaluation is that if the prediction model consists of de-centric targets involved in the same process even if complex and broad, then we should expect that the connectivity among the de-centric targets be lower than the connectivity among a set of randomly selected genes.

We defined the index of aggregation of a network [7] as the ratio of the size of the largest sub-network that exists in this network to the size of this network. Note that the size is calculated as the total number of genes within a given network/sub-network.

To test the hypothesis that the predicted targets are less connected than a randomly selected set of targets, we developed the null hypothesis test using the following re-sampling procedure :

1. Randomly select from the genetic regulators to targets pool, the same number of predicted targets generated from our method.
2. Retrieve the top 10 genetic regulators of each random target using |Ro – Rh| criteria.
3. Compute the index of aggregation of the superset.
4. Repeat steps 1 through 3 for 500 times to generate a distribution of the index of aggregation under random selection.
5. Compare the index of aggregation from our method with the distribution obtained in 4 and calculate the p-value.

2.5   Assessing Significance Testing of the De-Centric Genetic Regulatory Relationships

Our result is validated in the Yeast Fitness Database (http://chemogenomics.med. utoronto.ca/fitdb/fitdb.cgi). FitDB is a searchable database of quantitative chemical-genetic interactions based on data in Hillenmeyer [15]. A gene search allows viewing of the compounds that are most sensitive to the gene specified in a heterozygous and homozygous yeast deletion strain, including a view of yeast deletion strains that behave similarly to the gene of interest. Compounds can also be searched to identify heterozygous or homozygous deletion strains exhibiting hypersensitivity to compound, including a view of compounds that behave similarly to the compound of interest.

Hypergeometric test was introduced to see the significant of the top 10 high $|R_o - R_h|$ pairs in genetic regulator candidates. The validated number of top 10 high $|R_o - R_h|$ pairs in Yeast Fit database was k. The total number of the top 10 high $|R_o - R_h|$ pairs is n. The validated number of candidate pairs in Yeast Fit database is K. The total number of candidate pairs is N. The situation of hypergeometric test is illustrated by the following contingency table: (Form 2)

Form 2. Contingency table of hypergeometric test

	top 10	residue	total
validated	K	$K - k$	K
un-validated	$n - k$	$N + k - n - K$	$N - K$
total	n	$N - n$	N

The variable number of top 10 high $|R_o - R_h|$ pairs X follows the hypergeometric distribution by its probability mass function (pmf) is given by the formula below:

$$P(X = k) = \frac{\binom{K}{k}\binom{N-K}{n-k}}{\binom{N}{n}}$$

3   Results

3.1   Determination of De-Centric Genetic Regulated Targets from the Processed Data

The distribution of the number of genetic regulators for each target follows Gaussian distribution which means the number of genetic regulators is random variables independently drawn from the same distribution. 453 targets was selected, of which genetic regulator is equal or below 2 after clustering of gene expression profiles and transcription factor binding site pattern matching. (Fig 1) According to the formula: $2^N = k'+1$, the N was determined by the cluster k' and all targets was clustered into 8 groups corresponding k' varied from 0 to 7. We chose the left shade region separated by red line which displayed the target's genetic regulator equal or below 2 excluding 0.

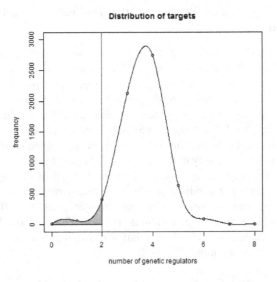

Fig. 1. Distribution of targets with different genetic regulators in the yeast data set. The red line stands for the genetic regulators equal 2 within step levels k' from 2 to 3. The shade area presents the genetic regulators equal or below 2 step levels excluding 0 within k' from 1 to 3.

The $2^N=k'+1$ is the key point to control the theoretical genetic regulator number of targets. Especially the selection of binary-state N determine the theoretical genetic regulator number directly. The reason we call N theoretical genetic regulator number is that in the formula $2^N=k'+1$, we cannot certain the number of N when several genetic regulator have same changing DNA expression level on their target giving merged step-levels. Assumed that a target t have n genetic regulators x_1, x_2 ... x_n. And $f(x_1, x_2 ... x_n)$ stands for step levels due to combine genetic regulatory effect on t. The states of x can be either 1 or 0 to indicate x is activate or not. The number of target's step-levels $f_t(x_1, x_2 ... x_n)$ is greater than n: $f_t(x_1, x_2...x_n) \geq n + 1$ For instance, if x_1, x_2, x_3 have similar regulation ability on t, the $f(1,0,0)$, $f(0,1,0)$, $f(0,0,1)$ will give a merged level. Similarly the $f(0,1,1)$, $f(1,0,1)$, $f(1,1,0)$ will give another merged level. In this case, the step levels of t is 4=n+1 that stands for four levels of $f(0,0,0)$, $f(1,0,0)$, $f(1,1,0)$ and $f(1,1,1)$. Hence, the N will be 2 means that our predicted targets of 2 genetic regulators will contain some targets of 3 genetic regulators in extreme case. And the mis-clustering will be more severe when applied for prediction of targets with more genetic regulators.

3.2   Identification of Activation/Inhibition Genetic Regulatory Relationship Details

According to the Pearson Correlation model of Activation/Inhibition, 89 targets were selected with 3483 activation pairs in initial 453 targets and 94 targets were selected with 2272 inhibition pairs in initial 453 targets. After applying the criteria of rejection, we found 83 targets with 3200 activation pairs and 93 targets with 2129 inhibition pairs. After ordering all 5329 regulation pairs in the genome for each target by finding max f(t, x): $|R_o - R_h|$ and choosing top 10 high pairs, 610 activation pairs and 494 inhibition pairs were selected. (Index 1) 176 targets candidates were queried in Yeast Fitness Database, 64 targets were identified afterwards. Among them, 33 targets with 126 activation pairs and 31 targets with 97 inhibition pairs were certificated. (Form 3)

Form 3. Validated de-centric targets and genetic regulatory relationship

	Activation			inhibition		
	Validated	total	rate	validated	total	rate
pairs	126	610	21%	97	494	20%
	(33)	(83)	(40%)	(31)	(93)	(33%)

3.3   Construction of De-Centric Target-Regulator Network

We constructed the network of the 610 activation pairs to targets and 494 inhibition pairs. 112 targets were found to be new candidates with its genetic regulator. The hub nodes with high-connectivity linked to the majority of nodes to formed a main structure of network. While the peripheral nodes with low-connectivity distributed around the hub nodes and formed relativity small sub-networks or even one-to-one model. The ratio of the size of the largest sub-network that exists in this network to the size of this network, we defined as index of aggregation, reflect hub-nodes weight. The index of aggregation in the activation genetic regulatory relationship is 64%. The index of aggregation in the inhibition genetic regulatory relationship is 62%.

The empirical distribution of the index of aggregation was obtained after 500 random re-samplings (Fig 2). Only 1 runs out of 500 resulted in an index of aggregation value greater than 99.8% in both de-centric genetic activation regulatory network and de-centric genetic inhibition regulatory network. Therefore, the p-value of the index of aggregation is 0.002. It is not surprising to observe such a significant result since the results are selected in a way that the theoretical genetic regulator is equal or below 2. Hence, the aggregation test confirmed result of clustering of gene expression profiles and transcription factor binding site pattern matching is significant in discovering de-centric genetic regulatory relationship.

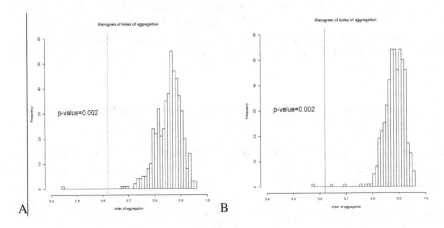

Fig. 2. A: Histogram of the index of aggregation distribution for the enrichments of sets of activation pairs (size=83) randomly selected from pairs. B: Histogram of the index of aggregation distribution for the enrichments of sets of inhibition pairs (size=93) randomly selected from pairs. The red line indicates the index of aggregation value for activation genetic regulatory pairs from de-centric network(activation=0.64/inhibition=0.62).

The contingency table of genetic regulatory relationship network is shown in Form 4. The total number of activation/inhibition pairs of genetic regulatory relationship predicted by Pearson Correlation model is 5329. 1104 genetic regulatory pairs were selected by top 10 high $|R_o - R_h|$ and 5224 were residue. After validation in Yeast Fit Database, the validated top 10 high $|R_o - R_h|$ is 223 and un-validated top 10 high $|R_o - R_h|$ is 881. The validated residue is 1196 and un-validated residue is 3029.

After applied the formula in the top10 high $|R_o - R_h|$ genetic regulatory pairs, the P-value of hypergeometric test is 3.46e-08, which indicates that the top10 high $|R_o - R_h|$ pairs are more significant in the pairs of genetic regulator candidates to their targets than random one.

$$P(X = k) = \frac{\binom{K}{k}\binom{N-K}{n-k}}{\binom{N}{n}}$$

Form 4. Contingency table of genetic regulatory relationship network

	top 10	bottom20%	total
validated	223	1196	1419
un-validated	881	3029	3910
total	1104	4225	5329

The 174 de-centric genetic regulatory targets (81 active targets, 91 inhibited targets, 2 in both active targets set and inhibited targets set) predicted were queried in the Yeastract Database (http://www.yeastract.com/index.php). The evidence of DNA binding and expression in database's documents covers 31 of 173 de-centric genetic regulatory targets. The 142 residue de-centric genetic regulatory targets haven't been discovered probably due to the limited experimental technique. The number of regulators of 31 de-centric genetic regulatory targets is small and not above 5. The ratio of the target's number which regulators' number equal or below 2 to the total target's number is 0.84 (Fig 3). Hence, the method is able to detect the de-centric genetic regulatory targets confirmed by the Yeastract Database.

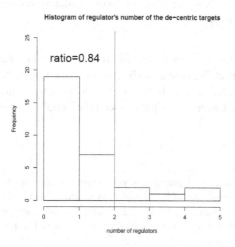

Fig. 3. Histogram of regulator's number of the de-centric targets. The x axis shows the number of regulators of the de-centric genetic regulatory target. Red line is drawn to indicate the ratio of the target's number which regulators' number equal or below 2 to the total target's number is 0.84.

4   Discussion

In this study, we developed a new computational technique to find the potential genetic regulators of none-hub targets. This method tested by empirical distribution p value. We exploited Pearson Correlation Model to explore the potential regulation relationship of genetic regulators to targets. The prediction pairs were identified by the validation of Yeast Fitness Database. This method contributes to de-centric genetic regulatory relationships identification, therefore highly complementary to other methods that tend to discover hub related genetic regulatory relationship. It can be also generalized to other biology genetic regulatory relationship mining areas.

In our method, we also explored 112 new targets. For each new target, we gave the top 10 high suspected genetic regulatory pair candidates (Index 1), although which one is the accurate genetic regulator we cannot affirm. This genetic regulatory relationship of regulators to targets can be more than the direct regulation and control. The genetic regulators could be in the upstream of its targets since we found the closest correlation instead of closest regulation in one protein deletion when applied with the microarray of 158 viable protein kinase/phosphatase deletions.

In summary, our method incorporated microarray of GSE25644 and helped us to successfully carry out genetic regulators to targets relationship. The computational results, which began with inputs that are not necessarily highly reliable, showed high biological relevance. Going down the ranked targets' genetic regulators, one may generate many new biological hypotheses about the new functions of genes in the regulation pairs network context beyond the scope of this work. We are also in the process of developing better and accurate prediction of genetic regulatory relationship and applying these methods to the study of other regulatory relationship in other data types more than microarray.

Acknowledgement. This work was supported in part by internal research funding obtained by Jake Chen at Wenzhou Medical University from China at Indiana University School of Informatics from Indianapolis, IN. We also thank the generous support by Capital Normal University, China to seed the initial implementation of ideas.

References

1. James, C.G., et al.: Microarray analyses of gene expression during chondrocyte differentiation identifies novel regulators of hypertrophy. Mol. Biol. Cell 16(11), 5316–5333 (2005)
2. Lorenz, W.W., et al.: Microarray analysis and scale-free gene networks identify candidate regulators in drought-stressed roots of loblolly pine (P. taeda L.). BMC Genomics 12, 264 (2011)
3. Snyder, M., Gallagher, J.E.: Systems biology from a yeast omics perspective. FEBS Lett. 583(24), 3895–3899 (2009)
4. Goodson, H.V., Anderson, B.L.: Synthetic Lethality Screen Identifies a Novel Yeast. The Journal of Cell Biology 133(6), 1277–1291 (1996)
5. Zhang, Y., et al.: Model-based analysis of ChIP-Seq (MACS). Genome Biol. 9(9), R137 (2008)

6. Barabasi, A.L., Gulbahce, N., Loscalzo, J.: Network medicine: a network-based approach to human disease. Nat. Rev. Genet. 12(1), 56–68 (2011)
7. Chen, J., Shen, C.: Mining Alzheimer Disease Relevant Proteins from Integrated Protein Interactome Data. In: Pacific Symposium on Biocomputing, vol. 11, pp. 367–378 (2006)
8. Yeh, R.F., Lim, L.P., Burge, C.B.: Computational inference of homologous gene structures in the human genome. Genome Res. 11(5), 803–816 (2001)
9. Audic, S., Claverie, J.-M.: The Significance of Digital Gene Expression Profiles. Genome Res. 7, 986–995 (1997)
10. Mehldau, G., Myers, G.: A system for pattern matching applications on biosequences. Comput. Appl. Biosci. 9(3), 299–314 (1993)
11. Huang, N., Jacobson, M.P.: Physics-based methods for studying protein-ligand interactions. Current Opinion in Drug Discovery & Development, coddd2007 10(3), 325–331 (2007)
12. Wang, W., et al.: A systematic approach to reconstructing transcription networks in Saccharomycescerevisiae. Proc. Natl. Acad. Sci. U. S. A. 99(26), 16893–16898 (2002)
13. Wan, P., et al.: Mechanisms of Radiation Resistance in Deinococcus radiodurans R1 Revealed by the Reconstruction of Gene Regulatory Network Using Bayesian Network Approach. J. Proteomics Bioinform. 6(7) (2013)
14. van Wageningen, S., et al.: Functional overlap and regulatory links shape genetic interactions between signaling pathways. Cell 143(6), 991–1004 (2010)
15. Hillenmeyer, M.E., et al.: The chemical genomic portrait of yeast: uncovering a phenotype for all genes. Science 320(5874), 362–365 (2008)

Classification of Mutations by Functional Impact Type: Gain of Function, Loss of Function, and Switch of Function

Mingming Liu, Layne T. Watson, and Liqing Zhang

Departments of Computer Science and Mathematics,
Virginia Tech, Blacksburg, VA, USA
{mingml,ltw,lqzhang}@vt.edu

Abstract. Genomic variations have been intensively studied since the development of high-throughput sequencing technologies. There are numerous tools and databases predicting and annotating the functional impact of genetic variants, such as determining whether a variant is neutral or deleterious to the functions of the corresponding protein. However, there is a need for methods that not only identify neutral or deleterious mutations but also provide fine grained prediction on the outcome resulting from mutations, such as gain, loss, or switch of function. This paper proposes the deployment of multiple hidden Markov models to computationally classify mutations by functional impact type.

Keywords: Genomic variants, gain of function, loss of function, switch of function, hidden Markov model.

1 Introduction

Mutations contribute to human evolution and disease development. Most variants are so-called "neutral", because either their impact on protein function is minor or the affected protein is not important for diseases or important traits. So it is worthwhile to computationally order the genetic variants in terms of their importance. With an increasing amount of genomic variability data, computational tools for prediction of the functional impacts of these variants on proteins have been developed, including SIFT SNP [1], Polyphen [2], GERP [3] and SCONE [4]. These methods compute a quantitative score to measure the degree of deleteriousness. However, they cannot predict the likely cellular outcome resulting from mutations, such as gain, loss, or switch of function. Biologically, loss of function mutations cause the gene product to have less or no function; gain of function mutations change the gene product to have a new and possibly abnormal function; switch of function is in the middle of loss of function and gain of function, which is a switch from one set to an alternative set of specific functions of the mutant protein [5]. Experimentally deciding the type of mutation is time consuming and expensive, thus computationally predicting the type of mutation is important. Computational classification of mutations by type as

M. Basu, Y. Pan, and J. Wang (Eds.): ISBRA 2014, LNBI 8492, pp. 236–242, 2014.

well as strength of the impact will contribute to a more complete understanding of functional alterations in a disease genome.

Earlier work addressed prediction of the functional type of variants[5–7] by trying to identify activating variants, but none provide a precise computational classification definition for all these types: loss, gain, and switch of function. This paper computationally classifies genomic variants into these three types on the basis of previous work on functional effects prediction of genetic variants using HMMvar[8], a method based on the principle of evolutionally conservation and hidden Markov models (HMMs).

2   Methods

Figure 1 shows the pipeline of the new proposed method. It starts with a set of homologous sequences of the target wild type protein sequence. Then a multiple sequence alignment (MSA) is generated as in HMMvar [8]. Then the K-means clustering algorithm is used to group the sequences into different subfamilies, including the target subfamily that contains the target wild type protein sequence (the dashed line shown in Figure 1), from which a set of hidden Markov models (one per subfamily) is built. These HMMs are then integrated into the decision rules to predict different types of mutations.

2.1   HMMvar

HMMvar [8] quantitatively predicts the functional effects of variants. It builds a hidden Markov model (HMM) based on a set of homologous sequences of the target query sequence. Then the wild type protein sequence and mutant type protein sequence are matched against the HMM respectively. HMM provides a score to measure the similarity between the sequence and the "protein family" represented by the hidden Markov model. If the mutant type sequence fits almost the same as the wild type sequence, the mutation has little effect on the protein function. To identify different types of mutations, homologous sequences are clustered and each cluster is viewed as a "subfamily", which captures specific functions. If a mutant sequence fits better than the corresponding wild type sequence in a "subfamily", then probably the variant causes the protein to obtain a new function. With this assumption, clustering the homologous sequences, including the query sequence, identifies "subfamilies".

First, the homologous sequences are aligned by the multiple sequence alignment algorithm MUSCLE [9], then K-means clustering is performed based on the MSA. The distance between sequences is the edit distance and the cluster prototype sequence is defined by the most frequent amino acid at each position at every iteration($k = 3$ clusters here). For each of the clusters, a HMM is built, which represents the "subfamily" or specific functions that differ from those of target group (cluster). Denote these "subfamilies" as $C_0, C_1, ..., C_{k-1}$, where C_0 is the target group that contains the wild type sequence, and the corresponding HMMs as $H_0, H_1, ..., H_{k-1}$.

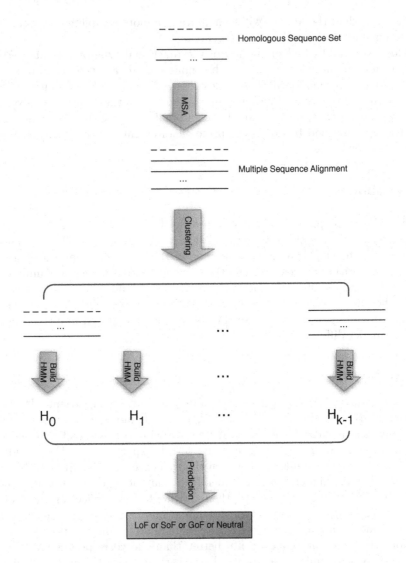

Fig. 1. Flowchat of the classification procedure. The dashed line represents the target query sequence.

2.2   Classification of Mutations

As in HMMvar [8], the HMMs are used to predict the degree of harm in the variants. The difference is that HMMvar uses only one HMM built from the MSA of all the homologous sequences, but here k HMMs are built. For a given variant v_i, let S_i^m ($0 \le m \le k-1$) denote the quantitative HMMvar score of variant v_i obtained from H_m. Note that H_0 is the HMM built from the target group C_0 where the wild type sequence clustered (the dashed line shown in Figure 1), thus S_i^0 is the score of variant v_i calculated from H_0.

The decision rule for classifying different types of mutations is shown in Figure 2, for a given variant v_i and predefined cutoff t ($t = 2$ from the experiments for HMMvar [8]). $S_i^0 > t$ indicates that in the target "subfamily", the wild type sequence fits better than the mutant type sequence, which is a sign that the variant is a loss of function. Further examined with other "subfamilies", if for all other "subfamilies" the wild type sequence fits better than the mutant type sequence, then classify this variant as loss of function (LoF). Otherwise, it is classified as switch of function (SoF) because although the variant probably causes the protein loss of function represented by C_0, it obtains the specific function represented by some C_m. On the other hand, if $S_i^0 \leq t$, the variant could be potentially gain of function (GoF). Then if there exists at least one other "subfamily" that the mutant type sequence fits better than the wild type sequence, the variant is classified as gain of function, otherwise, it is functionally neutral.

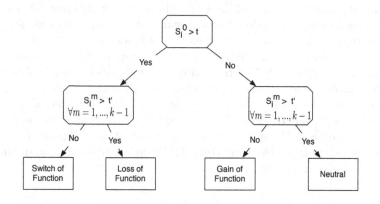

Fig. 2. Decision rules for the classification of mutations

3   Results

3.1   Datasets

Two data sets were used for the validation, TP53 SNPs and SWISS-PROT mutagenesis data [6]. A set of 2,565 SNP mutants and corresponding biological activity levels were obtained from the database IARC TP53 [10]. The mutants associated with TP53 were partitioned into four classes in terms of transactivity level: nonfunctional, partially functional, functional (wildtype), and supertrans (higher activity than wildtype) [11]. Transactivity level was measured by eight promoter-specific activity levels and the classification was made in terms of the median of these eight levels. Mutations are classified as "nonfunctional" if the median is ≤ 20, "partially functional" if the median is > 20 and ≤ 75, "functional" if the median is > 75 and ≤ 140, and "supertrans" if the median is > 140. The SWISS-PROT mutagenesis data consists of 6922 variants from 4061 proteins labeled by "Activating", "Neutral", and "Deleterious", which are extracted from the text descriptions by key words in the database. The final dataset used for analysis included 225 activating, 1222 neutral, and 5475 deleterious variants.

3.2 Predictions

In terms of the classification rules shown in Figure 2, the TP53 SNPs are classified into four classes as shown in Figure 3. Three dotted lines ($y = 20$, $y = 75$, and $y = 140$) separate the plot vertically into four regions, which represent "nonfunctional", "partially functional", "functional", and "supertrans" regions, respectively, from bottom to top. From Figure 3, the median of the transactivity level in the GoF group is the highest among these four groups as the GoF group is enriched by "functional" or "supertrans" variants. On contrast, the LoF group is dominated by "partially functional" or "nonfunctional" variants.

To validate the effectiveness of the scores obtained from different HMMs, these scores are used to train a random forest classifier. The "nonfunctional" and "partially functional" variants are treated as loss of function, and "functional" and "supertrans" variants are treated as neutral and gain of function, respectively. The number of trees used is 500. In random forest classifier, the out of bag error rate (OOB error) is an internal unbiased classification error estimator [12]. For each tree in the forest, one creates a training sample from the original sample by taking points with repetition, and what is left out can be used to compute the out of bag estimation. The results show that the out of bag estimate of error rate is 25.66%, which indicate the scores from HMMvar are reasonable to distinguish different classes of mutations.

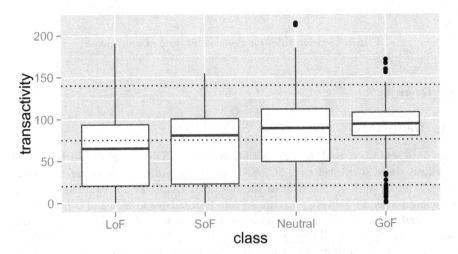

Fig. 3. Predictions based on the TP53 dataset

Around 7000 variants from the SWISS-PROT database were predicted. This dataset was labeled by three functional effects of mutations, "Activating", "Neutral", and "Deleterious". Figure 4 displays the prediction results. The deleterious variants constitute the most to the predicted loss of function group and this constitution decreases as it goes from LoF to SoF to neutral to GoF, while the activating variants constitute the least to the predicted loss of function group and this constitution increases as it goes from LoF to SoF to neutral to GoF. The neutral variants have the same trend as the activating variants. Although only 3.3% of the dataset is activating, 8.7% of the mutations in the gain of function group are activating. Compared to other groups, the activating variants are enriched in the gain of function group, and the deleterious variants are enriched in the loss of function group, similar to results for the TP53 mutations.

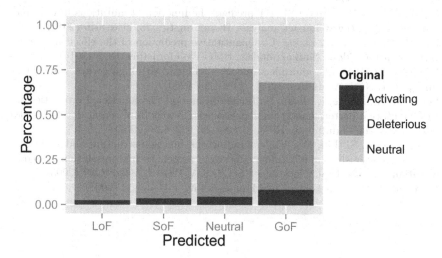

Fig. 4. Predictions based on SWISS-PROT mutagenesis dataset

4   Discussion and Conclusion

This paper proposed using multiple hidden Markov models to predict the fine grained functional impact of mutations on proteins. Using a decision rule for classifying the types of functional outcome and applying the prediction pipeline to two different data sets provides evidence that the pipeline is capable of identifying different types of mutations. The results suggest that the gain of function mutations tend to be activating variants and loss of function mutations tend to be deleterious variants. Worth noting is that with our computational pipeline, the fine grained prediction on functional outcome of mutations depends on two predefined cutoff values (t and t' shown in Figure 2, where $t = 2$ and $t' = 1$ achieve the best performance). As empirical functional essays of mutation outcomes become available in the future, it is desirable to include them to further improve the determination of these cutoffs which in turn strengthens the quality of the prediction.

References

1. Pauline, C., Henikoff, S.: Predicting Deleterious amino acid substitutions. Genome Res. 111, 863–874 (2001)
2. Ramensky, V., Bork, P., Sunyaev, S.: Human non-synonymous SNPs: server and survey. Nucleic Acids Res. 30(17), 3894–3900 (2002)
3. Cooper, G., Stone, E., Asimenos, G.: Distribution and intensity of constraint in mammalian genomic sequence. Genome Res. 15(7), 901–913 (2005)
4. Asthana, S., Roytberg, M., Stamatoyannopoulos, J.: Analysis of sequence conservation at nucleotide resolution. PLOS Comput. Biol. 3, e254 (2007)
5. Reva, B., Antipin, Y., Sander, C.: Predicting the functional impact of protein mutations: application to cancer genomics. Nucleic Acids Res. 39, e118 (2011)
6. Lee, W., et al.: Bi-directional SIFT predicts a subset of activating mutations. PLoS ONE 4, e8311 (2009)
7. Ng, S., et al.: PARADIGM-SHIFT predicts the function of mutations in multiple cancers using pathway impact analysis. Bioinformatics 28, i640–i646 (2012)
8. Liu, M., Watson, L.T., Zhang, L.: Quantitative prediction of the effect of genetic variation using hidden Markov models. BMC Bioinformatics 15, 5 (2014)
9. Edgar, R.C.: MUSCLE: multiple sequence alignment with high accuracy and high throughput. Nucl. Acids Res. 32(5), 1792–1797 (2004)
10. Petitjean, A., Mathe, E., Kato, S.: Impact of mutant p53 functional properties on TP53 mutation patterns and tumor phenotype: lessons from recent developments in the IARC TP53 database. Hum. Mutat. 28, 622–629 (2007)
11. Kato, S., Han, S., Liu, W.: Understanding the function-structure and function-mutation relationships of p53 tumor suppressor protein by high-resolution missense mutation analysis. Proc. Natl. Acad. Sci. U.S.A. 100(14), 8424–8429 (2003)
12. Breiman, L.: Random Forests. Machine Learning 45(1), 5–32 (2001)

Network Analysis of Human Disease Comorbidity Patterns Based on Large-Scale Data Mining

Yang Chen[1,2] and Rong Xu[2]

[1] Department of Electrical Engineering and Computer Science, School of Engineering
[2] Division of Medical Informatics, School of Medicine
Case Western Reserve University, Cleveland OH, 44106

Abstract. Disease comorbidity is an important aspect of phenotype associations and reflects overlapping pathogenesis between diseases. Existing comorbidity studies usually focused on specific diseases and patient populations. In this study, we systematically mined and analyzed disease comorbidity patterns without restricting disease types and patient populations. We presented a data mining approach and extracted comorbidity patterns from a patient-disease database in the drug adverse event reporting system. The database contains records of 3,354,043 patients. We first demonstrated that the data are not severely biased towards specific patient populations and valuable for comorbidity mining. Then we developed an automatic pipeline to process the data, and applied an association rule mining algorithm to mine comorbidity relationships among multiple diseases. Our approach extracted 8,576 comorbidity patterns for 613 diseases. We constructed a disease comorbidity network from these patterns and demonstrated that the comorbidity clusters reflect genetic associations between diseases. Different from previous studies based on relative risk, which tends to identify comorbidities for rare diseases, our approach extracted many patterns for common diseases. We applied the approach on colorectal cancer, and found interesting relationships between colorectal cancer and metabolic disorders, which may lead to promising pathogenesis discoveries.

1 Introduction

Phenotype similarity reflects overlapping pathogenesis (Brunner *et al.*, 2004; Tiffin *et al.*, 2009), thus has been used to predict associated genes (Wu *et al.*, 2008; Li *et al.*, 2010; Vanunu *et al.*, 2010), protein complexes (Lage *et al.*, 2007), and drug treatments (Iorio *et al.*, 2011) for human diseases. Disease comorbidity is an important aspect of the human phenome. Comorbidity patterns often lead to unexpected disease links (Oti *et al.*, 2008), and offer novel insights to explain disease pathogenesis. For example, Blair et al. used the comorbidity relationships between complex and Mendelian diseases as a lead and discovered the genetic variants for complex diseases from Mendelian variants (Blair *et al.*, 2013). Avery et al. started from the co-occurrence of dyslipidemia, obesity, and

M. Basu, Y. Pan, and J. Wang (Eds.): ISBRA 2014, LNBI 8492, pp. 243–254, 2014.

other metabolic dysregulations, and identified their unifying genetic mechanisms (Avery *et al.*, 2011). Recent researches also detected the genetic overlaps between cancers and their comorbidities (Joseph *et al.*, 2014; Toffanin *et al.*, 2010).

Several previous studies have systematic detected disease comorbidity patterns through data mining, but most of them focused on specific diseases and patient populations. Rzhetsky et al. (Rzhetsky *et al.*, 2007) developed a statistical model to analyze a database of medical records. They identified co-occurrence relationships among 160 diseases, and emphasized on psychiatry disorders. Park et al. (Park *et al.*, 2009) and Hidalgo et al. (Hidalgo *et al.*, 2009) extracted comorbidity patterns from Medicare claims with statistical measures of relative risk and ϕ-correlation. They included patients aged 65 or older into the analysis. Roque et al. mined disease comorbidities from the free text of medical records from a psychiatric hospital (Rogue *et al.*, 2011).

In this study, we systematically mined and analyzed disease comorbidity pattern without restricting disease types and patient populations. We propose to extract comorbidity patterns from FDA adverse event reporting system (FAERS), which contain records of 3,354,043 patients. FAERS has been extensively mined for detecting drug safety signals, but its use in mining disease comorbidities has not been explored yet. We analyzed the patient demographics and showed that the data are valuable for the comorbidity study.

We developed a data mining approach to extract comorbidity patterns with an association rule mining approach. Most previous studies mined disease comorbidities using relative risk and ϕ correlation, which are intrinsically biased towards rare diseases and exclusively detect pairwise relationships. Unlike previous studies, our approach identified comorbidity patterns for many common diseases, and can flexibly detect associations among multiple diseases. We constructed a human disease comorbidity network, and investigated the genetic coherence of the comorbidity clusters in the network to demonstrated its potential in predicting disease mechanisms. We applied the approach in mining comorbidity patterns for colorectal cancer. The patient population was stratified to eliminate the effects of age and gender. Our result comorbidity patterns are supported by literature evidences, and may indicate promising hypothesis for the genetic mechanisms of colorectal cancer.

2   Material and Methods

2.1   Data Sets

The adverse event reports contain records of 3,354,043 patients. Among all patients, 66% and 94% have their age and gender information available. Figure 1(a)-(b) show distributions of age and gender. Unlike the Medicare system, FAERS contains patients in of ages from one day to hundreds of years. The distributions are not severely inclined to particular gender or age levels.

The data represents the diseases that patients have by 10,122 indications of drugs that patients take. These indication terms include not only diseases, but also treatment procedures, such as surgery; common symptoms, such as pain; and

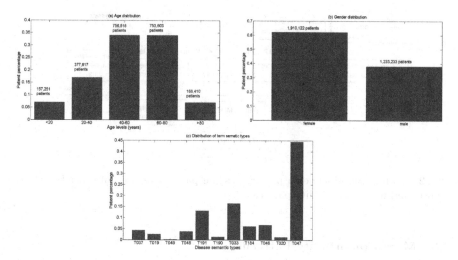

Fig. 1. (a) Age distribution of the patients in the adverse event reports. (b) Gender distribution. (c) Distribution of disease semantic types: T047, Disease or Syndrome; T020, Acquired Abnormality; T046, Pathologic Function; T184, Sign or Symptom; T033, Finding; T190, Anatomical Abnormality; T191, Neoplastic Process; T048, Mental or Behavioral Dysfunction; T049, Cell or Molecular Dysfunction; T019, Congenital Abnormality; T037, Injury or Poisoning.

ill-defined events, such as unevaluable events. We mapped the indication terms to the concept unique identifiers (CUIs) in Unified Medical Language System (UMLS) and extracted their semantic types. Figure 1(c) listed the distribution of eleven semantic types, in which the types such as "disease or syndromes," "neoplastic process," and "mental or behavioral dysfunction" contain disorder concepts. With the disease data for million of patients, we were able to conduct large-scale comorbidity mining and extract interesting disease associations.

2.2   Preprocess Data

We developed an automatic pipeline to preprocess the patient-indication pairs (Figure 2). We mapped all indication terms to CUIs and classified them by semantic types using the UMLS metathesaurus. Then We selected the identifiers of six semantic types: Mental or Behavioral Dysfunction, Neoplastic Process, Acquired Abnormality, Congenital Abnormality, Disease or Syndrome, and Anatomical Abnormality. We combined the synonyms among terms corresponding to these identifier and removed those only appearing once in the data, since rare diseases may lead to unstable association patterns. Finally, the data contains 3,033,368 links between 2,371,406 patients and 3,994 diseases.

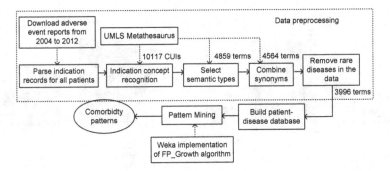

Fig. 2. Automatic pipeline to pre-process the patient-disease data in adverse event reports and mine comorbidity patterns

2.3   Mine Comorbidity Patterns

We explored comorbidity patterns among the 3,994 diseases with association rule mining. Due to the large number of patients and diseases in the adverse event reports, exhausting all possible association patterns is computationally impractical. We applied the frequent pattern growth algorithm, which uses an tree structure to compress the input and grow the patterns in a bottom-up manner (Han *et al.*, 2000). Previous effort (Han *et al.*, 2000) has demonstrated that this algorithm outperforms other popular pattern mining methods, such as the Apriori algorithm (Agrawal *et al.*, 1994). The frequent pattern growth algorithm has also been successfully applied in biomedical domain to extract drug adverse effects (Luo *et al.*, 2013).

We implemented the algorithm using the Weka java package (Hall *et al.*, 2009). The result of the algorithm is a set of patterns indicating how diseases are associated with each other. The pattern between two sets of diseases is represented in the form $X \Rightarrow Y$, where X is the pattern body and Y is the pattern head. For example, $[anxiety, amnesia] \Rightarrow [depression]$ indicates that when patients have anxiety and amnesia, are also likely to have depression. Note that though each pattern is directed with an arrow, they do not indicate causations between diseases, but represent co-occurrences. To avoid confusion, we currently ignored the directions of patterns, and considered all diseases in set X and Y associated.

The mining algorithm requires a few parameters: the minimum support was set to 0.0008%, which means at least 20 patients should have all the diseases in each pattern at the same time; the maximum number of diseases in each pattern was set at 3; and confidence was chosen to measure and rank the patterns. The confidence score of pattern $X \Rightarrow Y$ is defined as:

$$confidence(X \to Y) = |X \cup Y|/|X|, \tag{1}$$

where $|X \cup Y|$ is the number of patients who have diseases in both X and Y, and $|X|$ is the number of patients who have diseases in X.

2.4   Construct Disease Comorbidity Network

To obtain a global view of the comorbidity patterns, we constructed a disease comorbidity network based on the results of association rule mining. For each pattern X $\Rightarrow$ Y, we collected the diseases in set X and Y, and assumed they have phenotypic relationships between each other. Then we established an edge between each pair of diseases in $X \cup Y$. After processing all patterns in the same way, we constructed an unweighted network, containing all diseases in the patterns as the nodes, and representing phenotypic associations in the patterns with edges. In order to analyze the genetic coherence of the comorbidity network, we clustered the network with a widely-used community detection algorithm (Newman et al., 2004). This algorithm has successfully revealed natures of biology networks in previous studies (Palla et al., 2005; Ahn et al., 2010). We investigated the genetic associations among diseases in each comorbidity cluster to measure the genetic coherence of the comorbidity network.

2.5   Evaluate the Genetic Coherence of Comorbidity Clusters

A number of factors may determine the co-occurrence of diseases, some of which are environmental, genetic, or treatment-induced. In this study, we captured the contribution of genetic factors and estimated the genetic coherence of the comorbidity network. We evaluated whether diseases in the same comorbidity cluster shared significantly more associated genes, pathways, and gene ontology (GO) annotations than random clusters, assuming that these genetic factors directly reflected the underlying mechanisms of diseases. We obtained the disease-gene associations from the OMIM database (February 2014) (McKusick et al., 2007). Then we mapped the OMIM disease genes to their GO annotations (July 2012) (Ashburner et al., 2000) and pathways from the Molecular Signature Database (version 4.0) (Subramanian et al., 2005) to link diseases to related GO annotations and pathways. For all possible disease pairs in each cluster, we first normalized the number of overlapping genes, pathways and GO terms:

$$S_{ij} = |G_i \cap G_j| \, / \, |G_i \cup G_j| \qquad (2)$$

where G_i and G_j are the sets of genes, GO terms or pathways associate with disease i and j, and $|A|$ represents the number of elements in set A. Then we averaged the pairwise overlaps of genetic factors in the clusters, and averaged across all clusters to measure the genetic coherence of the entire comorbidity network. Similar genetic coherence measures have been used before (VanDriel et al., 2006; Park et al., 2009; Oti et al., 2009).

We compared the genetic coherence of the comorbidity network with 30 randomized disease networks. The random networks were generated by shuffling the nodes of the comorbidity network while keeping the network structure. Since we did not assume normal distributions for the genetic coherence, all statistically comparisons were done with two-sided Wilcoxon rank sum test implemented in the R statistical software package.

3   Results

We detected 8,576 disease association patterns with confidence larger than 10%. These patterns involve all kinds of diseases, such as cancers, mental disorders, and metabolic disorders. The comorbidity network based on these patterns contains 613 nodes and 3,277 edges. The community detection algorithm identified eight clusters from the giant component of the network.

3.1   Comorbidity Clusters Reflect Disease Genetic Associations

Figure 3 shows the genetic coherence of the comorbidity network compared with the random variants. The diseases in comorbidity clusters share significantly more disease associated genes ($p < e^{-8}$) than those in random clusters. The scores based on pathways and GO annotations are higher than that based on associated genes for the comorbidity network, since diseases may share functionally related genes, though they are not directly associated with the same genes. The chance for randomly selected diseases sharing GO annotations and pathways is also higher than sharing genes, because one GO annotation or pathway usually associates with many genes. However, the comorbidity network has significantly higher genetic coherence scores than random networks based on sharing GO annotations ($p < e^{-6}$) and pathways ($p < e^{-5}$). We also found that the genetic coherence has small magnitudes, which indicates that genetic factors can only partially explain the disease comorbidity relationships. Environmental factors, patient lifestyles, and the side effects of treatments may also contribute to the observed comorbidity patterns.

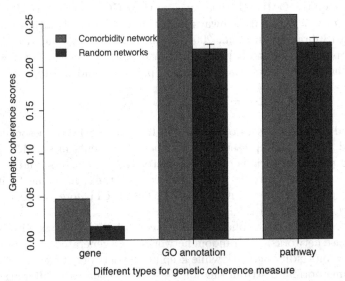

Fig. 3. Genetic coherence based on shared genes, GO annotations and pathways in clusters of the comorbidity network compared with randomized networks

We then show a few examples of comorbidity clusters. The cluster in Figure 4 (a) mostly consists of immunity-mediated diseases, such as Crohn's disease, ulcerative colitis, and ankylosing spondylitis. These diseases all involve chronic inflammatory. Crohn's disease and ulcerative colitis are two sub-types of the inflammatory bowel disease. The striking co-occurrence relationship between Crohn's disease and ankylosing spondylitis has been recognized for many years (Mielants et al., 1995), and recent studies (Elewaut et al., 2010) explored their genetic associations. Figure 4 (b) shows a cluster of mental disorders, including depression, bipolar and schizophrenia. Figure 4 (c) contains the sub-network of a bigger cluster, in which diabetes mellitus is connected with its major symptoms and complications, and cancers on different organ sites are interconnected with each other. Interestingly, the cluster links metabolic disorders like diabetes with cancers. Literature evidences over a decade support that metabolic abnormalities increase the risk of developing cancers (Vecchia et al., 1997; Schoen et al., 1999; Calle et al., 2003; Zoncu et al., 2011).

3.2 Comparison with the Relative Risk

The relative risk is a widely-used statistical measure to detect disease comorbidities. We compared our result with the disease comorbidity pairs extracted with the relative risk. We did not directly compare the genetic coherence of comorbidity patterns between the two methods, because they both identify interesting disease phenotypic relationships. Instead, we show that the two methods extract comorbidities for different diseases. We first calculated the relative risks of all possible disease pairs:

$$R_{ij} = C_{ij}N/P_iP_j, \tag{3}$$

where C_{ij} represents the co-occurrence counts of disease i and j, N is the total number of patients in the data, and P_i and P_j are the prevalence of the two diseases. Then we ranked the disease pairs, and extracted the top pairs until the number of diseases is the same with that in our comorbidity network. Last, we compared the prevalence of diseases in the top comorbidity patterns extracted through the relative risk and association rule mining.

Our result mostly contains common diseases, such as diabetes, hypertension, and depression. In contrast, the relative risk mostly identified comorbidity relationships between rare diseases among the population. For the relative risk approach, the diseases with the highest prevalence is rectal cancer stage IV, which appears in 622 (0.03%) patients. Half of the diseases in our result have higher prevalence than 0.03%. The average disease prevalence in the result of the relative risk approach is 55, significantly lower than 4486 for our approach ($p < e^{-9}$). Therefore, our comorbidity mining result is complementary with that of the relative risk.

3.3 Case Study on Colorectal Cancer

Colorectal cancer is deadly, complex, and common around the world (Haggar et al., 2009). Its multi-factorial pathogenesis has not been fully understood yet.

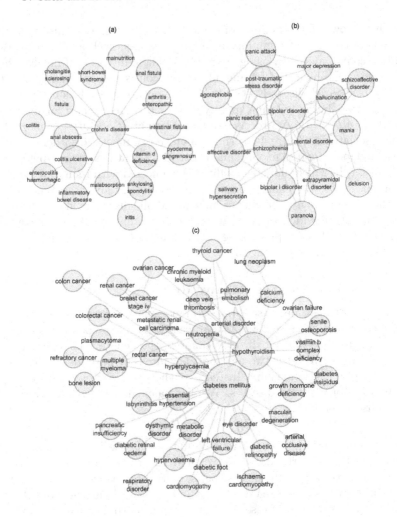

Fig. 4. Examples of comorbidity clusters based on the comorbidity patterns from association rule mining

Analyzing the phenotypic relationships of colorectal cancer may potentially introduces novel discoveries of its genetic mechanisms. We extracted the top 20 comorbidity patterns of colorectal cancer and listed them in Table 1. The patterns show that metabolic disorders, including hypercholesterolaemia, hyperlipidaemia, hypertension, and type 2 diabetes, frequently co-occur with colorectal cancer. Different from previous comorbidity studies, our result show associations among multiple disorders, instead of only disease pairs. Table 1 shows that metabolic syndrome components, such as hypercholesterolaemia, hypertension and hyperlipidaemia, also occur with each other in the top patterns. Since metabolic syndrome is diagnosed when these components present together, we hypothesized that metabolic syndrome is also associated with colorectal cancer.

Table 1. Top 20 comorbidity patterns involving colorectal cancer mined from the complete

Pattern body	Pattern head	Confidence
Hypercholesterolaemia, colorectal cancer metastatic	hypertension	1
Hyperlipidaemia, colorectal cancer metastatic	hypertension	1
colorectal cancer, neoplasm	neoplasm malignant	0.91
type 2 diabetes mellitus, colorectal cancer metastatic	hypertension	0.87
diabetes mellitus, colorectal cancer metastatic	hypertension	0.78
Hyperlipidaemia, colorectal cancer	hypertension	0.77
gastrooesophageal reflux disease, colorectal cancer metastatic	hypertension	0.67
colorectal cancer, neoplasm malignant	neoplasm	0.67
neoplasm malignant, neoplasm	colorectal cancer	0.59
colorectal cancer metastatic, rash	hypertension	0.55
Hyperlipidaemia, colorectal cancer	diabetes mellitus	0.54
gastritis, colorectal cancer	hypertension	0.47
diabetes mellitus, colorectal cancer	hypertension	0.46
diabetes mellitus, colorectal cancer	hyperlipidaemia	0.22
hypertension, colorectal cancer	diabetes mellitus	0.18
hypertension, colorectal cancer metastatic	type 2 diabetes mellitus	0.16
hypertension, colorectal cancer metastatic	hypercholesterolaemia	0.14
hypertension, colorectal cancer metastatic	diabetes mellitus	0.12
hypertension, colorectal cancer	hyperlipidaemia	0.12
hypertension, colorectal cancer metastatic	hyperlipidaemia	0.1

We stratified the patients by age and gender to reduce the effect of these confounding factors. The complete data were first divided into two age groups: 1,103,342 patients younger than 65, and 545,436 patients older than 65. The association rule mining algorithm extracted 69 and 347 patterns of colorectal cancer for the young and old strata, respectively. Among the younger patients, colorectal cancer is frequently associated with hypertension (confidence 100%), diabetes (confidence 83%), hyperlipidaemia (confidence 75%), and hypercholesterolaemia (confidence 50%). Among the older patients, colorectal cancer is associated with diabetes with a confidence of 95%, and the other three disorders with confidences of 100%. Also, the co-occurrence between metabolic syndrome components remains in the patterns. Similarly, the data were divided into 905,725 female and 1,346,709 male patients. The algorithm found 333 and 111 patterns for females and males. The four metabolic disorders associated with colorectal cancer with confidence higher than 70% among male patients and higher than 80% among female patients.

Both complete and stratified data suggest that colorectal cancer is associated metabolic disorders. Interestingly, a growing body of literature evidence support our result and investigated the reason for this disease co-occurrence. Previous studies show that type 2 diabetes and colorectal cancer patient shares similar lifestyles, such as high fat dietary and few exercises (Vecchia et al., 1997). Researches have also convincingly proved the association between insulin resistance, which contributes to metabolic syndrome and type 2 diabetes, and the development of colon cancer (Schoen et al., 1999; Komninou et al., 2003; Volkova et al., 2011). Therefore, we hypothesized that metabolic disorders and colorectal cancer may have underlying genetic associations. Further investigation of the hypothesis has the potential to help illuminate common genetic mechanisms of colorectal cancer and metabolic disorders.

4   Discussion

In this work, we extracted comorbidity patterns from a unique database–the patient disease records from the drug adverse events reporting system. We first demonstrated that the data is valuable for comorbidity mining, and then developed a data mining approach based on association rule mining. We demonstrated the genetic coherence of the comorbidity network based on the patterns extracted by our approach, and showed that we can obtain complementary results compared with the approach based on the relative risk. In a case study on colorectal cancer, we proved that our result has the potential to help elucidate genetic mechanisms of complex diseases.

Our study may be improved if the noises can be reduced in the data. Though we controlled the semantic types and removed non-disease terms, the remaining disease terms lie on multiple levels in the hierarchy of disease ontology. For example, diabetes mellitus, type 1 diabetes and type 2 diabetes all appear in our data, and we cannot identify the specific type of diabetes mellitus, though type 2 diabetes is more common than type 1. Yet, different ways to group the disease synonyms may affect the comorbidity mining result.

In addition, a number of other factors, such as side effects induced by drugs and treatments, environmental factors and patient lifestyles, may also determined whether diseases co-occur in one patient. We are currently developed approaches to include these factors and further elucidate the factors that cause the disease comorbidity relationships. However, more complicated computational framework and more detailed data may be required, and our current data have limitations. For example, we may predict whether the drugs treating colorectal cancer induce metabolic disorders, if the time series data are available to describe if the patients develop metabolic disorders before or after taking the drugs.

5   Conclusion

We demonstrated the feasibility of mining disease comorbidity patterns from the large-scale observational data form the drug adverse event reporting system. Unlike previous comorbidity studies, our approach based on association rule mining is able to extract associations both between disease pairs and among multiple diseases. We constructed a disease comorbidity network, and provided the initial promising evidence that the network genetic associations between diseases. When integrated with other phenotype and genome data, our comorbidity network has the potential to facilitate novel discoveries of disease mechanisms.

References

Brunner, H.G., Van Driel, A.: From syndrome families to functional genomics. Nat. Rev. Genet. 5, 545–551 (2004)

Tiffin, N., Andrade-Navarro, M.A., Perez-Iratxeta, C.: Linking genes to diseases: it's all in the data. Genome Med. 1(8), 77 (2009)

Houle, D., Govindaraju, R.D., Omholt, S.: Phenomics: the next challenge. Nat. Rev. Genet. 11(12), 855–866 (2010)

Wu, X., Jiang, R., Zhang, M.Q., Li, S.: Network-based global inference of human disease genes. Mol. Syst. Biol. 4, 189 (2008)

Li, Y., Patra, J.C.: Genome-wide inferring gene–phenotype relationship by walking on the heterogeneous network. Bioinformatics 26(9), 1219–1224 (2010)

Vanunu, O., Magger, O., Ruppin, E., Shlomi, T., Sharan, R.: Associating Genes and Protein Complexes with Disease via Network Propagation. PLoS Comput. Biol. 6(1), e1000641 (2010)

Lage, K., et al.: A human phenome-interactome network of protein complexes implicated in genetic disorders. Nature Biotechnology 25(3), 309–316 (2007)

Hwang, T., Atluri, G., Xie, M., et al.: Co-clustering phenome-genome for phenotype classification and disease gene discovery. Nucleic Acids Research 40(19), e146 (2012)

Iorio, F., Bosotti, R., Scacheri, E., et al.: Discovery of drug mode of action and drug repositioning from transcriptional responses. Proc. Nat. Acad. Sci. 107(33), 14621–14626 (2011)

van Driel, M.A., Bruggerman, J., Vriend, G., Brunner, H.G., Leunissen, J.A.: A text-mining analysis of the human phenome. Eur. J. Hum. Genet. 14, 535–542 (2006)

Oti, M., Huynen, M.A., Brunner, H.G.: Phenome connections. Trends Genet. 24(3), 103–106

Blair, D.R., Lyttle, C.S., Mortensen, J.M., et al.: A nondegenerate code of deleterious variants in mendelian Loci contributes to complex disease risk. Cell 155(1), 70–80 (2013)

Avery, C.L., He, Q., North, K.E., et al.: A phenomics-based strategy identifies loci on APOC1, BRAP, and PLCG1 associated with metabolic syndrome phenotype domains. PLoS Genetics 7(10), e1002322 (2011)

Joseph, C.G., Darrah, E., Shah, A.A., et al.: Association of the Autoimmune Disease Scleroderma with an Immunologic Response to Cancer. Science 343(6167), 152–157 (2014)

Toffanin, S., Friedman, S.L., Llovet, J.M.: Obesity, inflammatory signaling, and hepatocellular carcinoma-an enlarging link. Cancer Cell 17(2), 115–117 (2010)

Park, J., Lee, D.S., Christakis, N.A., Barabási, A.L.: The impact of cellular networks on disease comorbidity. Mol. Syst. Biol. 5, 262 (2009)

Hidalgo, C.A., Blumm, N., Barabási, A.L., Christakis, N.A.: A dynamic network approach for the study of human phenotypes. PLoS Comput. Biol. 5, e1000353 (2009)

Rogue, F.S., Jensen, P.B., Schmock, H., et al.: Using Electronic Patient Records to Discover Disease Correlations and Stratify Patient Cohorts. PLoS Comput. Biol. 7(8), e1002141 (2011)

Rzhetsky, A., Wajngurt, D., Park, N., Zheng, T.: Probing genetic overlap among complex human phenotypes. Proc. Natl. Acad. Sci. 104, 11694–11699 (2007)

Han, J., Pei, J., Yin, Y.: Mining Frequent Patterns without Candidate Generation. In: Proc. ACM SIGMOD Int. Conf. Manag. of Data, New York, NY, USA, pp. 1–12 (2000)

Agrawal, R., Srikant, R.: Fast Algorithms for Mining Association Rules in Large Databases. In: Proc. 20th Int. Conf. on VLDB, San Francisco, CA, USA, pp. 487–499

Luo, Z., Zhang, G.Q., Xu, R.: Mining Patterns of Adverse Events Using Aggregated Clinical Trial Results. In: AMIA Summits Transl. Sci. Proc., San Fransisco, CA, USA, pp. 112–116 (2013)

Newman, M.E., Girvan, M.: Finding and evaluating community structure in networks. Physical Review E 69(2), 026113 (2004)

Palla, G., Derényi, I., Farkas, I., Vicsek, T.: Uncovering the overlapping community structure of complex networks in nature and society. Nature 435(7043), 814–818 (2005)

Ahn, Y.Y., Bagrow, J.P., Lehmann, S.: Link communities reveal multiscale complexity in networks. Nature 466(7307), 761–764 (2010)

McKusick, V.A.: Mendelian Inheritance in Man and its online version, OMIM. American Journal of Human Genetics 80(4), 588 (2007)

Ashburner, M., Ball, C.A., Blake, et al.: Gene Ontology: tool for the unification of biology. Nature Genetics 25(1), 25–29 (2000)

Subramanian, A., Tamayo, P., Mootha, V.K., et al.: Gene set enrichment analysis: a knowledge-based approach for interpreting genome-wide expression profiles. Proceedings of the National Academy of Sciences of the United States of America 102(43), 15545–15550 (2005)

Oti, M., Huynen, M.A., Brunner, H.G.: The biological coherence of human phenome databases. The American Journal of Human Genetics 85(6), 801–808 (2009)

Hall, M., Frank, E., Holmes, G., Pfahringer, B., Reutemann, P., Witten, I.H.: The WEKA Data Mining Software: An Update. SIGKDD Explorations 111, 10–18 (2009)

Mielants, H., Veys, E.M., Cuvelier, C., et al.: The evolution of spondyloarthropathies in relation to gut histology II. Histological aspects. J. Rheumatol. 22, 2273–2278 (1995)

Elewaut, D.: Linking Crohn's disease and ankylosing spondylitis: it's all about genes! PLoS Genetics 6(12) (2010)

Vecchia, C.L., Negri, E., Decarli, A., Franceschi, S.: Diabetes mellitus and colorectal cancer risk. Cancer Epidemiol. Biomarkers Prev. 6(12), 1007–1010 (1997)

Schoen, R.E., Tangen, C.M., Kuller, L.H., et al.: Increased blood glucose and insulin, body size, and incident colorectal cancer. J. Natl. Cancer Inst. 91(13), 1147–1154 (1999)

Calle, E.E., Rodriguez, C., Walker-Thurmond, K., Thun, M.J.: Overweight, obesity, and mortality from cancer in a prospectively studied cohort of US adults. New England Journal of Medicine 348(17), 1625–1638 (2003)

Zoncu, R., Efeyan, A., Sabatini, D.M.: mTOR: from growth signal integration to cancer, diabetes and ageing. Nat. Revs. Mol. Cell Bio. 12(1), 21–35 (2011)

Haggar, F.A., Boushey, R.P.: Colorectal cancer epidemiology: incidence, mortality, survival, and risk factors. Clinics in Colon and Rectal Surgery 22(4), 191 (2009)

Komninou, D., Ayonote, A., Richie, J.P., Rigas, B.: Insulin resistance and its contribution to colon carcinogenesis. Experimental Biology and Medicine 228(4), 396–405 (2003)

Volkova, E., Willis, J.A., Wells, J.E., Robinson, B.A., Dachs, G.U., Currie, M.J.: Association of angiopoietin-2, C-reactive protein and markers of obesity and insulin resistance with survival outcome in colorectal cancer. British Journal of Cancer 104(1), 51–59 (2011)

Identification of Essential Proteins
by Using Complexes and Interaction Network*

Min Li[1], Yu Lu[1], Zhibei Niu[1], Fang-Xiang Wu[3], and Yi Pan[1,2]

[1] School of Information Science and Engineering,
Central South University, Changsha 410083, P.R. China
[2] Department of Computer Science,
Georgia State University, Atlanta, GA 30302-4110, USA
[3] Department of Mechanical Engineering and Division of Biomedical Engineering
University of Saskatchewan, Saskatoon, SK S7N 5A9, Canada
limin@mail.csu.edu.cn

Abstract. Essential proteins are indispensable in maintaining the cellular life. Identification of essential proteins can provide basis for drug target design, disease treatment as well as synthetic biology minimal genome. However, it is still time-consuming and expensive to identify essential protein based on experimental approaches. With the development of high-throughput experimental techniques in the post-genome era, a large number of PPI data and gene expression data can be obtained, which provide an unprecedented opportunity to study essential proteins at the network level. So far, many network topological methods have been proposed to identify the essential proteins. In this paper, we propose a new method, United complex Centrality(UC), to identify essential proteins by integrating protein complexes information and topological features of PPI network. By analysis of the relationship between protein complexes and essential proteins, we find that proteins appeared in multiple complexes are more inclined to be essential compared to these only appeared in a single complex. The experiment results show that protein complex information can help identify the essential proteins more accurate. Our method UC is obviously better than traditional centrality methods(DC, IC, EC, SC, BC, CC, NC) for identifying essential proteins. In addition, even compared with Harmonic Centricity which also used protein complexes information, it still has a great advantage.

Keywords: essential proteins, PPI network, protein complexes, traditional centrality methods.

1 Introduction

Essential proteins are the proteins that play an irreplaceable role in the cellular life, they are closely related to survival and reproduction of the organism[1-3].

---

* This work is supported in part by the National Natural Science Foundation of China under Grant No.61370024, No.61232001, and No.61379108, the Program for New Century Excellent Talents in University (NCET-12-0547).

M. Basu, Y. Pan, and J. Wang (Eds.): ISBRA 2014, LNBI 8492, pp. 255–265, 2014.

Removal of any one of essential proteins leads to a fatal defect on an organism[4,5].Compared with non-essential proteins, essential proteins tend to be more conservative in biological evolution[6-8], so that they are not easy to change. Meanwhile, it is proved that essential proteins are often disease genes in the human body cell[9,10]. Thus, the identification of essential proteins in lower organisms is of great importance to finding disease genes and drug targets. At the same time, it has important applications in disease diagnosis and drug design, etc.

Recent years, many researchers concerned more about the connections between topological properties in PPI networks and essential proteins[11-18]. Highly connected center nodes(hubs) in PPI networks are often the essential proteins, that is so-called centrality-lethality rule. After that, researchers successively proposed a series of centrality methods. For examples:Degree Centrality (DC)[19-21], Betweenness Centrality (BC)[22,23], Closeness Centrality (CC)[24], Subgraph Centrality (SC)[25], Eigenvector Centrality(EC)[26], Information Centrality (IC)[27], Neighborhood Centrality (NC)[28], Local Average Connectivity-based Method (LAC)[29]. These centrality methods are used to predict essential proteins based on topological features of PPI networks. Experimental analyses had shown that these centrality methods are much more effective than choosing essential proteins randomly. Specially, NC and LAC are better than six traditional centrality methods (DC, BC, CC, SC, EC, IC).

However, merely using PPI to predict essential protein is far less, the PPI data produced by current high throughput technology exists some false interactions(false positive)[33], these will effect the accuracy of essential protein predicting. Thus, researches hope to integrate multi-information to help identify the essential proteins more accurate[30-32,45]. Meanwhile, a large number of studies have shown that there exists close relationship between protein complex and essential proteins. Hart, etc.[34] pointed out that the essentiality is one of properties of protein complex. The essentiality of proteins are not merely decided by a single protein, often by functional protein complex. Zotenko, etc[14] proposed the concept of Essential Complex Biological Modules, a group of highly connected function modules which have the same or similar biological functions. They also consider that the reason why hub nodes are inclined to be essential proteins is there exists a lot in essential complex modules, so they joined in more biological functions.On this basis, Ren, etal.[45] introduced the information of protein complex and proposed a new centrality measuring method of proteins named Harmonic Centricity(HC).

In this paper, we propose a new method to identify essential proteins based on information of protein complexes and topology features of PPI networks. We first verify that proteins in complexes are more inclined to be essential proteins and sort the proteins of PPI network into two parts: proteins in complexes and proteins not in complexes. Secondly, after the classification, we analyze the proteins in complexes and find that proteins appeared in multiple complexes are more likely to be essential. So we classify the proteins of complexes into two parts: proteins appeared in a single complex and proteins appeared in multiple

complexes. Finally, we combine the classified proteins and the topology characteristics of PPI network to identify the essential proteins. The results show that
this method can improve the prediction precision of essential proteins compared
with other centrality measures. Even compared to Harmonic Centricity which
also used protein complexes information, it still has a great advantage.

2   Materials and Methods

Materials

PPI Network: PPI network data of *S.cerevisiae* are downloaded from DIP
dataset[37], we call it as YDIP. It contains 4950 proteins and 21788 interactions.

Protein Complex: Protein complex datasets are obtained from MIPS_216[36],
MIPS_408[36], krogan[38], Gavin[39], Ho[40], the map of integrated dataset in
YDIP is named YC_union. YC_union contains 1208 complexes and 2572 proteins.
We only wiped out the same complexes when integrate YC_union.

Essential Proteins: The essential proteins of *S.cerevisiae* are obtained from
the following databases:MIPS(Mammalian Protein-Protein Interaction Database)
[36], SGD(Saccharomyces Genome Database)[41], DEG(Database of Essential
Genes)[42] and SGDP(Saccharomyces Genome Deletion Project)[43]. A protein
in the yeast PPI network is considered as an essential protein if it can be matched
at least in one yeast essential database. Out of all the 4950 proteins in YDIP,
1151 proteins are essential, 3799 are non-essential if it can't be marked in our
essential datasets.

Methods

Protein complex is the basis of executing biological procedure, which generates
all kinds of molecular mechanisms to executing a large number of biological functions. Complexes consist of specific proteins which usually have constant structures and functions[35]. In order to investigate the connections between protein
complexes and essential proteins, we download the PPI network and complex information of *S.cerevisiae*, and list the number of proteins and essential proteins
in yeast PPI network and several complex datasets in Table 1. Among these,
PPI network is named YDIP. Different from usually that only concern the relationship between the single known complex and essential proteins, we integrate
five known complex datasets (MIPS_216, MIPS_408, krogan, Gavin, Ho) named
YC_union. As shown in Table 1, only 23.3% of the 4950 proteins in YDIP are essential proteins. While in the single protein complex datasets, the proportion of
essential proteins are 42.6%, 38.4%, 66.00%, 43.40%, 34.80%, respectively. And
out of the 2572 proteins in YC_union, 35.2% proteins are essential proteins.

In order to investigate that whether there exists the relationship between
protein complexes and essential proteins or not, we divide all the proteins in
YDIP into two part: proteins in YC_union and proteins not in YC_union. We also
count the number and the proportion of their essential proteins. The results are

Table 1. Numbers of proteins and essential proteins in PPI network and protein complexes

	dataset Name	Number of proteins	Number of essential proteins	Ratio of essential proteins
PPI network	YDIP	4950	1151	23.30%
Protein complex	MIPS_216	1184	504	42.60%
	MIPS_408	1835	711	38.70%
	Krogan	247	163	66.00%
	Gavin	1287	558	43.40%
	Ho	1228	427	34.80%
Union complex	YC_union	2572	906	35.20%

YC_union is a protein complex dataset which mapped in YDIP by integrating five known protein complex datasets: MIPS_216, MIPS_408, krogan, Gavin, Ho.

Table 2. The proportion of essential proteins in YC_union and not in YC_union

	proteins in YC_union	proteins not in YC_union
Number of proteins	2572	2378
Number of essential proteins	906	245
Ratio of essential protein	35.2%	10.3%

shown as Table 2. From Table 2 we can see that among the 4950 proteins in YDIP, the proportion of essential proteins in YC_union is 35.2% and the proportion of essential proteins not in YC_union is only 10.3%. Obviously, proteins in protein complexes are more likely to be essential.

It has been shown that many complexes exist a lot of overlapping part in PPI networks. That is to say, some proteins may belong to multiple protein complexes[36]. Obviously, if we wiped out these overlapping proteins, the protein complex will lost its function of this part. Thus, we divide the proteins in YC_union into two parts: proteins only appeared in one complex (overlap=1) and proteins appeared in multiple complexes(overlap>1). Here, given a protein u, its overlap denotes the number of complexes which protein u appeared in. We also count the number and the proportion of their essential proteins. As shown in Table 3, the proportion of essential proteins that appeared in multiple complexes is 43.0%, and the proportion of essential proteins that only appeared in one complex is 21.0%. Thus, essential proteins are more likely to appear in overlapping part of complexes.

A PPI network is described as an undirected graph $G = (V, E)$, which the nodes represent proteins, the edge represents an interaction between proteins. For an edge $E(i, j)$ connecting node i and node j, we pay attention to how many

Table 3. The proportion of essential proteins that only appeared in one complex and appeared in multiple complexes

	Proteins only appeared in one complex	Proteins appeared in multiple complexes
Number of proteins	910	1662
Number of essential proteins	191	715
Ratio of essential protein	21.0%	43.0%

other nodes that adjoin both i and j[28]. The edge clustering coefficient of $E(i,j)$ can be defined by the following formula:

$$ECC_{ij} = \frac{z_{i,j}}{min(k_i - 1, k_j - 1)} \tag{1}$$

where, $z_{i,j}$ is the number of triangles that include the edge actually in the network, k_i and k_j denotes the degree of node i and j, respectively, $min(k_i - 1, k_j - 1)$ is the number of triangles in which the edge $E(i,j)$ may possibly participate at most.

Then we combine the complex overlapping information and ECC, propose a new method to predict essential proteins, named UC(United complex Centrality). United complex centrality of protein i, $UC(i)$, is defined as:

$$UC(i) = \sum_{j=1}^{n} \left(\frac{o(j) + 1}{O + 1} \times ECC_{ij} \right) \tag{2}$$

where $o(j)$ is the overlap number of protein j which interacts with protein i in YC_union, O is the biggest number of complexes which protein appeared in, ECC_{ij} is the edge clustering coefficient between protein i and protein j.

In the equation(2), if the number of complexes which protein appeared in is higher, this edge is more like to be important. Considering that protein j which interacts with protein i is not in YC_union, O and $o(j)$ should add 1 when we count the UC value of protein i.

3   Experiments and Results

The Analysis of Essential Proteins in Complexes

In the method section, we count and analyze the proportion of essential proteins in PPI network and known protein complexes, as well as the proportion of essential proteins in YC_union which is a union complex mapped from YDIP. In order to further analyze the essentiality of proteins in complexes, we divide the proteins in YC_union which mapped from YDIP into 10 groups (overlap=1, overlap=2...overlap=9 and overlap≥10). Then we calculate the proportion of proteins and essential proteins in YC_union for each group. From Fig 1, we can see the proportion of proteins in YC_union is decreased, but the proportion of essential proteins is increased. Especially, the proportion of essential proteins is 69.9% when overlap=8. When overlap=1, the proportion of proteins in YC_union reach the highest, but the proportion of essential proteins reach the lowest, only 21.0%. Apart from that, when overlap ≥10, the number of proteins is too low, so we put them into one group. But the proportion of essential proteins are still 58.9%. In conclusion, the proteins in the complexes tend to be essential proteins, especially the proteins whose overlapping times are higher.

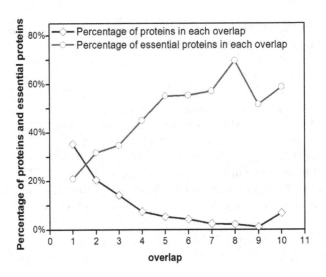

Fig. 1. Analysis about the proportion of proteins and essential proteins in YC_union for each overlap

For complex YC_union, we know it contains 2572 proteins and 906 essential proteins. In order to examine the relationship between YC_union and the essentiality of proteins is not random. We retain the size of every complex in YC_union and randomly choose proteins of YDIP to format new complex in 10 times. Then we calculate the proportion of essential proteins when proteins only appeared in one complex and proteins appeared in multiple complexes. From Fig 2, we can see that when proteins only appeared in one complex and proteins appeared in multiple complexes, the proportion of essential proteins in random complexes are all about 0.23. While the proportion is 0.43 in YC_union when proteins appeared in multiple complexes. This is higher than the case when proteins only appeared in one complex, also higher than randomly tests. Thus, there exists no random relationship between YC_union and the essentiality of proteins. Proteins are more likely to be essential proteins when they appeared in multiple complexes.

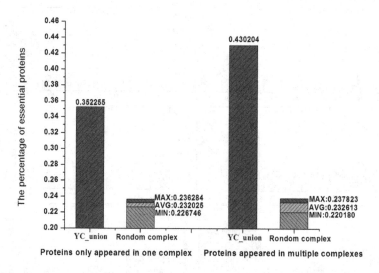

Fig. 2. The proportion of essential proteins when proteins only appeared in one complex and proteins appeared in multiple complexes. MAX means the maximum proportion of essential proteins in 10 times random complexes, AVG means the average proportion of essential proteins in 10 times random complexes, MIN means the minimum proportion of essential proteins in 10 times random complexes.

UC Compared with other Methods

In order to compare UC's performance with other methods in the prediction of essential proteins, we calculate the essentiality of every protein based on DC, IC, EC, SC, BC, CC, NC, HC and UC. As similar with previous experimental procedures[40], We do descending order to all the proteins according to the value and choose top 1%, 5%, 10%, 15%, 20%, 25% proteins as candidate essential proteins for each method. Then we calculate how many essential candidates are true essential proteins based on the above top percentage. As Fig 3 showed, UC performs better than other methods from top 1% to 25%. In top 1%(the number of proteins is 49), the true essential proteins number of every traditional centrality method is lower than 30, HC is also only 31. While the true essential proteins number of UC is 40, the identify ratio is reaching 81.6%. Especially, when we choose top 1% and 5% proteins as candidate essential proteins, the identifying performance of UC all increased 85% compared with EC and SC. With the increase of the number of the selected essential candidates, less improvement is obtained by UC compared with EC and SC. But even choose top 25% proteins as candidate essential proteins, the number of true essential proteins produced by UC still increased over 30% compared with EC and SC. Besides, comparing with the best result in above identifying measures from top 1% to top 25%, the result of UC still has great advantage in any top percentage.

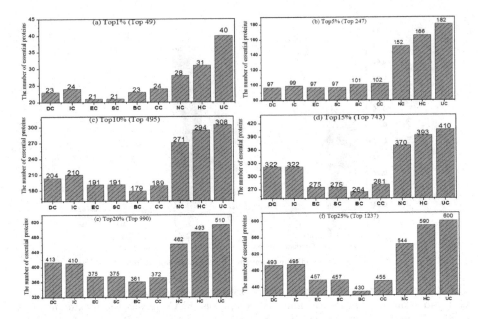

Fig. 3. The number of true essential proteins by UC, HC, DC, IC, EC, SC, BC, CC and NC when selecting the top 1%, top 5%, top 10%, top 15%, top 20%, top 25% proteins

4   Conclusion and Discussion

The identification of essential proteins based on PPI network is always a heat point in post-genome era, but the method usually concentrated on several properties of topology level, not that deeply in digging biological meaning and biological function. However, essential proteins are of great importance in biology functions. It is necessary to combine biology information in the procedure of essential proteins identifying. Previous research has shown that there are close relationship between protein complexes and essential proteins[4,14]. In this paper, we propose a new method to identify essential proteins based on information of protein complexes and topology features of PPI networks. The results show that proteins in overlapping part of complexes are more inclined to be essential proteins. Apart from that, this method improved a lot in performance of predicting essential proteins compared with traditional centrality methods(DC, BC, CC, SC, EC, IC, and NC).

In this paper, we only combine information of protein complexes and network topology characteristics to identify essential proteins. However, there are still some biological characteristics such as protein function information, domain information, gene expression information, etc closely related to essential proteins. In future, we will further analyze the essentiality of proteins by using these biological information.

References

1. Pál, C., Papp, B., Hurst, L.D.: Genomic function (communication arising): rate of evolution and gene dispensability. Nature 421(6922), 496–497 (2003)
2. Zhang, J., He, X.: Significant impact of protein dispensability on the instantaneous rate of protein evolution. Molecular Biology and Evolution 22(4), 1147–1155 (2005)
3. Liao, B.Y., Scott, N.M., Zhang, J.: Impacts of gene essentiality, expression pattern, and gene compactness on the evolutionary rate of mammalian proteins. Molecular Biology and Evolution 23(11), 2072–2080 (2006)
4. Winzeler, E.A., Shoemaker, D.D., Astromoff, A., et al.: Functional characterization of the S. cerevisiae genome by gene deletion and parallel analysis. Science 285(5429), 901–906 (1999)
5. Kamath, R.S., Fraser, A.G., Dong, Y., et al.: Systematic functional analysis of the Caenorhabditis elegans genome using RNAi. Nature 421(6920), 231–237 (2003)
6. Kondrashov, F.A., Ogurtsov, A.Y., Kondrashov, A.S.: Bioinformatical assay of human gene morbidity. Nucleic Acids Research 32(5), 1731–1737 (2004)
7. Furney, S.J., Albá, M.M., López-Bigas, N.: Differences in the evolutionary history of disease genes affected by dominant or recessive mutations. BMC Genomics 7(1), 165 (2006)
8. Fraser, H.B., Hirsh, A.E., Steinmetz, L.M., et al.: Evolutionary rate in the protein interaction network. Science 296(5568), 750–752 (2002)
9. Xu, J., Li, Y.: Discovering disease-genes by topological features in human protein - protein interaction network. Bioinformatics 22(22), 2800–2805 (2006)
10. Park, D., Park, J., Park, S.G., et al.: Analysis of human disease genes in the context of gene essentiality. Genomics 92(6), 414–418 (2008)
11. Jeong, H., Mason, S.P., Barabási, A.L., et al.: Lethality and centrality in protein networks. Nature 411(6833), 41–42 (2001)
12. Estrada, E.: Virtual identification of essential proteins within the protein interaction network of yeast. Proteomics 6(1), 35–40 (2006)
13. He, X., Zhang, J.: Why do hubs tend to be essential in protein networks? PLoS Genetics 2(6), e88 (2006)
14. Zotenko, E., Mestre, J., O'Leary, D.P., et al.: Why do hubs in the yeast protein interaction network tend to be essential: reexamining the connection between the network topology and essentiality. PLoS Computational Biology 4(8), e1000140 (2008)
15. Chua, H.N., Tew, K.L., Li, X.L., et al.: A unified scoring scheme for detecting essential proteins in protein interaction networks. In: 20th IEEE International Conference on Tools with Artificial Intelligence, ICTAI 2008, vol. 2, pp. 66–73. IEEE (2008)
16. Batada, N.N., Hurst, L.D., Tyers, M.: Evolutionary and physiological importance of hub proteins. PLoS Computational Biology 2(7), e88 (2006)
17. Seo, C.H., Kim, J.R., Kim, M.S., et al.: Hub genes with positive feedbacks function as master switches in developmental gene regulatory networks. Bioinformatics 25(15), 1898–1904 (2009)
18. Acencio, M.L., Lemke, N.: Towards the prediction of essential genes by integration of network topology, cellular localization and biological process information. BMC Bioinformatics 10(1), 290 (2009)
19. Vallabhajosyula, R.R., Chakravarti, D., Lutfeali, S., et al.: Identifying hubs in protein interaction networks. PLoS One 4(4), e5344 (2009)

20. Pang, K., Sheng, H., Ma, X.: Understanding gene essentiality by finely characterizing hubs in the yeast protein interaction network. Biochemical and Biophysical Research Communications 401(1), 112–116 (2010)
21. Ning, K., Ng, H.K., Srihari, S., et al.: Examination of the relationship between essential genes in PPI network and hub proteins in reverse nearest neighbor topology. BMC Bioinformatics 11(1), 505 (2010)
22. Freeman, L.C.: A set of measures of centrality based on betweenness. Sociometry, 35–41 (1977)
23. Joy, M.P., Brock, A., Ingber, D.E., et al.: High-betweenness proteins in the yeast protein interaction network. BioMed Research International 2005(2), 96–103 (2005)
24. Wuchty, S., Stadler, P.F.: Centers of complex networks. Journal of Theoretical Biology 223(1), 45–53 (2003)
25. Estrada, E., Rodriguez-Velazquez, J.A.: Subgraph centrality in complex networks. Physical Review E 71(5), 056103 (2005)
26. Bonacich, P.: Power and centrality: A family of measures. American Journal of Sociology, 1170–1182 (1987)
27. Stephenson, K., Zelen, M.: Rethinking centrality: Methods and examples. Social Networks 11(1), 1–37 (1989)
28. Wang, H., Li, M., Wang, J., Pan, Y.: A new method for identifying essential proteins based on edge clustering coefficient. In: Chen, J., Wang, J., Zelikovsky, A. (eds.) ISBRA 2011. LNCS (LNBI), vol. 6674, pp. 87–98. Springer, Heidelberg (2011)
29. Li, M., Wang, J., Chen, X., et al.: A local average connectivity-based method for identifying essential proteins from the network level. Computational Biology and Chemistry 35(3), 143–150 (2011)
30. Tang, X., Wang, J., Zhong, J., Pan, Y.: Predicting essential proteins based on weighted degree centrality. IEEE/ACM Transactions on Computational Biology and Bioinformatics (2014)
31. Li, M., Zheng, R., Zhang, H., Wang, J., Pan, Y.: Effective identification of essential proteins based on priori knowledge, network topology and gene expressions. Methods (2014)
32. Kim, W.: Prediction of essential proteins using topological properties in GO-pruned PPI network based on machine learning methods. Tsinghua Science and Technology 17(6), 645–658 (2012)
33. Sprinzak, E., Sattath, S., Margalit, H.: How reliable are experimental protein - protein interaction data? Journal of Molecular Biology 327(5), 919–923 (2003)
34. Hart, G.T., Lee, I., Marcotte, E.M.: A high-accuracy consensus map of yeast protein complexes reveals modular nature of gene essentiality. BMC Bioinformatics 8(1), 236 (2007)
35. Spirin, V., Mirny, L.A.: Protein complexes and functional modules in molecular networks. Proceedings of the National Academy of Sciences 100(21), 12128–12128 (2003)
36. Mewes, H.W., Amid, C., Arnold, R., et al.: MIPS: analysis and annotation of proteins from whole genomes. Nucleic Acids Research 32(suppl. 1), D41–D44 (2004)
37. Xenarios, I., Rice, D.W., Salwinski, L., et al.: DIP: the database of interacting proteins. Nucleic Acids Research 28(1), 289–291 (2000)
38. Gavin, A.C., Aloy, P., Grandi, P., et al.: Proteome survey reveals modularity of the yeast cell machinery. Nature 440(7084), 631–636 (2006)
39. Krogan, N.J., Cagney, G., Yu, H., et al.: Global landscape of protein complexes in the yeast Saccharomyces cerevisiae. Nature 440(7084), 637–643 (2006)

40. Ho, Y., Gruhler, A., Heilbut, A., et al.: Systematic identification of protein complexes in Saccharomyces cerevisiae by mass spectrometry. Nature 415(6868), 180–183 (2002)
41. Issel-Tarver, L., Christie, K.R., Dolinski, K., et al.: Saccharomyces Genome Database. Methods in Enzymology 350, 329 (2002)
42. Zhang, R., Lin, Y.: DEG 5.0, a database of essential genes in both prokaryotes and eukaryotes. Nucleic Acids Research 37(suppl. 1), D455–D458 (2009)
43. http://www-sequence.stanford.edu/group/yeast_deletion_project (Saccharomyces Genome Deletion Project)
44. Li, M., Wang, J., Wang, H., Pan, Y.: Essential proteins discovery from weighted protein interaction networks. In: Borodovsky, M., Gogarten, J.P., Przytycka, T.M., Rajasekaran, S. (eds.) ISBRA 2010. LNCS, vol. 6053, pp. 89–100. Springer, Heidelberg (2010)
45. Ren, J., Wang, J., Li, M., Wang, H., Liu, B.: Prediction of essential proteins by integration of PPI network topology and protein complexes information. In: Chen, J., Wang, J., Zelikovsky, A. (eds.) ISBRA 2011. LNCS (LNAI), vol. 6674, pp. 12–24. Springer, Heidelberg (2011)

GenoScan: Genomic Scanner for Putative miRNA Precursors

Benjamin Ulfenborg, Karin Klinga-Levan, and Björn Olsson

Systems Biology Research Centre, School of Bioscience
University of Skövde
Skövde, Sweden
bjorn.olsson@his.se

Abstract. The significance of miRNAs has been clarified over the last decade as thousands of these small non-coding RNAs have been found in a wide variety of species. By binding to specific target mRNAs, miRNAs act as negative regulators of gene expression in many different biological processes. Computational approaches for discovery of miRNAs in genomes usually take the form of an algorithm that scans sequences for miRNA-characteristic hairpins, followed by classification of those hairpins as miRNAs or non-miRNAs. In this study, two new approaches to genome-scale miRNA discovery are presented and evaluated. These methods, one ensemble-based and one using logistic regression, have been designed to detect miRNA candidates without relying on conservation or transcriptome data, and to achieve high-confidence predictions in reasonable computational time. GenoScan achieves high accuracy with a good balance between sensitivity and specificity. In a benchmark evaluation including 15 previously published methods, the regression-based approach in GenoScan achieved the highest classification accuracy.

Keywords: miRNA discovery, machine learning, hairpin classification.

1 Introduction

MicroRNAs are small non-coding RNAs (ncRNAs) that regulate protein expression post-transcriptionally. Currently, miRBase contains more than 30 000 mature miRNAs from more than 200 species [1]. MicroRNAs have been implicated in the regulation of many biological processes, such as proliferation, differentiation, apoptosis and metabolism [2], [3], [4]. The biogenesis and regulatory action of miRNAs includes several steps. Following transcription, the pri-miRNA forms a hairpin secondary structure, which is cleaved by the endoribonuclease Drosha into a precursor miRNA (pre-miRNA). Exportin 5 transports the pre-miRNA to the cytoplasm, where it is cleaved by Dicer to remove the hairpin loop and release the mature miRNA. The mature miRNA is incorporated into the RNA-induced silencing complex, which inhibits translation by binding to specific target mRNAs [5]. A recent estimate proclaims that more than 60% of human protein-coding genes are regulated by miRNAs [6].

M. Basu, Y. Pan, and J. Wang (Eds.): ISBRA 2014, LNBI 8492, pp. 266–277, 2014.
© Springer International Publishing Switzerland 2014

A multitude of computational methods have been developed to facilitate the discovery of novel miRNAs in genome sequences. This process can generally be divided into three steps. First, the set of sequences in which to search for novel miRNAs is identified. This can be an entire genome [7], [8], conserved regions [9], [10], transcribed regions [11], [12] or regions able to bind mRNAs [13], [14]. Restricting the search to regions that are conserved or transcribed reduces the risk of false positives. This is important since the non-protein-coding regions of the human genome contain ~11 million sequences that can fold into miRNA-like hairpins [15]. The drawback is that the search is not able to detect miRNAs that are poorly conserved or not expressed in the samples from which the transcripts are derived. Second, each candidate sequence is folded *in silico* by secondary structure prediction software [16], [17] to determine if it forms a hairpin shape. Thirdly, the sequences with hairpin structures are evaluated to determine if they are miRNAs or not.

The earliest algorithms employed filter-based approaches consisting of thresholds for various properties of the hairpin, such as minimum free energy (MFE), the number of base pairs, or the nucleotide composition [7], [9], [18]. Later algorithms have used machine learning by training on known miRNAs and non-miRNAs and classifying hairpins based on their similarity to the training data. Examples include support vector machines (SVM) [19], [20], [21], hidden Markov models (HMM) [22], [23], [24], random forests [25], [26], artificial neural networks [27] and decision trees [28]. Recent algorithms have incorporated ensemble approaches where several machine learning classifiers are combined to improve performance [29], [30]. Some tools also implement methods to facilitate miRNA discovery from next-generation sequencing (NGS) data [12], [31], [32], [33].

A shortcoming in the current literature on computational discovery of miRNAs is the lack of benchmarking. Although most methods have been evaluated by classifying a test set of known miRNAs and non-miRNAs, it is unclear which methods are most suitable when working with whole genomes or large genomic regions. Common measures of evaluation include sensitivity, specificity, accuracy and Matthews's correlation coefficient. To determine the usefulness of computational methods on the genome scale, it is also helpful to take into account measures such as false discovery rate, genome prediction rate (the fraction of hairpin-forming sequences in a genome classified as miRNAs) and computational time.

This study presents novel genome-scale miRNA discovery software referred to as *GenoScan*, which allows scanning of an entire genome or to focus on a set of genomic regions. GenoScan includes two methods for hairpin classification. The starting point of developing the first method was to benchmark published hairpin classification methods to evaluate their suitability for genome-wide scans. Based on the results, algorithms were selected and integrated into an ensemble classifier for improved performance. The second method is based on a logistic regression model trained on the datasets used during benchmarking. The software implementation allows processing of whole genomes, filtering and targeting specific regions of interest.

2   Methods

2.1   Dataset Generation

Three datasets were generated: the positive, the negative, and the genomic dataset. The positive dataset was extracted from the 1600 *H. sapiens* pre-miRNA hairpin sequences in miRBase ver 19 [34]. First, all pairs of sequences were aligned with BLAST [35] and one sequence removed for every pair with >= 95% identity. Second, every hairpin not fulfilling any of the following criteria, according to miRBase annotation, was removed: (1) not supported by cloning, (2) supported by fewer than three experiments, and (3) supported by fewer than ten NGS reads. This removes e.g. degradation products from NGS experiments that have been wrongly annotated as miRNAs. Third, the hairpin sequences were folded by RNAfold [16] and sequences containing more than one terminal loop were removed. The remaining 1200 hairpins constitute the positive dataset.

The negative dataset was based on permutations of known human pre-miRNAs, hairpins extracted from CDS loci, and hairpins extracted from ncRNA loci (excluding miRNAs). Permutations were generated by the uShuffle tool [36], using $k = 2$, which preserves the k-mer frequencies of the original sequence. The H. sapiens genome build hs_ref_GRCH37.p10 (hg19) was downloaded from NCBI Refseq [37], followed by extraction of CDS and ncRNA sequences based on NCBI Refseq annotation. These sequences were folded with RNAfold followed by hairpin extraction. All pairs in the negative dataset were aligned with BLAST, and one sequence removed from every pair with identity >= 95%. Finally, hairpins with more than one terminal loop, and those with a sequence identity >= 95% to known human pre-miRNAs were also removed. The remaining hairpins were merged into a single dataset, such that 1/3 comes from permutations, 1/3 from CDS-derived hairpins and 1/3 from ncRNA-derived hairpins. This resulted in 1361 hairpins in the negative dataset.

The genomic dataset was generated by moving a 200 nt window with a step size of 20 nt across the genome. The resulting sequences were folded with RNAfold, followed by hairpin extraction from loci not overlapping with CDS regions. Next, BLAST was used to generate pairwise alignments, and one sequence was removed for every pair with identity >= 95%. Hairpins containing more than one terminal loop were also removed. From the remaining hairpins, 10 000 sequences were randomly selected and placed in the genomic dataset.

The following hairpin extraction algorithm was used when generating the negative and genomic datasets. For each hairpin loop in a sequence, the algorithm extends the stem of the hairpin until the length is 120 bp, or a hairpin junction is seen. Each hairpin was retained if it fulfilled all of the following criteria: (1) sequence length (min: 40), (2) GC-content (min: 0.2, max 0.8), (3) triplet-repeat score (max: 5), (4) stem length (min: 20), (5) loop size (max: 20), (6) base pair propensity (min: 0.5), (7) unpaired stretch propensity (max: 0.35), (8) normalized unpaired stretch length (max: 0.35). The triplet-repeat score (3) is calculated as the total number of triplets divided by the number of unique triplets. Base pair propensity (6) is the number of base pairs in the hairpin stem divided by the length of the stem. Unpaired stretch propensity (7) is the number of unpaired base stretches in the stem divided by stem length. An unpaired base stretch is a symmetric/asymmetric loop or bulge in the hairpin stem. The normalized unpaired stretch length of an unpaired stretch is calculated as the number of bases in the unpaired stretch divided by the stem length.

2.2   Benchmark Measures

Benchmarking of the hairpin classification algorithms was based on sensitivity (Sn), specificity (Sp), Matthew's correlation coefficient (MCC), false discovery rate (FDR), genome prediction rate (GPR) and computational time (CT). Sensitivity and specificity are defined as the fraction of miRNA hairpins classified as miRNAs and the fraction of non-miRNA hairpins classified as non-miRNAs, respectively. Matthew's correlation coefficient ranges from -1 to 1, where -1 indicates complete disagreement between assigned class and real class, and 1 indicates perfect agreement. False discovery rate is the fraction of non-miRNA hairpins classified as miRNAs. Genome prediction rate is the fraction of hairpins in a genome (N_{genome}) that are classified as miRNAs ($N_{classified}$). Computational time CT is the average number of seconds required to classify a single hairpin as either miRNA or non-miRNA, where T is the number of CPU seconds taken to classify all hairpins in a dataset and D_{size} is the number of sequences of the dataset.

$$GPR = \frac{N_{classified}}{N_{genome}} \tag{1}$$

$$CT = \frac{T}{D_{size}} \tag{2}$$

2.3   Hairpin Classification Methods

When surveying the literature three criteria were used for including hairpin classification methods in the benchmarking: i) the method must be able to classify hairpins as miRNAs or non-miRNAs; ii) the method must take nucleotide sequences as input; iii) there must be a stand-alone executable implementation, or a web-server with batch functionality. The survey covered 57 published methods, many of which lacked executable implementation. Of the 15 methods included in the study, four are based on SVMs (Triplet-SVM [19], microPred [20], MiRPara ver 4.1 [21], Mirident [39]), three on HMMs (ProMir ver 1.0 [22], HMMMir [8], CSHMM [24]), two on algorithm-specific hairpin criteria (MIRcheck [38], MIReNA ver 2.0 [31]), two on ensemble methods (miR-BAG ver 1.0 [29], HeteroMirPred [30]), two on random forests (MiPred [25], HuntMi [26]), one on decision trees (CID-miRNA [28]), and one on fixed-order Markov models (FOMmir [40]). All algorithms were run with default parameters, except MiRPara for which the "animal 15" model was used. For miR-BAG, organism was set to *H. sapiens*.

For the ensemble classifier in GenoScan, algorithms were selected based on the results from benchmarking and divided into primary and secondary algorithms. Primary algorithms are well suited for generating an initial set of predictions from genomic sequences, based on high sensitivity and low computational time. Secondary algorithms are suitable for increasing the confidence in the predictions and to remove false positives, due to their high specificity and low false discovery rates. The ensemble classifier was evaluated on the positive, negative and genomic datasets.

2.4   Logistic Regression Modeling

In addition to the ensemble classifier, a novel hairpin classification algorithm based on a logistic regression model was implemented in R using the *glm* function. The initial model contained the following covariates: (1) hairpin stem length, (2) terminal loop size, (3) hairpin minimum free energy divided by stem length, (4) number of symmetric loop positions, (5) number of asymmetric loop positions, (6) number of gap positions, (7) number of bulge positions, (8) number of wobbles and (9) proportion of GC nucleotides in the stem, and (10-17) the number of each of the eight possible secondary structure triplets, i.e. combinations of "(" and ".", for the whole hairpin divided by stem length. During model simplification by iteratively removing the least significant covariate until only significant ones remained ($p =< 0.05$), the following covariates were removed: symmetric loop positions, asymmetric loop positions, and the secondary structure triplets "(.." and ".(.". The final model had a residual deviance of 1505.3 as compared to 3549.3 in the null model. Its performance was evaluated by leave-one-out cross validation on the positive and negative datasets.

3   Results and Discussion

3.1   Benchmarking Results

The benchmarking results are presented in Table 1. Sn, Sp, MCC and FDR are estimated from the positive and negative datasets, while GPR is based on the genomic dataset. Benchmarking was carried out in Ubuntu 13 and Fedora 18 using an Intel Core i5 2300 processor (2.8 GHz, four cores) and 12 GB DDR3 memory. Most methods show a performance ranging from 0.5 to 0.7 MCC (mean 0.48). This indicates fairly good classification accuracy, although with some room for improvement. Highest MCC was achieved by MIReNA (0.68), followed by CID-miRNA (0.66) and FOMmir (0.65). HuntMi achieved the highest Sn of 100%, but low specificity (50%). The three other methods with high Sn, MiPred (96%), Mirident (90%) and microPred (90%), all had low Sp (45%, 70% and 47%, respectively). ProMir had the highest Sp of 98%, but the second lowest Sn. There were generally large gaps between Sn and Sp, with an average of 29 percentage units.

Dramatic differences could be observed with respect to computational time of the stand-alone implementations. MIRcheck, Triplet-SVM, ProMir and MIReNA required only milliseconds per hairpin, while MiPred and CID-miRNA took 6s and microPred and HuntMi more than 11s. The slowest method was miR-BAG, with 171s per hairpin. This has enormous implications for scanning a whole genome sequence.

Table 1. Results from benchmarking hairpin classification methods

Algorithm	Sn (%)	Sp (%)	MCC	FDR (%)	GPR (%)	CT (s)
MIRcheck	5.6	96	0.05	41	2	0.002
Triplet-SVM	67	92	0.62	12	6	0.003
ProMir	54	98	0.59	4	2	0.004
MiPred	96	45	0.47	39	38	6.067
CID-miRNA	69	94	0.66	9	6	6.425
HHMMiR	60	89	0.51	18	17	0.054
microPred	90	47	0.41	40	48	11.310
CSHMM	86	74	0.60	26	31	1.986
MIReNA	76	90	0.68	12	5	0.002
MiRPara	66	83	0.51	22	8	0.248
Mirident	90	70	0.61	27	34	0.269
miR-BAG	81	47	0.25	43	45	171.200
FOMmir	88	77	0.65	23	25	-
HeteroMirPred	83	77	0.60	24	35	2.430
HuntMi	100	50	0.56	36	35	11.890

A potential source of bias in benchmarking is to include some items in the test set that were used in the training set, leading to overestimation of performance. There are three reasons why this was difficult to avoid in this study. First, most methods cannot be retrained on new datasets. Two notable exceptions are CSHMM and HuntMi. Second, the datasets used by most methods are not available, making it difficult to deduce which hairpins have been used as positive and negative training sets. Third, the known human miRNAs is fairly few, so excluding all miRNAs used by all methods would probably result in exclusion of most known miRNAs.

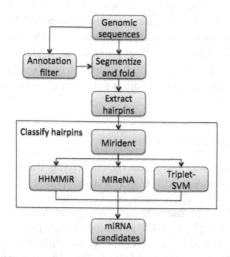

Fig. 1. Outline of the ensemble classifier method

Table 2. Classification performance of the ensemble classifier

Confidence	Sn (%)	Sp (%)	MCC	FDR (%)	GPR (%)	CT (s)
1	82	84	0.66	18	17	0.31
2	68	94	0.65	9	7	0.31
3	51	99	0.58	2	0.5	0.31

3.2   The GenoScan Hairpin Classification Methods

The GenoScan classification methods were designed with four goals in mind: high performance (Sn and Sp), reasonable GPR, low FDR and reasonable computational time. GenoScan takes genomic sequences as input, followed by optional filtering based on supplied sequence annotation. Sequences that pass the filter are segmented into 200 nt windows with 20 nt overlaps and folded with RNAfold. The resulting hairpins are passed to the Classify hairpins step. Two approaches to hairpin classification were evaluated. The first is an ensemble classifier (Fig. 1) that was created by integrating published hairpin classification algorithms to provide accurate and confident hairpin classification. Initial miRNA candidates are generated by Mirident (primary algorithm), which combines high Sn with low computational time. All hairpins are subsequently passed to HHMMiR, MIReNA and Triplet-SVM (secondary algorithms), which achieve high Sp and low FDR. Hairpins passing HHMMiR, MIReNA or Triplet-SVM are merged into a single set and filtered based on the confidence parameter $C = \{1, 2, 3\}$, which indicates how many secondary algorithms a hairpin must pass in order to qualify as a miRNA candidate. The C parameter allows adjustment of the balance between Sn and Sp, and to control the number of candidates reported.

The second method is a logistic regression model that was trained on the positive and negative datasets. The model calculates the probability that a given hairpin is a miRNA, and if the probability is equal to or greater than the threshold P the hairpin is reported as a miRNA candidate. The purpose of evaluating this method is to investigate the usefulness of logistic regression for miRNA discovery, and to see if single algorithms can achieve classification performance comparable to the ensemble approach.

3.3   GenoScan Performance Evaluation

The ensemble classifier was evaluated by classifying the hairpins in the positive, negative and genomic datasets with $C = 1, 2$ and 3 (Table 2). Increasing C results in better Sp, FDR and GPR, but reduces Sn. In comparison to the individual hairpin classification methods, the ensemble achieves better balance between Sn and Sp (Fig. 2) for $C = 1$, although the GPR is relatively high (17 %). For $C = 3$ the GPR drops to 0.5 %,, thus surpassing the best GPR for the individual classification methods, which ranged from 2 % for ProMir and MIRcheck to 48 % for microPred. The cost is a dramatic reduction of Sn to 51 %, however with an Sp of 99 %. For the individual methods with low GPR (2 to 8 %) the sensitivity ranges from 5.6 to 76 %. Individual methods with higher sensitivity have considerably higher GPR. When scanning whole genomes, the number of hairpins to be classified potentially number in the millions, and this demands a method with very low FDR and GPR, and reasonable computation time. The lower GPR of the ensemble classifier makes it more suitable for whole-genome miRNA discovery than the individual hairpin classification methods. For the human genome, the hairpin extraction algorithm described in the method section identifies ~27 million single terminal loop hairpins, of which ~19 million are extracted.

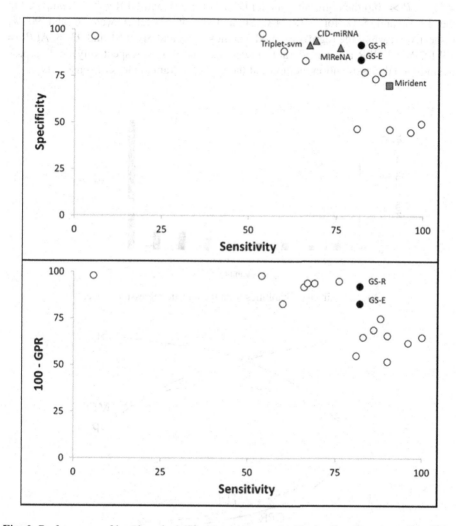

Fig. 2. Performance of benchmarked algorithms compared with the GenoScan ensemble (GS-E) and regression (GS-R) classifiers, with $C = 1$ for GS-E and $P >= 0.5$ for GS-R. *Upper panel*: Sensitivity vs. specificity. Algorithms included in the ensemble classifier are represented by filled gray markers. *Lower panel*: Sensitivity vs. 100 – GRP (%).

The logistic regression model was evaluated with different values for the probability threshold $P = \{0, 0.1, ..., 1.0\}$, where hairpins with a probability of P or higher are classified as miRNAs. The probabilities assigned to the left-out positive and negative hairpins during leave-one-out cross validation are shown in Fig. 3. The Sn and Sp of the model peaks at $P >= 0.5$ and 0.6, with MCC = 0.75 (Fig. 4). This is higher than the MCC of all the benchmarked methods, where MIReNA had the highest (0.68). It was also higher than the MCC of the ensemble classifier (0.66 for $C = 1$). At $P >= 0.6$ the regression model has FDR = 9 % and GPR = 5 %, compared to the 12 % and 5 % for MIReNA. Sn and Sp of MIReNA was 76 and 90 %, respectively, while the regression model achieved Sn and Sp of 81 and 93 %. At $P >= 0.9$ FDR and GPR are considerably lower (2.5 and 0.6 %, respectively) and surpass the individual classification methods, at the cost of a dramatic loss of sensitivity.

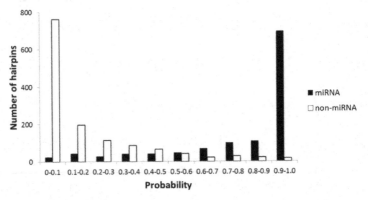

Fig. 3. Hairpin probabilities from the logistic regression model

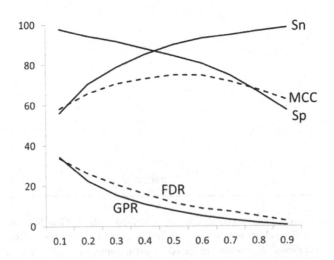

Fig. 4. Performance of the regression model classifier at $P = \{0.1,..., 0.9\}$. All performance measures are shown in percentages, except MCC, which is shown as the original MCC value multiplied by 100. (The peak MCC value at $P = 0.5$ and 0.6 is 0.75.)

In summary, the performance evaluation shows that both methods in GenoScan give highly accurate classifications results with a good balance between sensitivity and specificity. It was also shown that a regression model with only 13 covariates achieves higher classification accuracy than previously published algorithms.

3.4   Availability

GenoScan is available as downloadable software from the Comprehensive Perl Archive Network (CPAN) in the form of the Software::GenoScan module. It was developed and tested in the Linux Ubuntu environment.

3.5   Conclusions

In this study the miRNA discovery software GenoScan was presented. The performance evaluation indicates that high accuracy can be achieved by combining existing algorithms in an ensemble classifier. Furthermore, the evaluation of the logistic regression model shows that the performance of a classifier is not necessarily proportional to its complexity. GenoScan does not rely on conservation or transcriptome data for generating candidate miRNAs and supports user-defined annotation filters to exclude or target specific regions. This makes GenoScan a flexible tool for genome-scale miRNA discovery. The main limitation of the current version of the software is its low sensitivity with stringent settings of the confidence parameter and probability threshold. Future developments of GenoScan include training the logistic regression model on multiple organisms, improving sensitivity without increasing genomic prediction rate, and implementing support for miRNA discovery in next-generation sequencing data.

References

1. Griffiths-Jones, S., Saini, H.K., van Dongen, S., Enright, A.J.: miRBase: tools for microRNA genomics. Nucleic Acids Research 36(Database issue), D154–D158 (2008)
2. Huang, Y., Zou, Q., Wang, S.P., Tang, S.M., Zhang, G.Z., Shen, X.J.: The discovery approaches and detection methods of microRNAs. Molecular Biology Reports 38(6), 4125–4135 (2011)
3. Bartel, D.P.: MicroRNAs: target recognition and regulatory functions. Cell 136(2), 215–233 (2009)
4. Ghildiyal, M., Zamore, P.D.: Small silencing RNAs: an expanding universe. Nature Reviews. Genetics 10(2), 94–108 (2009)
5. He, L., Hannon, G.J.: MicroRNAs: small RNAs with a big role in gene regulation. Nature Reviews. Genetics 5(7), 522–531 (2004)
6. Friedman, R.C., Farh, K.K., Burge, C.B., Bartel, D.P.: Most mammalian mRNAs are conserved targets of microRNAs. Genome Research 19(1), 92–105 (2009)
7. Lim, L.P., Lau, N.C., Weinstein, E.G., Abdelhakim, A., Yekta, S., Rhoades, M.W., Burge, C.B., Bartel, D.P.: The microRNAs of Caenorhabditis elegans. Genes & Development 17(8), 991–1008 (2003)

8. Kadri, S., Hinman, V., Benos, P.V.: HHMMiR: efficient de novo prediction of microRNAs using hierarchical hidden Markov models. BMC Bioinformatics 10(suppl. 1), S35 (2009)

9. Lai, E.C., Tomancak, P., Williams, R.W., Rubin, G.M.: Computational identification of Drosophila microRNA genes. Genome Biology 4(7), R42 (2003)

10. Huang, T., Fan, B., Rothschild, M.F., Hu, Z., Li, K., Zhao, S.: MiRFinder: an improved approach and software implementation for genome-wide fast microRNA precursor scans. BMC Bioinformatics 8, 341 (2007)

11. Adai, A., Johnson, C., Mlotshwa, S., Archer-Evans, S., Manocha, V., Vance, V., Sundaresan, V.: Computational prediction of miRNAs in Arabidopsis thaliana. Genome Research 15(1), 78–91 (2005)

12. Friedländer, M.R., Chen, W., Adamidi, C., Maaskola, J., Einspanier, R., Knespel, S., Rajewsky, N.: Discovering microRNAs from deep sequencing data using miRDeep. Nature Biotechnology 26(4), 407–415 (2008)

13. Lindow, M., Jacobsen, A., Nygaard, S., Mang, Y., Krogh, A.: Intragenomic matching reveals a huge potential for miRNA-mediated regulation in plants. PLoS Computational Biology 3(11), e238 (2007)

14. Thieme, C.J., Gramzow, L., Lobbes, D., Theissen, G.: SplamiR–prediction of spliced miRNAs in plants. Bioinformatics 27(9), 1215–1223 (2011)

15. Bentwich, I.: Prediction and validation of microRNAs and their targets. FEBS Letters 579(26), 5904–5910 (2005)

16. Lorenz, R., Bernhart, S.H., Höner, C., Tafer, H., Flamm, C., Stadler, P.F., Hofacker, I.L., Tafer, H., Flamm, C., Stadler, P.F., Hofacker, I.L.: ViennaRNA Package 2.0. Algorithms for Molecular Biology: AMB 6, 26 (2011)

17. Zuker, M.: Mfold web server for nucleic acid folding and hybridization prediction. Nucleic Acids Research 31(13), 3406–3415 (2003)

18. Ohler, U., Yekta, S., Lim, L.P., Bartel, D.P., Burge, C.B.: Patterns of flanking sequence conservation and a characteristic upstream motif for microRNA gene identification. RNA 10(9), 1309–1322 (2004)

19. Xue, C., Li, F., He, T., Liu, G.-P., Li, Y., Zhang, X.: Classification of real and pseudo microRNA precursors using local structure-sequence features and support vector machine. BMC Bioinformatics 6, 310 (2005)

20. Batuwita, R., Palade, V.: microPred: effective classification of pre-miRNAs for human miRNA gene prediction. Bioinformatics 25(8), 989–995 (2009)

21. Wu, Y., Wei, B., Liu, H., Li, T., Rayner, S.: MiRPara: a SVM-based software tool for prediction of most probable microRNA coding regions in genome scale sequences. BMC Bioinformatics 12(1), 107 (2011)

22. Nam, J., Shin, K., Han, J., Lee, Y., Kim, V.N., Zhang, B.: Human microRNA prediction through a probabilistic co-learning model of sequence and structure. Nucleic Acids Research 33(11), 3570–3581 (2005)

23. Terai, G., Komori, T., Asai, K., Kin, T.: miRRim: a novel system to find conserved miRNAs with high sensitivity and specificity. RNA 13(12), 2081–2090 (2007)

24. Agarwal, S., Vaz, C., Bhattacharya, A., Srinivasan, A.: Prediction of novel precursor miRNAs using a context-sensitive hidden Markov model (CSHMM). BMC Bioinformatics 11(suppl. 1), S29 (2010)

25. Jiang, P., Wu, H., Wang, W., Ma, W., Sun, X., Lu, Z.: MiPred: classification of real and pseudo microRNA precursors using random forest prediction model with combined features. Nucleic Acids Research 35(Web Server issue), W339–W344 (2007)

26. Gudyś, A., Szcześniak, M.W., Sikora, M., Makałowska, I.: HuntMi: an efficient and taxon-specific approach in pre-miRNA identification. BMC Bioinformatics 14, 83 (2013)

27. Rahman, M.E., Islam, R., Islam, S., Mondal, S.I., Amin, M.R.: MiRANN: a reliable approach for improved classification of precursor microRNA using Artificial Neural Network model. Genomics 99(4), 189–194 (2012)
28. Tyagi, S., Vaz, C., Gupta, V., Bhatia, R., Maheshwari, S., Srinivasan, A., Bhattacharya, A.: CID-miRNA: a web server for prediction of novel miRNA precursors in human genome. Biochemical and Biophysical Research Communications 372(4), 831–834 (2008)
29. Jha, A., Chauhan, R., Mehra, M., Singh, H.R., Shankar, R.: miR-BAG: Bagging Based Identification of MicroRNA Precursors. PloS One 7(9), e45782 (2012)
30. Lertampaiporn, S., Thammarongtham, C., Nukoolkit, C., Kaewkamnerdpong, B., Ruengjitchatchawalya, M.: Heterogeneous ensemble approach with discriminative features and modified-SMOTEbagging for pre-miRNA classification. Nucleic Acids Research 41(1), e21 (2012)
31. Mathelier, A., Carbone, A.: MIReNA: finding microRNAs with high accuracy and no learning at genome scale and from deep sequencing data. Bioinformatics 26(18), 2226–2234 (2010)
32. Hackenberg, M., Rodríguez-Ezpeleta, N., Aransay, A.M.: miRanalyzer: an update on the detection and analysis of microRNAs in high-throughput sequencing experiments. Nucleic Acids Research 8(Web Server issue), W132–W138 (2011)
33. Guan, D.-G., Liao, J.-Y., Qu, Z.-H., Zhang, Y., Qu, L.-H.: mirExplorer: detecting microRNAs from genome and next generation sequencing data using the AdaBoost method with transition probability matrix and combined features. RNA Biology 8(5), 922–934 (2011)
34. Kozomara, A., Griffiths-Jones, S.: miRBase: integrating microRNA annotation and deep-sequencing data. Nucleic Acids Research 39(Database issue), D152–D157 (2011)
35. Altschul, S.F., Gish, W., Miller, W., Myers, E.W., Lipman, D.J.: Basic local alignment search tool. Journal of Molecular Biology 215(3), 403–410 (1990)
36. Jiang, M., Anderson, J., Gillespie, J., Mayne, M.: uShuffle: a useful tool for shuffling biological sequences while preserving the k-let counts. BMC Bioinformatics 9(i), 192 (2008)
37. Pruitt, K.D., Tatusova, T., Brown, G.R., Maglott, D.R.: NCBI Reference Sequences (RefSeq): current status, new features and genome annotation policy. Nucleic Acids Research 40(Database issue), D130–D135 (2012)
38. Jones-Rhoades, M.W., Bartel, D.P.: Computational identification of plant microRNAs and their targets, including a stress-induced miRNA. Molecular Cell 14(6), 787–799 (2004)
39. Liu, X., He, S., Skogerbø, G., Gong, F., Chen, R.: Integrated sequence-structure motifs suffice to identify microRNA precursors. PloS One 7(3), e32797 (2012)
40. Shen, W., Chen, M., Wei, G., Li, Y.: MicroRNA Prediction Using a Fixed-Order Markov Model Based on the Secondary Structure Pattern. PloS One 7(10), e48236 (2012)

Searching SNP Combinations Related to Evolutionary Information of Human Populations on HapMap Data[*]

Xiaojun Ding[1], Haihua Gu[2], Zhen Zhang[1], Min Li[1], and Fangxiang Wu[3]

[1] School of Information Science and Engineering,
Central South University, Changsha 410083, P.R. China
feathersky@gmail.com
[2] School of Computer and Software,
Nanjing College of Information Technology, Nanjing 210023, P.R. China
guhh@njcit.cn
[3] Division of Biomedical Engineering,
University of Saskatchewan, Saskatoon, SK, S7N 5A9, Canada

Abstract. The International HapMap Project is a partnership of scientists and funding agencies from different countries to develop a public resource that will help researchers find genes associated with human disease and response to pharmaceuticals. The project has collected large amounts of SNP(single-nucleotide polymorphism) data of individuals of different human populations. Many researchers have revealed evolution information from the SNP data. But how to find all the SNPs related to human evolution is still a hard work. At most time, these SNPs work together which leads to the differences between different human populations. The number of SNP combinations is very large, thus it is impossible to check all the combinations. In this paper, a novel algorithm is proposed to find the SNP combinatorial patterns whose frequencies are quite different in two different populations. The numbers of the multi-SNP combinations are regarded as the differences between each paired human populations, then a hierarchical clustering algorithm is used to construct the evolution trees for human populations. The trees from 4 chromosomes are consistent and the result can be validated by other literatures, which indicates that evolutionary information is well mined. The multi-SNP combinations found by our method can be studied further in many aspects.

Keywords: Multi-SNP combination, SNP-SNP interaction, Evolution tree.

1 Introduction

For different human beings, 99.9% of the bases of the genome are the same, while the remaining 0.1% makes a person unique, they code the different attributes,

---

[*] This work is supported in part by the National Natural Science Foundation of China under grant NO.61232001, NO.61379108 and NO.61370172, the Program for New Century Excellent Talents in University (NCET-12-0547)

M. Basu, Y. Pan, and J. Wang (Eds.): ISBRA 2014, LNBI 8492, pp. 278–288, 2014.

characteristics and traits of a person. A single-nucleotide polymorphism (SNP) is a DNA sequence variation occurring when a single nucleotide - A, T, C or G - in the genome (or other shared sequence) differs between members of a biological species or paired chromosomes. SNPs are distributed highly non-randomly in the human genome through a variety of processes from ascertainment biases to the action of mutation hotspots and natural selection [1].

Many information can be inferred from SNP data [2–5]. For example, Patterson et al. used SNP data to analyze population mixture for 53 diverse populations [6]. Gattepaille et al. reconstructed the historical variation of population size from sequence and SNP data [7]. Gutenkunst et al. inferred the joint demographic history of multiple populations from multidimensional SNP frequency data [8].

Proteins are often bound together to perform some functions [9]. The protein-protein interactions can be inferred from SNP-SNP interactions. To find these SNP-SNP interactions, researchers have proposed many methods to search the SNP combinatorial patterns whose frequencies in two different groups are quite different. For example, Mao et al. presented an optimum random forest algorithm which first sorts all SNPs, finds out the most disease associated SNPs for a given threshold, then generates many random trees, and selects the best trees [10]. But how to select the most disease associated SNPs is a problem. Brinza et al. presented a solution named CS algorithm [11]. The algorithm finds the best p-value of the exposed-closure of each single-SNP, after that it searches for the best p-value among exposed-closure of all 2-SNPs combinations and so on. However some single-SNP with a higher p-value but they actually are parta of some significant multi-SNP combinations. Brinza also presented a Complimentary Greedy Search algorithm [12], while the algorithm can only find case-free or control-free factors. Chuang et al. presented a PSO algorithm to find the significant MSCs [13]. Zhang presented a Bayesian graphical model for genome-wide multi-SNP association mapping [14]. However, for huge data, the above methods will be very slow or miss many important high order SNP combinations.

To overcome the shortage, we propose a new method which can deal with huge data and apply it on the HapMap data to mine evolutionary information. First, the SNP combinations which lead to the difference between each paired human populations are searched . Secondly, a hierarchical clustering algorithm is used to construct the evolution trees. Four trees are obtained from the SNP data of chromosome X, chromosome Y, chromosome 1 and chromosome 2. It is found that these four evolution trees are highly similar. The conclusions of papers [15], [16] are also consistent with the evolution trees in our result. It indicates that our method can efficiently reveal the SNPs which are related to the evolutionary information of human populations. Studies on these SNPs will help us understand human evolutionary history and promote the population-specific disease studies. Our software is available at http://files.cnblogs.com/feathersky3000/MSCD.zip.

2   Methods

First, we will search the SNPs related to human evolutionary history. Secondly, evolution trees will be constructed based on these SNPs information to validate our result.

Candidate SNP Loci Detecting

In the paper, the SNP data of one individual is represented by a vector $v = [v_1, v_2, ..., v_n]^T$. Here n is the number of SNPs and v_i is number of minor alleles of the associated sites on the chromosome, v_i can be 0, 1 or 2. The SNP datasets of a population can be represented by $D = [d_1, d_2, ..., d_r]$ and the SNP datasets of another population can be represented by $H = [h_1, h_2, ..., h_t]$. Here d_i and h_i are SNP vectors of two different individuals. r and t are the numbers of individuals in each population. In order to let 0, 1 and 2 play the same important roles, each SNP vector is transformed from n dimension vector to $3 \times n$ vector. 0, 1 and 2 are transformed into 001, 010 and 100 respectively.

We suppose that the two populations have some common features which are irrelevant to the population difference and some features which lead to the population differences. There might be several SNP which are related to one feature. For example, the features that one individual is fat or tall are not the population difference. There may exist a new coordinate system under which the SNP vector of individuals in the two populations can be represented as following.

$$d_i = a_{i1}X_1 + a_{i2}X_2 + ... + p_{i1}Y_1 + p_{i2}Y_2 + ... + 0Z_1 + 0Z_2 + ...$$
$$h_i = b_{i1}X_1 + b_{i2}X_2 + ... + 0Y_1 + 0Y_2 + ... + q_{i1}Z_1 + q_{i2}Z_2 + ... \tag{1}$$

Where $X_1, X_2, ..., Y_1, Y_2, ..., Z_1, Z_2, ...$ are not only axes but also features related to SNP sets, any two of the axes are orthogonal. The distributions of the two population should be different on these axes. To find the new coordinate system, firstly, a measurement is needed which does not change under linear coordinate transform and can reflect the difference between two populations. If the mean of all individuals is at the origin, the variance will not change under linear coordinate transform. The variance is regarded as energy. Another good property of energy is that the energy equals to the sum of component energy on each axis. $Energy = energy_{X_1} + energy_{X_2} + ... + energy_{Y_1} + energy_{Y_2} + ... + energy_{Z_1} + energy_{Z_2} + ...$ Where $energy_{\alpha-axis}$ is component energy on the α axis.

D and H are merged into a matrix T. For each row of T, the mean is calculated and is subtracted from each entry in the row. At last matrix T is separated to get the new D and H.

If α-axis approximates an axis in Y of a discriminative coordinate system in the formula (1) . Most individuals of the first population have energy on α-axis, so $E_\alpha(population_1) = \alpha'DD'\alpha/(r - 1)$ is great. Most individuals of the second population have no energy on α-axis, so $E_\alpha(population_2) = \alpha'HH'\alpha/(t-1) \approx 0$.

Similarly, if α-axis approximates an axis in Z, $E_\alpha(population_1) = \alpha'DD'\alpha/(r-1) \approx 0$, and $E_\alpha(population_2) = \alpha'HH'\alpha/(t-1)$ is great enough. Therefore, it is a great possibility that an axis α maximizing the difference between $E_\alpha(population_1)$ and $E_\alpha(population_2)$ is similar to an axis in Y or in Z.

To approximate Y and Z, we search for α such that

$$\alpha = \underset{\alpha}{\operatorname{argmax}} \left| \alpha' \frac{HH'}{t-1}\alpha - \alpha' \frac{DD'}{r-1}\alpha \right|, \tag{2}$$

and

$$\| \alpha \| = 1. \tag{3}$$

Let

$$K(\alpha) = \alpha' \frac{HH'}{t-1}\alpha - \alpha' \frac{DD'}{r-1}\alpha. \tag{4}$$

We use the gradient method to get an α maximizing $K(\alpha)$ to approximate the axis Z and an α minimizing $K(\alpha)$ to approximate the axis Y.

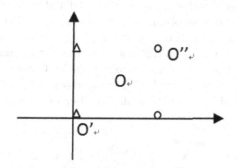

Fig. 1. The distribution of individuals. individuals of *population*₁ are denoted by triangle and individuals of *population*₂ are denoted by circle.

On some axes of X shown in Equation (1), the projections of both populations are not zero, but their value distributions may be different. For example, in Figure 1, if the point O is set as the origin, most values of individuals in the first population on axis X are less than 0 and most values of the individuals in the second population on axis X are greater than 0. Under the situation, whatever Y or Z is, the energy distribution difference of the two populations is always 0 because of its symmetry. However, we can set the point $O' = [0,0,0,...0]_{3n}$ and $O'' = [1,1,1,...1]_{3n}$ as the origin respectively to destroy the symmetry, then compute the axis Y_1, Z_1 for origin O' and Y_2, Z_2 for origin O''. But at here, another rule can be used to filter out more SNPs firstly. We find the axis α^* which maximizes the mean difference between the cases and the controls. The axis α^* satisfies

$$\alpha^* = \underset{\alpha}{\operatorname{argmax}} \left| \frac{1}{t} \sum_{i=1}^{t} G_{h_i}\alpha - \frac{1}{r} \sum_{i=1}^{r} G_{d_i}\alpha \right|. \tag{5}$$

The solution of α^* is shown as Formula 6.

$$\alpha^* = \text{normalize}\left(\frac{1}{t}\sum_{i=1}^{t}h_i - \frac{1}{r}\sum_{i=1}^{r}d_i\right) \tag{6}$$

After obtaining the axis α^*, the SNPs whose corresponding component absolute values are smaller than the average absolute value of the component values of α^* are removed. Then we set O' and O'' as the origin respectively to compute the axes Y_1, Z_1, Y_2 and Z_2.

Searching Multi-SNP Combinations

In the following, we use MSC to denote a multiple SNP loci combination and its value. We select candidate SNP loci from the axes Y, Z, Y_1, Z_1, Y_2 and Z_2. Each component of these vectors corresponds a SNP locus, and the absolute value of the component indicates how important the corresponding SNP locus is in making difference between the two populations. From each axis, top 28 components sorted by absolute values are selected and the corresponding loci are candidate SNP loci. Thus we obtain six sets of candidate SNP loci with each set from each vector, and we will use a pruning tree-search strategy to significant MSCs from each set. Given a MSC, we use num_i to denote the number of individuals in $population_i$ whose values match the MSC. $totalN = num_1 + num_2$. Its p-value is computed by Formula (9), where r is number of individuals in the first population individuals and t is the number of individuals in the second population.

$$p_1(MSC) = \sum_{i=num_1}^{totalNum} \binom{totalN}{i}(\frac{r}{r+t})^i(\frac{t}{r+t})^{totalN-i} \tag{7}$$

$$p_2(MSC) = \sum_{i=0}^{num_1} \binom{totalN}{i}(\frac{r}{r+t})^i(\frac{t}{r+t})^{totalN-i} \tag{8}$$

$$p-value(MSC) = Min(p_1, p_2) \tag{9}$$

The p-value threshold is often traditionally set 0.05. However, for multiple hypothesis testing, correction is needed. The Bonferroni [17] correction adjustment is often used for this purpose. In our experiment, the MSC with p-value below the Bonferroni corrected p-value threshold is regarded as a significant MSC. Once the significant MSCs are found by the pruning tree-search strategy, we remove the SNPs at the loci existing in at least one significant MSC from H and D, and the above processing (i.e. new Y, Z, Y_1, Z_1, Y_2 and Z_2 computing, new six sets of candidate SNP loci selection and significant MSCs searching) is repeated until there is no new significant MSC.

After all significant MSCs are found, the numbers of significant MSCs between the two different human populations are regarded as their difference. Then the difference matrix for human populations can be obtained. Next we will use the matrix to construct its corresponding evolution tree.

Constructing Evolution Trees

The numbers of significant MSCs between two human populations are regarded as their differences. The larger the number is, the bigger difference is between the two human populations. There are many algorithms to construct evolution tree. While in our experiments, the entries of the matrix do not satisfy the distance criterion. When more than two human populations are combined, it is hard to find one ancestry population to represent them. So we adopt a hierarchical clustering algorithm to construct evolution tree. First, each cluster contains one human population. We define the difference between $cluster_p$ and $cluster_q$ is defined as formula (10).

$$diff(cluster_p, cluster_q) = \frac{2 \times \sum\limits_{(i,j)} difference(population_i, population_j)}{K(K-1)} \quad (10)$$

Here, $population_i \in cluster_p \cup cluster_q$, $population_j \in cluster_p \cup cluster_q$, $i \neq j$, $K = |cluster_p \cup cluster_q|$. The two clusters with the smallest difference are combined into one cluster. This procedure will be repeated until there is only one cluster. The evolution tree is then constructed according to the procedure.

3 Results

The HapMap is a catalog of common genetic variants that occur in human beings. It describes what these variants are, where they occur in our DNA, and how they are distributed among people within populations and among populations in different parts of the world [18]. The data is obtained by sampling in 11 human populations. Abbreviations of the human populations are listed as Table 1 .

The SNP data on chromosome Y, chromosome X, chromosome 1 and chromosome 2 are used for the experiment. Since the SNP loci of different populations are not entirely identical, we select the common SNP loci in all of these human populations and search significant MSCs between each paired human populations. Part of the MSCs are listed in Table 2. From the table, we can find that the frequencies of these MSCs are quite different in two human populations. They can reflect the evolutionary history of human populations. For each chromosome, we get evolution trees as shown in Figure 2, 3, 4 and 5.

The evolution trees are not identical, but they are very similar. The human populations CHB, CHD and JPT are clustered together. The human populations ASW, LWK, MKK and YRI are clustered together. The human populations CEU, TSI, GIH and MEX are clustered together. We suppose these four evolution trees are randomly constructed. The probability of the event that on the four evolution trees the populations are clustered into the above three clusters is about 3.89×10^{-14}. The probability is so small which indicates these four trees are not constructed randomly, but constructed according to evolutionary information. In our experiments, these trees have a little differences. For example, on chromosome Y data, human population CHB and JPT are clustered first and

Table 1. The information of the eleven human populations

Label	Population sample
ASW	African ancestry in Southwest USA
CEU	Utah residents with Northern and Western European ancestry from the CEPH collection
CHB	Han Chinese in Beijing, China
CHD	Chinese in Metropolitan Denver, Colorado
GIH	Gujarati Indians in Houston, Texas
JPT	Japanese in Tokyo, Japan
LWK	Luhya in Webuye, Kenya
MEX	Mexican ancestry in Los Angeles, California
MKK	Maasai in Kinyawa, Kenya
TSI	Toscans in Italy
YRI	Yoruba in Ibadan, Nigeria

Table 2. Part of significant MSCs between different human populations on different chromosomes

p_1	p_2	SNPs and values	num_1	num_2	f_1	f_2	chr	pvalue
CEU	YRI	(rs17849631,rs28562204,rs6641979)=000	1	193	0.01	0.92	Y	1.51E-49
TSI	YRI	(rs28562204,rs4892831,rs6641979)=020	0	192	0	0.92	Y	7.21E-34
CEU	MKK	(rs1569562,rs17389,rs6418925)=200	1	154	0.01	0.84	X	2.31E-43
CHB	MKK	(rs12391615,rs1266324)=00	133	0	0.96	0	X	1.98E-49
CEU	JPT	(rs10157120,rs10218696)=20	0	100	0	0.86	1	1.61E-40
CEU	LWK	(rs1036001,rs10465612)=00	0	101	0	0.92	1	2.49E-42
CEU	YRI	(rs10198899,rs10207082)=02	158	0	0.91	0	2	7.27E-55
CHB	MKK	(rs10187416,rs1113669)=22	135	0	0.97	0	2	3.67E-50

p_i is the abbreviation of $population_i$. num_i is the number of individuals in $population_i$ whose values match the MSC. f_i is the frequency of individuals among $population_i$ whose values match the MSC. chr is the chromosome.

then combined with CHD. But on chromosome X data human population CHB and CHD are clustered first and then combined with JPT. The reason for this phenomenon may be that the three human populations have a common ancestry while the binary tree can not represent this kind of relationship. So the following evolution tree as shown in Figure 6 is obtained by comprehensive evaluation.

In the paper [16], the authors said that African = {ASW, LWK, MKK, YRI}, East Asian = {CHB, CHD, JPT}, European = {CEU, TSI}, and GIH and MEX are two independent groups.

In the paper [19], the authors measured the population differences in terms of fixation index (Fst), the SNPs with Fst values over 0.5 were defined as highly differentiated SNPs. In their result, ASW, LWK, MKK and YRI are cataloged in African. CHB, CHD and JPT are cataloged in Asian. CEU and TSI are cataloged in European. GIH occupies one catalog and MEX occupies one catalog.

The conclusions of above papers are in accordance with our results. So the SNPs combinations related to the evolutionary information are actually identified. These SNPs combinations will be very useful for other studies.

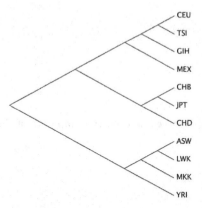

Fig. 2. Evolution tree based on HapMap data of chromosome Y

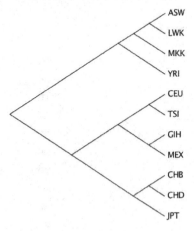

Fig. 3. Evolution tree based on HapMap data of chromosome X

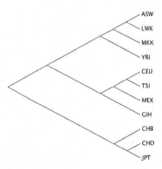

Fig. 4. Evolution tree based on HapMap data of chromosome 1

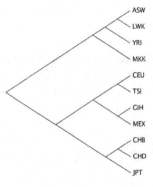

Fig. 5. Evolution tree based on HapMap data of chromosome 2

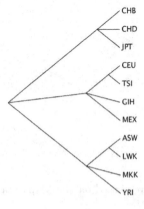

Fig. 6. Evolution tree by comprehensive evaluation

4 Conclusions

In this paper, we propose a novel method to search the MSCs between two different populations. Many multi-SNP combinations can be found and the speed of our method is very fast. Based on the information of the population-specific MSCs, we construct evolution trees of human populations and the trees are consistent on different chromosomes, which indicates that the SNPs related to evolution history of human populations are well mined. These SNPs combinations can help us understand human evolutionary history and promote the population-specific disease studies.

References

1. Amos, W.: Even small snp clusters are non-randomly distributed: is this evidence of mutational non-independence? Proceedings of the Royal Society B: Biological Sciences 277(1686), 1443–1449 (2010)
2. Cai, Z., Sabaa, H., Wang, Y., Goebel, R., Wang, Z., Xu, J., Stothard, P., Lin, G.: Most parsimonious haplotype allele sharing determination. BMC Bioinformatics 10(1), 115 (2009)
3. Sabaa, H., Cai, Z., Wang, Y., Goebel, R., Moore, S., Lin, G.: Whole genome identity-by-descent determination. Journal of Bioinformatics and Computational Biology 11(02) (2013)
4. Wang, Y., Cai, Z., Stothard, P., Moore, S., Goebel, R., Wang, L., Lin, G.: Fast accurate missing snp genotype local imputation. BMC Research Notes 5(1), 404 (2012)
5. Cheng, Y., Sabaa, H., Cai, Z., Goebel, R., Lin, G.: Efficient haplotype inference algorithms in one whole genome scan for pedigree data with non-genotyped founders. Acta Mathematicae Applicatae Sinica, English Series 25(3), 477–488 (2009)
6. Patterson, N., Moorjani, P., Luo, Y., Mallick, S., Rohland, N., Zhan, Y., Genschoreck, T., Webster, T., Reich, D.: Ancient admixture in human history. Genetics 192(3), 1065–1093 (2012)
7. Gattepaille, L., Jakobsson, M., Blum, M.G.: Inferring population size changes with sequence and snp data: lessons from human bottlenecks. Heredity (2013)
8. Gutenkunst, R.N., Hernandez, R.D., Williamson, S.H., Bustamante, C.D.: Inferring the joint demographic history of multiple populations from multidimensional snp frequency data. PLoS Genetics 5(10), e1000695 (2009)
9. Ding, X., Wang, W., Peng, X., Wang, J.: Mining protein complexes from ppi networks using the minimum vertex cut. Tsinghua Science and Technology 17(6), 674–681 (2012)
10. Mao, W., Lee, J.: A combinatorial analysis of genetic data for crohn's disease. In: The 1st International Conference on Bioinformatics and Biomedical Engineering, ICBBE 2007, pp. 1031–1034. IEEE (2007)
11. Brinza, D., Zelikovsky, A.: Combinatorial methods for disease association search and susceptibility prediction. In: Bücher, P., Moret, B.M.E. (eds.) WABI 2006. LNCS (LNBI), vol. 4175, pp. 286–297. Springer, Heidelberg (2006)
12. Brinza, D.: Discrete algorithms for analysis of genotype data. Computer Science Dissertations, 19 (2007)

13. Chuang, L.Y., Lin, M.C., Chang, H.W., Yang, C.H.: Analysis of snp interaction combinations to determine breast cancer risk with pso. In: 2011 IEEE 11th International Conference on Bioinformatics and Bioengineering (BIBE), pp. 291–294. IEEE (2011)
14. Zhang, Y.: A novel bayesian graphical model for genome-wide multi-snp association mapping. Genetic Epidemiology (2012)
15. Farheen, S., Basu, A., Majumder, P.P.: Haplotype variation in the ace gene in global populations, with special reference to India, and an alternative model of evolution of haplotypes. The HUGO Journal 5(1-4), 35–45 (2011)
16. Xue, C., Liu, X., Gong, Y., Zhao, Y., Fu, Y.X., et al.: Significantly fewer protein functional changing variants for lipid metabolism in Africans than in Europeans. Journal of Translational Medicine 11(1), 67 (2013)
17. Dewey, M., Seneta, E.: Carlo emilio bonferroni. In: Statisticians of the Centuries, pp. 411–414. Springer (2001)
18. The hapmap project homepage,
    http://hapmap.ncbi.nlm.nih.gov/whatishapmap.html.en
19. Duan, S., Zhang, W., Cox, N.J., Dolan, M.E.: Fstsnp-hapmap3: a database of snps with high population differentiation for hapmap3. Bioinformation 3(3), 139 (2008)

2D Pharmacophore Query Generation

David Hoksza and Petr Škoda

[1] Charles University in Prague, FMP, Department of Software Engineering,
Malostranské nám. 25, 118 00, Prague, Czech Republic
hoksza@ksi.mff.cuni.cz
http://siret.cz/hoksza
[2] Charles University in Prague, FMP, Department of Software Engineering,
Malostranské nám. 25, 118 00, Prague, Czech Republic
skoda@ksi.mff.cuni.cz

Abstract. Using pharmacophores in virtual screening of large chemical compound libraries proved to be a valuable concept in computer-aided drug design. Traditionally, pharmacophore-based screening is performed in 3D space where crystallized or predicted structures of ligands are superposed and where pharmacophore features are identified and compiled into a 3D pharmacophore model. However, in many cases the structures of the ligands are not known which results in using a 2D pharmacophore model.

We introduce a method capable of automatic generation of 2D pharmacophore models given previous knowledge about the biological target of interest. The knowledge comprises of a set of known active and inactive molecules with respect to the target. From the set of active and inactive molecules 2D pharmacophore features are extracted using pharmacophore fingerprints. Then a statistical procedure is applied to identify features separating the active from the inactive molecules and these features are then used to build a pharmacophore model. Finally, a similarity measure utilizing the model is applied for virtual screening.

The method was tested on multiple state of the art datasets and compared to several virtual screening methods. Our approach seems to exceed the existing methods in most cases. We believe that the presented methodology forms a valuable addition to the set of tools available for the early stage drug discovery process.

Keywords: 2D pharmacophores, pharmacophore modeling, virtual screening.

1 Introduction

The main procedure to identify new leads in the drug discovery process has traditionally been medium or high-throughput screening (HTS). However, using high-throughput virtual screening (HTVS) for prioritization of chemical compounds in large chemical libraries became a common early-stage drug discovery practice. Usefulness of complementing HTS with HTVS has been supported by

M. Basu, Y. Pan, and J. Wang (Eds.): ISBRA 2014, LNBI 8492, pp. 289–300, 2014.
© Springer International Publishing Switzerland 2014

several studies [5,27]. Moreover, without HTVS it is impossible to screen activities of compounds in large real or virtual compound libraries which can contain up to tens of millions of compounds. Therefore, to minimize costs of the drug discovery campaigns it is typical to narrow down the number of compounds to several thousands using in silico methods, i.e. virtual screening.

Based on the information at hand about active compounds and/or the target, we distinguish different virtual screening approaches [8,25]. If we know the three-dimensional structure of the biological target, we can use methods classified as structure-based virtual screening (SBVS) [20,21]. If only the information about the bioactive ligands is available, but not about the protein-ligand complexes, we can utilize the so-called ligand-based virtual screening (LBVS) [19].

The structure-based approach is based on docking and includes two steps: positioning a ligand into the target active site (docking) and scoring the pose. Since SBVS is not the central topic of this paper we only refer to several reviews concerning SBVS [21,16,3].

The LBVS is built around the concept that similar structures carry out similar functions more often than dissimilar ones. This assumption is based on the shape complementary of the ligand and the target called key-and-lock principle [15] or simply similar property principle [14]. Thus, in the first step of LBVS the 2D or 3D structure of a reference ligand(s) is used to obtain a computer-based representation of the molecule(s). In the second step the representations are either aggregated into a query or the individual representations are used directly for similarity search in compound databases. Finally, the database is sorted with respect to the extent of similarity to the query ligand(s). It is assumed that the high-scoring database compounds have a higher probability to bind to the target due to the similarity principle.

The popularity and value of LBVS can be demonstrated by many success stories [19] and also by the fact that all the leading commercial drug discovery software toolboxes include a tool realizing tge LBVS [16,25].

Taboureau at al. [25] classify the LBVS approaches into five classes based on the utilized features: alignment-based, descriptor-based, graph-based, shape-based and pharmacophore-based. While, for example, graph-based methods utilize directly the 2D graph by identifying common substructures between query and database compounds, other methods require some preprocessing including a feature extraction process. The type of extracted features include 3D shape, physico-chemical features or pharmacophores.

According to IUPAC a pharmacophore is defined as "the ensemble of steric and electronic features that is necessary to ensure the optimal supramolecular interactions with a specific biological target structure and to trigger (or to block) its biological response.". However, much more straightforward description of the pharmacophore concept dates back to 1909 when Erlich defined it as "a molecular framework that carries (phoros) the essential features responsible for a drugs (pharmacon) biological activity" [4]. The features usually include hydrogen bond donor/acceptor, charge, hydrophobicity and aromaticity. In the context of protein-ligand binding, the pharmacophore can be viewed as a specialization of the (maybe too simple) lock-and-key principle which focuses only on shape complementarity.

A pharmacophore model can be built either from a set of known ligands (ligand-based) or from a target (structure-based). Ligand-based pharmacophore model is formed from the superposition of 3D poses of the ligands and identification of positions (including a tolerance radius) of common physico-chemical features. If the target or target-ligand complex is known, a pharmacophore can be built by probing the active site and focusing on the target-ligand interaction points.

When a pharmacophore model is prepared we can use it as a template for similarity searching in compound databases to find ligands which comply with the model. The procedure is known as pharmacophore-based virtual screening [11,24].

The pharmacophore concept is traditionally (and intuitively) imagined in three dimensions. However, the ligand's three dimensional structure is not always known and its prediction is tricky due to the ligand's flexibility. In [23], Schneider et al. introduced the concept of topological pharmacophore where only topological distances (shortest paths between atoms in 2D graph) are used instead of the 3D distances.

2 Method Outline

In this paper we introduce a method for building a 2D pharmacophore model from the topological pharmacophores and show how to utilize this model in the 2D pharmacophore-based virtual screening. Our method consists of three main parts: 1) Extraction of 2D pharmacophores from a set of active and inactive molecules, 2) building a pharmacophore model from the available pharmacophores, 3) definition of a similarity measure used in virtual screening using the built pharmacophore model.

Our method differs in that the pharmacophore model is purely topological, thus it does not require any structural information about the existing ligands or target. However, unlike most of virtual screening methods, our pharmacophore model is built on a statistical analysis which needs both positive and negative examples, i.e., both known active and inactive molecules for the target. In most cases, the information about inactive compounds is available from the existing biochemical screens. If not, we believe it can be replaced by random sampling of large publicly available compound databases as we justify in the following section.

3 Model Building

The input to the model building process include two sets of molecules - known actives and inactives with respect to given biological target. These sets are the output of a previously carried out screening against the specific target. In the optimal case, the source is a confirmatory screen which contains less false positives in comparison to a primary screen. However, we can often find ourselves in the situation when we only know the active molecules. In such a case, we can randomly select inactive molecules from a large database of publicly available molecules such as ZINC [13], Pubchem [28] or ChEMBl [7]. We can do so relatively safely because the ratio between active and inactive molecules when considering large database is extremely low. Thus, when randomly choosing molecules the probability that we pick multiple active molecules by chance is very low.

The active and inactive molecules are then fed into the pharmacophore model building process which consists of the following steps:

1. 2D pharmacophore fingerprint generation for each molecule.
2. Running a statistical analysis to identify bits corresponding to the presence of individual pharmacophores which discriminate an active molecule from an inactive one.
3. Using the information about discriminative pharmacophores to build a bit string representing the resulting pharmacophore model.

In the following sections we discuss each of the steps in more detail.

3.1   2D Pharmacophore Fingerprint Generation

A 2D pharmacophore is a specific distance distribution of pharmacophore features in a molecular graph. The graph distance of two atoms in a molecular graph corresponds to the topological distance of their respective chemical features. The pharmacophore fingerprint is a bit string carrying the information about the presence or absence of every combination of features and distances.

In our implementation we used the 2D pharmacophore fingerprints implemented in RDKit [1]. The version of 2D fingerprints available in RDKit v. 2013.09.2 supports only 2-3 points pharmacophore where n point pharmacophore describes the mutual distances of n pharmacophore features. Therefore, each bit of the bit string representation captures the information about presence or absence of a specific topological distribution of two or three chemical features.

Thus, the input to the pharmacophore generation phase are two sets of molecules (actives and inactives) resulting in two sets of 2D pharmacophore fingerprints.

3.2   Statistical Analysis

After we identify the pharmacophores present in each molecule in the actives and inactives datasets we need to single out those which separate the two groups. To do so, we run a statistical test for every single pharmacophore.

Let us encode the presence of a pharmacophore by a 1 and the absence by a 0. We are therefore interested in such pharmacophores where the numbers of ones in the sets of active and inactive molecules significantly differ.

Since we are interested in the difference in the proportions of two independent samples, the utilization of the two-sample proportion z-test would be the obvious choice. However, there are three assumptions to use the proportion test: 1) independence within the groups, 2) independence between the groups and 3) normality which basically means that there should be at least 5 expected successes and 5 expected failures in the sample used from inference. The conditions 1 and 2 are fulfilled since the molecules are mutually independent (condition 1) and there is no dependence between active and inactive molecules (condition 2). The problem arises with the third condition because of our sample size. Generally, it means that the proportion test should be used only for cases when there are more than 10 observations which is often not the case with the active set. Therefore, we used Fisher's exact test [6] for testing the null hypothesis of independence of rows and columns in a contingency table (available in R [18]) where the condition on the sample size is relaxed. The null hypothesis is that there is no significant difference in the two proportions. The alternative hypothesis is that the two proportions significantly differ. The significance level is a parameter of the approach, but in general the standard 95% confidence level seems to work well (see 5).

3.3   Pharmacophore Model Construction

We use the knowledge about which pharmacophores are responsible (with the given confidence level) for the difference in activity to build the pharmacophore model. In our method we represent the model as a 2D pharmacophore fingerprint. We simply build a bit string corresponding to a virtual molecule having all the significant pharmacophores. That is, in the bit string we turn on the respective bits. This bit string is the resulting pharmacophore model carrying the information obtained from the known active and inactive compounds.

4   Virtual Screening

The idea behind the 3D pharmacophore-based virtual screening is to identify compounds being subject to specific spatio-chemical template defined by the 3D pharmacophore model. But 2D pharmacophores only approximate the spatial arrangement of properties. We can easily imagine two atoms being far apart in the molecular graph, but close in the 3D space. Therefore, we use the 2D model not for filtering, but for prioritization in the same way the standard ligand-based virtual screening does. To do so, we exploit the fact that the model is equivalent to a 2D pharmacophore fingerprint. Therefore, after defining a similarity measure, we can prioritize the database compounds based on the similarity to the fingerprint/model.

The most straightforward similarity measure between a pharmacophore and a compound would be to count pharmacophores in the model which are also present in the database compound. That can be simply achieved by counting number of common one-bits in the bit strings of the pharmacophore model and the 2D pharmacophore fingerprint of a database molecule. However, the size of such measure's domain can not be bigger than the number of pharmacophores in the model. That is, if the pharmacophore model contains 30 pharmacophores, which the statistical evaluation labeled as significant, then the similarity measure can take only 30 different values. That is impractical when screening libraries containing tens or hundreds of thousands of compounds. To tackle this problem we decided to tweak the similarity measure by taking into account also the p-value from the pharmacophore building process. So in the resulting measure each common one-bit is weighted by the p-value of its respective pharmacophore.

The following equations formally introduce the similarity measure:

$$s_{pp}(m_{pp}, m_{db}) = \sum_{i=1}^{m_{pp}.size} (m_{pp}[i] \times m_{db}[i]) \times w(i) \tag{1}$$

$$w(j) = 1 - \frac{pval[j] \times 0.5}{conf} \tag{2}$$

The m_{pp} and m_{db} are the bit vectors of the pharmacophore model and a database molecule for which the similarity is being computed, $m[i]$ represents the i-th bit of m, $w(i)$ is the weight corresponding to the i-th pharmacophore, $pval[i]$ its p-value and $conf$ is the confidence level used for the hypothesis testing. The purpose of the weighting function is to map the p-value to the interval $< 0.5; 1 >$. Such mapping uplifts the pharmacophores with lower p-values to emphasize the pharmacophores for which we have more evidence that they separate actives from inactives. In other words, the less evidence we have that a pharmacophore separates actives from inactives the less weight we assign to it in the similarity measure.

When using too strict confidence interval, the pharmacophore model will contain only few or even no pharmacophores at all. Therefore we decided to introduce a modification of the purely pharmacophore-based screening. It consists in combining the s_{pp} score with the MAX-rule data fusion strategy [9,10] commonly employed in LBVS when information about multiple bioactive ligands is present:

$$s_{pp-lb}(m_{pp}, m_{db}) = s_{pp} \times \max_{a \in A}(s_{lb}(m_{db}, a)) \tag{3}$$

s_{pp-lb} is an arbitrary similarity method (we use MACCS fingerprints [17] and Tanimoto coefficient [26] in our experiments), A is the set of known active compounds and the maximum function implements the MAX-rule strategy.

5   Experimental Evaluation

For experimental evaluation of our method we focused on choosing public and well established data sets. The choice of established data sets makes it easier to compare our method to others.

We decided to use the Maximum Unbiased Validation (MUV) data sets [22]. The MUV methodology was designed to select an unbiased subset of active and inactive ligands from a biochemical assay for testing ligand based virtual screening methods. Each MUV data set contains a spatially random distribution of actives and decoys in a simple descriptor space. The term simple descriptors is defined in [22] as "a vectorized form of the respective counts of all atoms, heavy atoms, boron, bromine, carbon, chlorine, fluorine, iodine, nitrogen, oxygen, phosphorus and sulfur atoms in each molecule as well as the number of H-bond acceptors, H-bond donors, the logP, the number of chiral centers, and the number of the ring system". Each data set in MUV contains exactly 30 known active and 15000 inactive molecules (decoys).

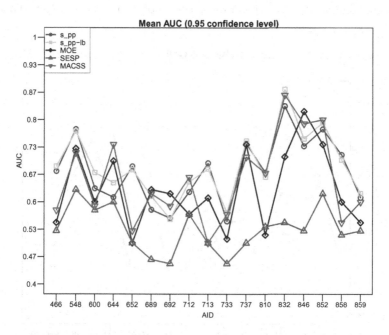

Fig. 1. The comparison of the 2D pharmacophore-based VS with other VS methods. The 95% confidence interval was used in the model building process. See section 4 for the description of our s_{pp} and s_{pp-lb} methods.

In [22], 17 MUV datasets were used to test the quality of several LBVS approaches. Specifically, MOE molecular properties descriptors [12], MACCS structural keys [17] and SESP, a class of versatile, alignment independent 2D topological indices based on atom pairs [2], were compared. The performance of the approaches was tested using the MAX-rule data fusion technique [9,10]. For each dataset, 10 active ligands were used as known actives to prioritize a database pooled from the the 15,000 decoy and remaining 20 active molecules. Then, the LBVS was performed, the ROC curve was built based on the positions of the 20 active molecules in the result set. Finally, the AUC (area under the curve) was computed. This process was repeated 100 times with different, randomly chosen, set of 10 actives, the AUCs were averaged and reported as the aggregated quality measure of a method.

To compare our approach to the above mentioned methods, we used a very similar strategy. We used the 10 active compounds as learning examples to build the pharmacophore model and used it with the s_{pp} similarity measure to sort the database and get the AUC value. This process was repeated 100 times. However, unlike the methods above, to build the pharmacophore model we need not only the known active compounds, but also known inactive compounds. Since we used one third (10 compounds) of the dataset for learning we used the same proportion for the decoys. Hence we used 10 active and 5000 inactive compounds for learning and the test set therefore consisted of 20 actives and 10,000 decoys.

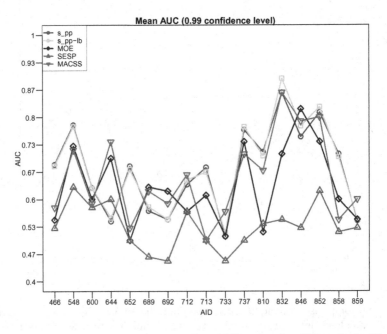

Fig. 2. Comparison of 2D pharmacophore-based VS with other VS methods. The 99% confidence interval was used in the model building process. See section 4 for the description of our s_{pp} and s_{pp-lb} methods.

5.1   2D Pharmacophore-Based Virtual Screening

The pharmacophore model building process is the core component influencing the performance of the virtual screening. Its crucial part is the identification of pharamcophores which constitute the resulting model. The identification process is based on the statistical test which can be influenced mainly by the the confidence level used. Higher confidence level results in less pharmacophores identified, but these have higher chance of being the real discriminators. We present results of our method for 95% (Fig. 1) and 99% (Fig. 2) confidence intervals. In both cases the method on most datasets outperforms all the other tested methods. It seems that our method outperforms the other approaches irrespective of whether we used the purely pharmacophore-based approach (s_{pp}) or the MAX-rule data fusion strategy modification (s_{pp-lb}). Specifically, on the 95% confidence level our method scores best on 11 out of the 17 datasets while for the 99% confidence level we score best in 10 cases. From the simple counts it might seem that higher confidence level does not bring any advantage. But closer inspection reveals that when the AUC is high enough a higher confidence level leads to higher AUCs and vice versa.

Table 1. The number of pharamcophores (PP count) in a pharmacophore model and the respective AUC for each dataset for different confidence levels. The gray rows show datasets with very low number of pharmacophores in the model.

	95%		99%	
AID	PP count	Mean AUC	PP count	Mean AUC
466	14.7	0.67	5.1	0.68
548	43.6	0.78	22.7	0.78
600	15.3	0.63	1.9	0.63
644	11.5	0.61	1.3	0.55
652	29.3	0.69	11.5	0.68
689	21.4	0.58	3.9	0.57
692	7.3	0.56	1.3	0.55
712	27.4	0.62	8.4	0.64
713	18.3	0.69	3.1	0.68
734	3.6	0.55	0.5	0.51
737	39.0	0.74	14.8	0.77
810	22.4	0.67	7.2	0.71
832	46.8	0.83	30.0	0.86
846	20.7	0.74	6.8	0.75
852	51.8	0.78	25.9	0.81
858	38.5	0.71	14.5	0.71
859	9.4	0.61	2.3	0.55

5.2   Relation of AUC and Size of the Pharmacophore Model

The extremes in AUC values for the higher confidence level are caused by the number of pharamcophores in the model. If the confidence level is high, it migh happen that not many pharamcophores will be considered as discriminative and thus the resulting pharmacophore model does not have enough power. To support this claim we compiled Tab. 1 which shows the average number of pharmacophores in the model and the respective AUC value. We can notice that higher confidence levels correspond to lower smaller models (the PP count columns). The gray lines highlight datasets where the model building process yielded an extremely low number of pharamcophores. These lines apparently correspond to datasets where the 99% confidence level performs considerably worse than the 95% level. The relation between the number of pharmacophores in a model and the AUC can be even more clearly observed in Fig. 3 where we brought together results for the two confidence levels. We can see that the lower the AUC is, the more the 95% confidence level outperforms the 99% level and vice versa.

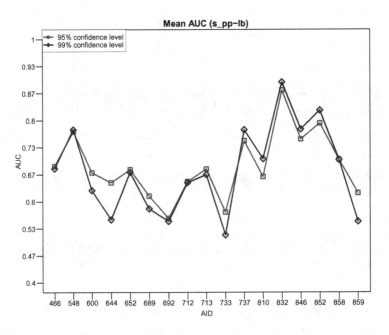

Fig. 3. Comparison of different confidence levels for pharmacophore-based screening with MAX-rule data fusion strategy. See section 4 for the description of the s_{pp-lb} method.

6   Conclusion

In this paper, we have proposed a novel method for pharmacophore model building based on the topological information and the information about active and inactive molecules with respect to a target of biological interest. We showed its value in virtual screening experiments where it outperformed other ligand-based screening methods.

In the future, we would like to perform an extensive experimental evaluation of the method to find its optimal (dataset-dependent) parametrization. Most importantly we would like to do an extensive comparison with other ligand-based virtual screening approaches and 3D pharmacophore models. Finally, we plan to develop a GUI application providing a user-friendly access to the presented method.

Acknowledgments. This work was supported by the Czech Science Foundation (GAČR) [projects Nr. 14-29032P and P202/11/0968], by Grant Agency of Charles University [project Nr. 154613] and by project SVV-2014-260100.

References

1. RDKit, http://www.rdkit.org/ (accessed: April 15, 2014)
2. Baumann, K.: An alignment-independent versatile structure descriptor for qsar and qspr based on the distribution of molecular features. Journal of Chemical Information and Computer Sciences 42(1), 26–35 (2002)
3. Cheng, T., Li, Q., Zhou, Z., Wang, Y., Bryant, S.H.: Structure-based virtual screening for drug discovery: a problem-centric review. AAPS J. 14(1), 133–141 (2012)
4. Ehrlich, P.: Über den jetzigen Stand der Chemotherapie. Verlag der Chemiker-Zeitung Otto v. Halem (1908)
5. Ferreira, R.S., Simeonov, A., Jadhav, A., Eidam, O., Mott, B.T., Keiser, M.J., McKerrow, J.H., Maloney, D.J., Irwin, J.J., Shoichet, B.K.: Complementarity between a docking and a high-throughput screen in discovering new cruzain inhibitors. J. Med. Chem. 53(13), 4891–4905 (2010)
6. Fisher, R.A.: On the Interpretation of 2 from Contingency Tables, and the Calculation of P. Journal of the Royal Statistical Society 85(1), 87–94 (1922)
7. Gaulton, A., Bellis, L.J., Bento, A.P., Chambers, J., Davies, M., Hersey, A., Light, Y., McGlinchey, S., Michalovich, D., Al-Lazikani, B., Overington, J.P.: ChEMBL: a large-scale bioactivity database for drug discovery. Nucleic Acids Research 40(D1), D1100–D1107 (2011)
8. Heikamp, K., Bajorath, J.: The future of virtual compound screening. Chem. Biol. Drug Des. 81(1), 33–40 (2013)
9. Hert, J., Willett, P., Wilton, D.J., Acklin, P., Azzaoui, K., Jacoby, E., Schuffenhauer, A.: Comparison of fingerprint-based methods for virtual screening using multiple bioactive reference structures. Journal of Chemical Information and Computer Sciences 44(3), 1177–1185 (2004)
10. Hert, J., Willett, P., Wilton, D.J., Acklin, P., Azzaoui, K., Jacoby, E., Schuffenhauer, A.: Comparison of topological descriptors for similarity-based virtual screening using multiple bioactive reference structures. Org. Biomol. Chem. 2, 3256–3266 (2004)
11. Horvath, D.: Pharmacophore-based virtual screening. Methods Mol. Biol. 672, 261–298 (2011)
12. Chemical Computing Group Inc.: Molecular Operating Environment (MOE), 2013.08 (2013)

13. Irwin, J.J., Sterling, T., Mysinger, M.M., Bolstad, E.S., Coleman, R.G.: ZINC: A Free Tool to Discover Chemistry for Biology. J. Chem. Inf. Model. 52(7), 1757–1768 (2012)

14. Johnson, M.A., Maggiora, G.M.: Concepts and Applications of Molecular Similarity. Wiley-Interscience (1990)

15. Jorgensen, W.L.: Rusting of the lock and key model for protein-ligand binding. Science 254(5034), 954–955 (1991)

16. Liao, C., Sitzmann, M., Pugliese, A., Nicklaus, M.C.: Software and resources for computational medicinal chemistry. Future Med. Chem. 3(8), 1057–1085 (2011)

17. McGregor, M.J., Pallai, P.V.: Clustering of large databases of compounds: using the mdl keys as structural descriptors. Journal of Chemical Information and Computer Sciences 37(3), 443–448 (1997)

18. R Development Core Team. R: A Language and Environment for Statistical Computing. R Foundation for Statistical Computing, Vienna, Austria (2008) ISBN 3-900051-07-0

19. Ripphausen, P., Nisius, B., Bajorath, J.: State-of-the-art in ligand-based virtual screening. Drug Discov. Today 16(9-10), 372–376 (2011)

20. Ripphausen, P., Nisius, B., Peltason, L., Bajorath, J.: Quo Vadis, Virtual Screening? A Comprehensive Survey of Prospective Applications. Journal of Medicinal Chemistry 53(24), 8461–8467 (2010)

21. Ripphausen, P., Stumpfe, D., Bajorath, J.: Analysis of structure-based virtual screening studies and characterization of identified active compounds. Future Med. Chem. 4(5), 603–613 (2012)

22. Rohrer, S.G., Baumann, K.: Maximum unbiased validation (muv) data sets for virtual screening based on pubchem bioactivity data. Journal of Chemical Information and Modeling 49(2), 169–184 (2009)

23. Schneider, G., Neidhart, W., Giller, T., Schmid, G.: Scaffold-hopping by topological pharmacophore search: A contribution to virtual screening. Angewandte Chemie International Edition 38(19), 2894–2896 (1999)

24. Sun, H.: Pharmacophore-Based Virtual Screening. Current Medicinal Chemistry 15(10), 1018–1024 (2008)

25. Taboureau, O., Baell, J.B., Fernandez-Recio, J., Villoutreix, B.O.: Established and emerging trends in computational drug discovery in the structural genomics era. Chem. Biol. 19(1), 29–41 (2012)

26. Tanimoto, T.: IBM Internal Report 17th November (1957)

27. Vidler, L.R., Filippakopoulos, P., Fedorov, O., Picaud, S., Martin, S., Tomsett, M., Woodward, H., Brown, N., Knapp, S., Hoelder, S.: Discovery of novel small-molecule inhibitors of BRD4 using structure-based virtual screening. J. Med. Chem. 56(20), 8073–8088 (2013)

28. Wang, Y., Xiao, J., Suzek, T.O., Zhang, J., Wang, J., Bryant, S.H.: Pubchem: a public information system for analyzing bioactivities of small molecules. Nucleic Acids Research 37(Web-Server-Issue), 623–633 (2009)

Structure-Based Analysis of Protein Binding Pockets Using Von Neumann Entropy

Negin Forouzesh, Mohammad Reza Kazemi, and Ali Mohades

Laboratory of Algorithm and Computational Geometry, Department of Mathematics
and Computer Science, Amirkabir University of Technology, Tehran, Iran
{n.forouzesh,mr.kazemi,mohades}@aut.ac.ir

Abstract. Protein binding sites are regions where interactions between
a protein and ligand take place. Identification of binding sites is a functional issue especially in structure-based drug design. This paper aims
to present a novel feature of protein binding pockets based on the complexity of corresponding weighted Delaunay triangulation. The results
demonstrate that candidate binding pockets obtain less relative Von Neumann entropy which means more random scattering of voids inside them.

Keywords: Protein binding site, Delaunay triangulation, Von Neumann
entropy.

1 Introduction

Proteins mainly accomplish their biological activities in interaction with other
molecules. Meanwhile these interactions, only some of surface atoms of the protein get involved. Thus identifying these regions, binding sites, help scientists to
study the mechanism of interactions and protein performance very well. Moreover, identification of binding sites is known as the basis of structure-based drug
design [1,2].

In recent years, various computational methods have been introduced and
developed for the purpose of finding protein pockets which results in predicting protein binding sites. Generally these methods can be classified into two
types: energy-based and geometry-based methods. Geometry-based algorithms
themselves are classified to grid-based [3,4,5], sphere-based [6,7] and alpha-shape
based [8,9] types. Besides, consensus methods [10,11] have been proposed recently in which some previous pocket detecting methods are combined together
in order to improve the prediction success rate entirely.

Usually many pockets are found for each protein and definitely not all of
them can be regarded as real binding sites. Therefore, it is necessary to evaluate
those pockets according to a ranking method and report the top-rated cases.
Pocket size, the number of atoms forming the pocket, is a widely used ranking
criterion; however, past studies [12,4] elucidated that some real binding sites
are disregarded when protein pockets are evaluated only by their size, especially
when top ranked candidates have tiny diversity in size.

M. Basu, Y. Pan, and J. Wang (Eds.): ISBRA 2014, LNBI 8492, pp. 301–309, 2014.

Comprehensive studies have been performed recently to extract novel features of protein binding pockets beside their size to improve the past prediction results. In [4] the degree of the conservation of involved surface residues in pockets was studied to report TOP1 pocket among TOP3 largest cases. In addition to pocket size, distance from the protein centroid, sequence conservation and the number of hydrophobic residues are chosen as the ranking criteria both in combination with each other and solely in [13].

The most substantial issue is that although shape of pockets and related features have been examined in previous works, up to our knowledge, the arrangement of atoms consisting the protein pockets have not been considerably discussed before. The main contribution of this paper is examining the complexity of protein binding pockets in comparison with other pockets. This leads to achieve a novel feature of binding sites which finally help us to predict them. To accomplish that, in this study we make use of weighted Delaunay triangulation of protein atoms which results in a geometric graph sensitive to the location of each atom. Afterward, the complexity of those graphs are analyzed by a useful complexity measure, Von Neumann entropy. Results show that candidate binding pockets usually obtain less relative Von Neumann entropy and consequently more disorderliness in their structure.

2   Preliminaries

In this section, basic concepts that are necessary for next part are introduced. First, some computational geometry tools both for bare and weighted points are reviewed. Secondly, matrix representation of graphs and related definitions in linear algebra are discussed.

Given a set of finite points $P = \{p_1, p_2, \ldots, p_n\}$ in the space, called sites, the Voronoi diagram is the set of cells, V_i, $1 \leq i \leq n$, defined by:

$$V_i = \{p|\ |p - p_i| \leq |p - p_j|\ ,\ 1 \leq j\ \leq n\}.$$

In other word, Voronoi diagram is a subdivision of the space into n cells. Each cell in this diagram corresponds to a site in P, under the condition that all points in cell V_i are closer to their corresponding site p_i rather than any other sites. If the sites lie in general position meaning that no three sites on a line, no four sites on a circle and no five sites on a sphere, then the dual graph of the Voronoi diagram results in a unique geometric graph called Delaunay triangulation in which sites are considered as vertices and edges are drawn between any two vertices whose corresponding cells are adjacent.

Let $P^w = \{p_1^w, p_2^w, \ldots, p_n^w\}$ be a set of weighted points where point p_i^w can be denoted as a spherical ball $b_i = b(z_i, r_i)$ with center $z_i \in \mathbb{R}^3$ and radius r_i . The distance of a point $x \in \mathbb{R}^3$ and a ball $b = b(z, r)$ is formulated as:

$$\pi_b(x) = |z - x|^2 - r^2.$$

Now the weighted Voronoi diagram (Power diagram) is defined by:

$$V_{b_i}(p) = \{p \in \mathbb{R}^3 \mid \pi_{b_i}(p) \leq \pi_{b_j}(p),\ 1 \leq j\ \leq n\}.$$

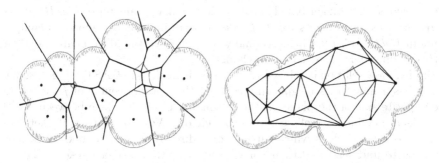

Fig. 1. Left: Weighted Voronoi diagram for a set of balls. Right: Weighted Delaunay triangulation for the same set of balls.

Such as the previous case, weighted Delaunay triangulation (Regular triangulation) can be obtained by the dual shape of weighted Voronoi diagram (see Fig 1).

Let $G(E, V)$ be a simple graph with n vertices and m edges. Adjacency matrix $A(G)$ is an $n \times n$ matrix in which $A_{uv} = A_{vu} = 1$ if two vertices u and v are adjacent and $A_{uv} = A_{uv} = 0$ otherwise. Degree matrix $D(G)$ is an $n \times n$ matrix in which diagonal element D_{uu} resembles the degree of vertex u and other elements are zero. Laplacian matrix $L(G)$ is defined as:

$$L(G) = D(G) - A(G).$$

Density matrix (G) is a normalized form of Laplacian matrix and is defined by the following relation:

$$\rho(G) = \frac{1}{tr[L(G)]} L(G)$$

where $tr[L(G)]$ is the sum of elements on the main diagonal of matrix $L(G)$. The trace of Laplacian matrix for each graph is equaled to the sum of all the vertices degree. Thus the previous relation is equaled to the following relation:

$$\rho(G) = \frac{1}{2m} L(G).$$

3   Methods and Materials

3.1   Protein Pocket Structure

We used CASTp [14] to detect protein pockets. In this method, protein atoms are modeled as spherical balls (weighted points). Weighted Voronoi diagram is computed for this set based on the concepts explained in preliminaries. Next, $ResB$ is defined as $ResB = \{V_b \cap b | b \in B\}$ (Fig 2.left). Like Delaunay triangulation, by connecting the centers of neighboring regions in $ResB$ a graph called $CpxB$ is acquired (Fig 2.middle). Obviously $CpxB \subseteq DelB$. Afterwards, spherical balls simultaneously get bigger based on the variation of a parameter α. $CpxB_\alpha$ grows as α increases until it gets to $DelB$. Pockets are informally defined as components in $DelB - CpxB$ which become voids before getting disappeared as their corresponding balls grow based on changes in α.

Briefly, Flow relation is utilized to find protein pockets in CASTp (Fig 2.right). Cell ρ has a flow to its neighboring cell σ if the Voronoi center of ρ locates in the opposite side of the plane passing through the common face between ρ and σ. Sinks are defined as cells containing their own Voronoi centers. Pockets are defined as a set of cells which directly or indirectly flow to a sink. Thus it is sufficient to find sinks and their corresponding flows to detect pockets. Fig 2 is adopted from [8].

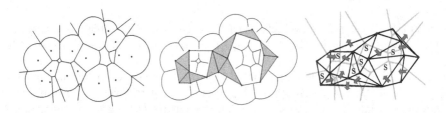

Fig. 2. Left:*Res B*, Middle: *Cpx B*, Right: Flow Relation (sinks are shown by S)

For each protein pocket we consider its graph consisting of vertices and edges corresponding to the centers of balls and Delaunay edges respectively. Then the adjacency matrix of these graphs is easily computed according to the definitions in preliminaries section.

3.2   Von Neumann Entropy

Information theoretical methods are common to compare complexity of networks. Shannon entropy is one of the representative network complexities which is defined by:

$$H\left(X\right) = H\left(p_1,\ p_2,\ \ldots,\ p_n\right) = -\sum_i p_i \log p_i$$

where X is a random variable with probability distribution $p_1,\ p_2,\ \ldots,\ p_n$. Entropy is widely used in various spheres, such as biology and chemistry, to measure complexity of graphs [15,16]. One approach to accomplish that is partitioning graph vertices to some classes $\{X_i\}$ such that the ratio of partitions size to graph vertices, $\frac{|X_i|}{|X|}$, results in a probability distribution. Shannon entropy of graphs is calculated for that probability distribution by:

$$H\left(G\right) = -\sum_i \frac{|X_i|}{|X|} \log\left(\frac{|X_i|}{|X|}\right).$$

However, there is no unique form to partition graphs and consequently different entropic measures can be assigned to them. Von Neumann entropy which is calculated by eigenvalues of a graph Laplacian matrix is a useful complexity measure while studying graphs [17]. Formally, Von Neumann entropy for any density matrix ρ is defined by:

$$S\left(\rho\right) = -tr\left(\rho \log \rho\right) = -\sum_i \lambda_i log \lambda_i.$$

Density matrix can be computed for graphs by the definition provided in preliminaries. Based on [17], Von Neumann entropy for an arbitrary graph G increases as edges of the graph scatter more randomly. In a study [17] of four different kinds of graphs with maximum twenty vertices, the smallest entropy is obtained for complete graph then it increases for star, random and perfect matching respectively. By Von Neumann entropy the complexity of graphs can be computed directly from their structures.

While examining the complexity of pockets, it is important to ignore any other parameters, size of corresponding graphs for instance. On the other hand, Von Neumann entropy of graphs increases as they grow in size and this also happens when studying different pockets with different sizes. To avoid this problem, relative Von Neumann entropy is utilized instead of its primary form as:

$$S\left(\rho \parallel \sigma\right) = \sum_i p_i \left(\log p_i - \sum_j P_{ij} \log q_j \right)$$

in which $\{p_i\}$ and $\{q_j\}$ are the eigenvalues of matrices ρ and σ and $P_{ij} = \langle X_i, Y_j \rangle^2$ where $\{X_i\}$ and $\{Y_j\}$ are the eigenvectors of matrices ρ and σ respectively. The corresponding graph entropy relative to a matrix with same dimension n and maximum entropy, $\frac{1}{n}I_n$, is chosen to examine the pocket complexity independent from its size:

$$S\left(\rho \parallel \frac{1}{n}I_n\right) = \log n - S\left(\rho\right).$$

Obviously when the graph entropy gets closer to the entropy of $\frac{1}{n}I_n$, the relative entropy decreases. Therefore, less relative Von Neumann entropy means more complexity in the corresponding graph of pocket and consequently more disorderliness in its structure.

To demonstrate that relative Von Neumann entropy does not change by pocket size variation, we use the following result from [18] for almost every graph G,

$$S\left(G\right) = \left(1 + o\left(1\right)\right) \log n.$$

This consequently results in:

$$S\left(G \parallel \frac{1}{n}I_n\right) = o\left(1\right) \log n.$$

By the definition as n grows, $o(1)$ approaches zero. Thus, the grow rate of relative Von Neumann entropy is almost less than the logarithm of its size, which grows slightly.

3.3   Test Dataset

A dataset of 48 bound/unbound structures first introduced in [4] is used to have a comprehensive test over both ligand-bound and unbound structures. A widely used method to check whether a pocket is the real binding pocket, is to measure if its geometry center is within $4A°$ of the ligand atoms. Whenever more than one pocket meet the condition, the one which is closer to the ligand is reported. This method was firstly used in [4].

4   Results and Discussion

We used CASTp website [1] to detect protein pockets. Although it provides comprehensive information about pocket size, atoms and mouths, it does not give simplices of each pocket. Therefore, to have Delaunay edges of pockets more than their vertices reported by CASTp, we constructed those graphs by the use of Computational Geometry Algorithms Library (CGAL)[2] which provides access to efficient geometric algorithms in the form of C++ library.

It is worth mentioning that weighted Delaunay triangulation is utilized as the basic graph according to its two convenient features for the purpose of predicting binding sites. First, since we want to examine the arrangement of points and its effect on forming binding sites, it is necessary to select a geometric graph. Delaunay triangulation is a geometric graph and is sensitive to the location of vertices. Second, there is an appropriate relation between Delaunay triangulation and void spheres in pockets. More precisely, every four vertices form a tetrahedron in Delaunay triangulation if the sphere passing through them is empty of any other points. Furthermore, these voids correspond to the regions in which the ligand atoms probably stand. Therefore, it is worthwhile to examine the distribution of them among protein surface when we want to predict the binding pockets. Indeed, Von Neumann entropy of Delaunay triangulation represents the distribution of voids on pocket surface. In a more comprehensive study, it is better to make use of dual complex of each pocket instead of its Delaunay triangulation to achieve a more accurate graph for each pocket.

Although we apply relative version of Von Neumann to ignore the effects of size, there are some small pockets with high entropy which are not eligible for being binding sites regarding their tiny available surface to interact with ligands. Figure 3 illustrates such examples.

To avoid taking those undesirable pockets to account as binding pockets, we narrowed down the list of pockets to 10 largest pockets of each sample in 48 bound/unbound dataset. The results are shown in Table 1.

---

[1]   http://cast.engr.uic.edu
[2]   http://www.cgal.org

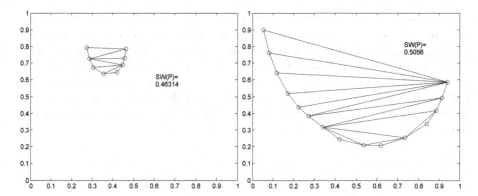

Fig. 3. Left: Extremely small pocket that has small relative Von Neumann entropy. Right: A more desired pocket.

Table 1. Prediction success rate of size and relative Von Neumann entropy

Ranking feature	Bound		Unbound	
	TOP1	TOP3	TOP1	TOP3
Pocket Size	67%	83%	58%	75%
Relative Von Neumann entropy	49%	77%	38%	83%

Table 2 shows the prediction success rate presented by different ranking methods adopted from [13] which is implemented on a same dataset. This is one of the recent studies about examining pockets properties and makes use of [19] for detecting protein pockets.

Table 2. Prediction success rate of different ranking features adopted from [13]

Methods	Unbound/Bound	
	TOP1	TOP3
Conservation score	57%	72%
Distance	56%	70%
Volume	44%	59%
Hidrophobic residues	30%	48%

We remind that pocket size is still the most successful feature to find binding pockets. Exploring novel features, preferably independent from size, can improve previous results. Pockets complexity measured by relative Von Neumann entropy can predict TOP3 pockets very well especially for unbound samples where the success rate even precedes previous results. In further studies, a hybrid criterion consisting of both size and entropy can be investigated.

Based on [17], Von Neumann entropy increases as graph edges scatter more randomly. Hence, for fixed number of edges, perfect matching and complete graph get maximum and minimum amount respectively. In particular, we have found that according to Table 1 in candidate binding pockets the edges of Delaunay triangulation and equivalently voids are scattered more uniformly than other pockets. In figure 4, two pockets with different edge distributions are shown. According to results, binding pockets are more likely to have a shape similar to left side figure rather than right side one.

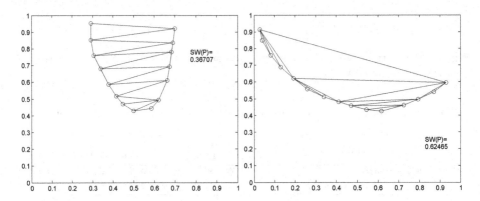

Fig. 4. Left:A pocket with uniform distribution of edges Right: A pocket with irregular distribution of edges

5   Conclusion

In this study we examined a novel feature in protein binding pockets based on the complexity of corresponding geometric graphs. CASTp was utilized for detecting pockets and weighted Delaunay triangulation was considered as pockets graphs. It was illustrated that binding pockets usually acquire less relative Von Neumann entropy which means more regular distribution of Delaunay edges and consequently uniform scattering of voids. Regarding small dependency of relative Von Neumann to the size of graphs, one can merge them in future studies to propose a comprehensive scoring criterion.

Acknowledgments. We would like to thank Dr. Fatemeh Zare-Mirakabad for her advice and assistance in analyzing biological aspects of this study. Besides, we appreciate Ms. Kamelia Jamaati for her valuable technical support in implementing C++ code.

References

1. Seco, J., Luque, J., Barril, X.: Binding Site Detection and Druggability Index from First Principles. Journal of Medicinal Chemistry 52, 2363–2371 (2009)
2. Pérot, S., Sperandio, O., Miteva, M.A., Camproux, A., Villoutreix, B.O.: Druggable pockets and binding site centric chemical space: a paradigm shift in drug discovery. Drug Discovery Today 15, 656–667 (2010)
3. Hendlich, M., Rippmann, F., Barnickel, G.: LIGSITE: Automatic and efficient detection of potential small molecule-binding sites in proteins. J. Mol. Graph. Model. 15, 359–363 (1997)
4. Huang, B., Schroeder, M.: LIGSITEcsc: predicting ligand binding sites using the Connolly surface and degree of conservation. BMC Structural Biology 6, 19–29 (2006)
5. Weisel, M., Proschak, E., Schneider, G.: PocketPicker: analysis of ligand binding-sites with shape descriptors. Chemistry Central Journal 1 (2007)
6. Laskowsk, R.A.: SURFNET: a program for visualizing molecular surfaces, cavities, and intermolecular interactions. J. Mol. Graph. 13, 323–330, 307–308 (1995)
7. Brady, G.P., Stouten, P.F.: Fast prediction and visualization of protein binding pockets with PASS. J. Comput. Aided Mol. Des. 14, 383–401 (2000)
8. Edelsbrunner, H., Facello, M., Liang, J.: On the definition and the construction of pockets in macromolecules. Descrete Applied Mathematics 88, 83–102 (1998)
9. Le Guilloux, V., Schmidtke, P., Tuffery, P.: Fpocket: an open source platform for ligand pocket detection. BMC Bioinformatics 10 (2009)
10. Haung, B.: MetaPocket: a meta approach to improve protein ligand binding site prediction. OMICS 13, 325–330 (2009)
11. Zhang, Z., Li, Y., Lin, B., Schroeder, M., Huang, B.: Identification of cavities on protein surface using multiple computational approaches for drug binding site prediction. Bioinformatics 27, 2083–2088 (2011)
12. Laskowski, R.A., Luscombe, N.M., Swindless, M.B., Thornton, J.M.: Protein clefts in molecular recognition and function. Protein Science 5, 2438–2452 (1996)
13. Gao, J., Liu, Q., Kang, H., Cao, Z., Zhu, R.: Comparison of Different Ranking Methods in Protein-Ligand Binding Site Prediction. International Journal of Molecular Science 13, 8752–8761 (2012)
14. Dundas, J., Ouyang, Z., Tseng, J., Binkowski, A., Turpaz, Y., Liang, J.: CASTp: computed atlas of surface topography of proteins with structural and topographical mapping of functionally annotated residues. Nucleic Acids Res. 34, W116–W118 (2006)
15. Dehmer, M., Barbarini, N., Varmuza, K., Graber, A.: A large scale analysis of information-theoretic network complexity measures using chemical structures. PLoS One 4, e8057 (2009)
16. Dehmer, M., Mowshowitz, A.: A history of graph entropy measures. Information Science 181, 57–78 (2011)
17. Passerini, F., Severini, S.: Quantifying complexity in networks: The Von Neumann entropy. IJATS 4, 58–67 (2009)
18. Du, W., Li, X., Li, Y., Severini, S.: A note on the von Neumann entropy of random graphs. Linear Algebra and its Application (2010)
19. Dai, T., Liu, Q., Gao, J., Cao, Z., Zhu, R.: A new protein-ligand binding sites prediction method based on the integration of protein sequence conservation information. BMC Bioinformatics 12(suppl. 14), S9 (2011)

A New Mathematical Model for Inbreeding Depression in Large Populations

Shuhao Sun, Fima Klebaner, and Tianhai Tian

School of Mathematical Sciences, Monash University
Melbourne VIC 3800 Australias
{Shuhao.Sun,Fima.Klebaner,Tianhai.Tian}@monash.edu

Abstract. It has been widely recognized that inbreeding mating results in increased homozygosity which generally leads to a decreased fitness of population. This conclusion was supported by a large number of experimental observations in natural populations. However, a theoretical analysis of this phenomenon is still lacking. Here we present a theoretic proof showing that for most natural populations, inbreeding mating does reduce the mean fitness of populations. It also suggests that inbreeding depression depends on not only the mating system but also the structure of population. As a consequence, we conclude that, for a natural inbreeding population without any inbreeding depression, most genotypes should be additive or co-dominant. This result gives an explanation to the question why hermaphroditism populations do not show severe inbreeding depression. Another major result of this research is that, for a large inbreeding population with directional relative genotype fitnesses, the mean fitness increases monotonically for any value of inbreeding coefficient. This result may provide a reason to explain the frequent occurrence of self-fertilization populations. We also characterize pseudo-overdominance for single locus, which suggests that there are many pseudo-overdominance populations among the class of overdominance populations.

Keywords: Inbreeding, Mathematical model, Evolution, Fitness.

1 Introduction

Inbreeding is reproduction from the mating of parents who are closely related genetically. It has been widely accepted that inbreeding results in increased homozygosity and increases the chances of offspring being affected by recessive or deleterious traits. Inbreeding generally leads to a decreased fitness of the population [3], [4], [6], [12], and thus it is usually called inbreeding depression. An individual who results from inbreeding is referred to as an inbred. The more closely related breeding pair is, the more homozygous deleterious genes the offspring may have. For alleles that confer an advantage in the heterozygosity and/or homozygous-dominant state, the fitness of the homozygous-recessive state may even be zero, which means meaning sterile or unviable offspring. Another mechanism responsible for inbreeding depression is overdominance of heterozygous

M. Basu, Y. Pan, and J. Wang (Eds.): ISBRA 2014, LNBI 8492, pp. 310–321, 2014.

alleles, which can lead to reduced fitness of a population with many homozygous genotypes, even if they are not deleterious. Here even the dominant alleles result in reduced fitness if present homozygously. Thus there are two distinct ways in which increased homozygosity can lower fitness: namely increasing homozygosity for partially recessive detrimental mutations and increasing homozygosity for alleles at loci with heterozygote advantage (overdominance). Deleterious alleles will generally be present in populations at low frequencies (mutation-selection balance), whereas overdominant alleles at a locus are maintained at intermediate frequencies by balancing selection. Currently it is still not known which one of these two mechanisms is more prevalent in nature [11], [14], [19], [20], [25].

For practical applications such as livestock breeding, the former mechanism may be more significant – it may yield completely unviable offspring. However, the latter can only lead to relatively reduced fitness. The partial dominance theory was generally considered to be the major cause of observed inbreeding depression [1], [5], [6], [10], but there are evidences that also support the overdominance hypothesis [9], [18], [20]. It was believed that purging was only achieved when inbreeding depression is caused by deleterious recessive. A fitness rebound in inbred populations provides support to the hypothesis of partial dominance mechanism [21]. Purging effects have been confirmed experimentally in a number of cases [1], [23]. But overall, the evidence for purging in plant and animal populations is limited and this has led to a question for the role of purging in restoring fitness. In this paper, we will present an explicit formula for calculating the purging rate.

According to Wright's formula [16], [26], at one autosomal locus with two alleles A and a with frequency x and y $(x + y = 1)$, the three diploid genotypes AA, Aa and aa have frequency $x^2 + fxy, 2xy - fxy, y^2 + fxy$, respectively, where f is the inbreeding coefficient. It was accepted that this formula gives a good approximation to the observed phenomena in nature. For example, for the harmful recessive condition albinism in humans, the observed affected individuals is about $y^2 = 1/20000$; leading to a recessive allele frequency $y = 1/141$. Using the first cousin mating $(f = 1/16)$ by Wright's formula, the expected frequency of affected individuals is $y^2 + fy(1 - y) = 1/2000$ which is about 10-fold higher than that in the total population. A statistic report on albinism population in both Japan and Europe verified this prediction [2]. In Japan, 56% in the offsprings of the first cousin marriages took up to 5% of the total population; while 20% in the offsprings of the first cousin marriages in Europe took up to 2% of the total population.

Based on Wright's formula above, we assume that the relative genotype fitnesses are $1, 1 - hs$ and $1 - s$, where s is the selection coefficient, and h the degree of dominance. In this work, we will prove that, for a large inbreeding mating population, the mean fitness increases monotonically for any inbreeding population with directional selection (i.e., $0 \leq h \leq 1$). However, it is not true for overdominant populations. For multiple loci, we will prove that inbreeding mating reduces the mean fitness of the population with the mixed selection types for many nature populations by assuming multiplicative interaction of genes.

Furthermore, our research shows that inbreeding depression depends on both inbreeding coefficient and structure of populations. As an application, we present an additional reason for the frequent occurrence of self-fertilization populations which loss 50% heterozygous each generation because most genotypes of these organisms are additive.

2   Theory of Inbreeding at a Single Locus

Inbreeding results in increased homozygosity of alleles that are identical by descent (IBD). Wright defined the inbreeding coefficient f as the probability that two homologous alleles in an individual are IBD [26], [16], [13]. Let x represent the frequency of allele A. The frequency and fitnesses of AA, Aa and aa are

$$x^2 + fx(1-x), \quad 2x(1-x) - fx(1-x), \quad (1-x)^2 + fx(1-x)$$

and W_{11}, W_{12}, W_{22}, respectively. Then the change of frequency is given by

$$\Delta x = x' - x$$
$$= \frac{x(1-x)}{\bar{w}}[(1-f)(w_{11} + w_{22} - 2w_{12})x + f(w_{11} - w_{12}) + w_{12} - w_{22}]$$

where x' is the frequency of A in the next generation, and the mean fitness [13] is defined by

$$\bar{w} = w_{11}x^2 + 2w_{12}x(1-x) + w_{22}(1-x)^2 + fx(1-x)(w_{11} + w_{22} - 2w_{12}).$$

To calculate the difference Δw between the mean fitnesses of two successive generations, we write it as

$$J = (1-f)(w_{11} + w_{22} - 2w_{12})x + f(w_{11} - w_{12}) + w_{12} - w_{22}$$

and w' for the mean fitness of next generation. Thus we have

$$\Delta x = x' - x = x(1-x)\frac{J}{\bar{w}},$$
$$\Delta w = \bar{w}' - \bar{w}$$
$$= \Delta x[(w_{11} - 2w_{12} + w_{22})(x' + x) + 2w_{12} - 2w_{22}$$
$$+ f(w_{11} - 2w_{12} + w_{22})(1 - x' - x)]$$
$$= \Delta x[J + (1-f)(w_{11} + w_{22} - 2w_{12})x' - f(w_{12} - w_{22}) + w_{12} - w_{22}]$$
$$= \frac{x(1-x)J^2}{\bar{w}^2}[2\bar{w} + (1-f)(w_{11} + w_{22} - 2w_{12})x(1-x) - \bar{w}f(w_{11} - w_{22})/J]$$
$$= \frac{x(1-x)J^2}{\bar{w}^2}[w_{11}x^2 + w_{22}(1-x)^2 + x(1-x)(w_{11} + w_{22})$$
$$+ \bar{w}(1 - f(w_{11} - w_{22})/J)].$$

Finally, the difference between the mean fitnesses of two successive generations is given by

$$\Delta w = \frac{x(1-x)J^2}{\bar{w}^2}\left[w_{11}x + w_{22}(1-x) + \bar{w}\left(1 - \frac{f(w_{11} - w_{22})}{J}\right)\right] \qquad (1)$$

2.1   Monotonical Increase of Mean Fitness

From equation (1), it is clear that a sufficient condition for $\Delta w > 0$ is

$$1 - \frac{f(w_{11} - w_{22})}{J} \geq 0.$$

Thus, if $f = 0$ (i.e., random mating), the mean fitness increases monotonically, which is a well-known result [13]. We also conclude that for an inbreeding population with $w_{11} = w_{22}$ and $f \neq 1$, the mean fitness increases monotonically.

The following result has potential application to self-fertilization plants. The proof of this theorem will be provided in the expanded version of this paper.

Theorem 2.1 For an inbreeding population of which the relative genotype fitness are $1, 1 - hs$ and $1 - s$ with $0 \leq h \leq 1$, the mean fitness increases monotonically for any inbreeding coefficient $0 \leq f \leq 1$.

Clearly this result partially generalizes the well-known result that the mean fitness increases monotonically for a random-mating population with constant relative fitnessess [13]. This theorem can be used to explain the reason why self-fertilization (self-pollination) plants, the most extreme form of inbreeding, reduce heterozygosity by 50% each generation, but do not extinct. Note that the majority of genotypes of a self-fertilization plant is either partial dominance or partial recessive [3], [4], [5], [8], [13]. However, Theorem 2.1 can not be generalized to the case with $h < -\frac{f}{1-f}$ (overdominance) or the case with $h > 1$ (underdominance). We rewrite the fitness as

$$w_{11} = 1 - s, \quad w_{12} = 1, \quad w_{22} = 1 - t,$$

where $0 < s, t \leq 1$. Furthermore, if $f \neq 0$ and either $s \leq t \leq s/f$ or $t \leq s \leq t/f$ hold, we can show that $\Delta w(x)$ is not necessarily positive. In fact, in this case, we have

$$J = (1 - f)(-s - t)x + f(-s) + t.$$

Let

$$x^* = \frac{fs - t}{-(s + t)(1 - f)}.$$

Then $0 < x^* < 1$ holds. In fact, it is equivalent to $s \leq t \leq s/f$ or $t \leq s \leq t/f$. Thus we have the following counterexamples:

(1) **Pseudo-Overdominance:** If $f \leq 0.3$, the allele's fitnesses are $0.8, 1$ and 0, and x is sufficient close to unity, then $\Delta w(x) < 0$ holds.

(2) **Overdominance:** If $s > t \geq 0$ and $x^* - x$ is positive and sufficient small, then $\Delta w(x) < 0$. However, if $t > s \geq 0$ and $x - x^*$ is positive and sufficient small, then $\Delta w(x) < 0$.

(3) **Underdominance:** If $0 \geq s > t$ and $x^* - x$ is positive and sufficient small, then $\Delta w(x) < 0$.

If we consider the whole history of an allele A, whose frequency may change from very low (a mutant) to very common (fixation), the following quantity is useful to describe whether a population may be extinct. Note that according to definition, the mean fitness $\bar{w}$ of a population depends on the initial frequency. To emphasize the importance of initial frequency x, we re-write the mean fitness as $\bar{w}(x)$. Theorem 2.1 shows that the mean fitness increases monotonically for an inbreeding population with directional selection coefficients (i.e., $0 \leq h \leq 1$), which is equivalent to $\Delta w(x) > 0$ for $x \neq 0, 1$. We define the *average mean fitness* (AMF) of a population as the integral

$$\text{AMF} = \int_0^1 \bar{w}(x)dx.$$

Thus for an inbreeding population with $0 \leq h \leq 1$, by Theorem 2.1, AMF is positive since an integral of a non-negative and non-constant function is positive. It is interesting to point out that, although the mean fitness may not increases monotonically for an overdominance inbreeding population, we still can prove the following theorem.

Theorem 2.2 The average mean fitness of an overdominance inbreeding population is positive, that is, $\int_0^1 \Delta w(x) > 0$.

The proof will be provided in the expanded version of this paper. Since the mean fitness increases monotonically only for single locus, Theorem 2.2 may have further potential applications.

2.2   Inbreeding Depression

Inbreeding is expected to play an important but complicated role in evolution. The next question is whether inbreeding reduce the mean fitness of a population. Based on Wright's model mentioned above, this question is equivalent to show whether $w_{11} + w_{22} - 2w_{12} < 0$ holds. The following result shows that at a single locus, inbreeding may or may not reduce the mean fitness of population. Here we show in Theorem 3.1 below that inbreeding does reduce the mean fitness at multi loci and mixed genotype fitnesses. Recall that genotype fitnesses (w_{11}, w_{12}, w_{22}) is called *underdominance* if both $w_{11} > w_{12}$ and $w_{22} > w_{12}$. However, it is called *overdominance* if both $w_{11} < w_{12}$ and $w_{22} < w_{12}$. The following results are an extension of some published results.

Proposition 2.3. (1) For the populations with underdominance or with partial recessive ($h \geq 1/2$), inbreeding mating increases the mean fitness of population.

(2) For the populations with overdominance or partial dominance ($h \leq 1/2$), inbreeding mating reduces the mean fitness of population.

In natural populations, the occurrence of alleles with underdominance is rare comparing with the ones of overdominance. In addition, the occurrence of partial dominant alleles is less often than that of partial recessive ones. The following result is more important in application. Part of the results has been described in [16]. But a result for multiple-genes is given in Theorem 3.1 below.

Proposition 2.4 At a single locus, a population possesses only additive genotypes if and only if no inbreeding depression occurs.

For a selfing population whose relative genotype fitness are $1, 1 - hs$ and $1 - s$ with $0 \leq h \leq 1/2$, inbreeding depression is $\delta = 1 - w_s/w_0$, where w_s (w_0) is the mean fitness of the inbreeding (random mating) population. Thus we have

$$\delta = \frac{-fxys(2h - 1)}{1 - 2shxy - sy^2} = \frac{0.5sx(1 - x)(1 - 2h)}{1 - 2x(1 - x)sh - (1 - x)^2 s}. \tag{2}$$

For example, if $s = 1$, $x = 0.9$ and $h = 0$, inbreeding depression at a single locus is $\delta = 0.045$. However, a complete recessive mutant usually will cost around 30% inbreeding depression [4], [5], [19]. We will discuss this issue in Section 3.1 below.

2.3   Pseudo-overdominance

Pseudo-overdominance describes recessive deleterious mutations at closely linked loci [4], [7]. It has overdominance phenotype but is determined by recessive deleterious mutants. It was expected that among the class of overdominance populations there might be many pseudo-overdominance individous. To classify it, we call the subclass of overdominance populations satisfying $(f + h - fh) > 0$ as *pseudo-dominance* but still call others (that is, $(f + h - fh) \leq 0$) as *overdominance*. If the overdominance coefficients are of $1 - s, 1$ and $1 - t$ with $s, t \geq 0$, a population is of pseudo-overdominance if and only if $(s < t < s/f)$ holds.

Note that a population with overdominance has an internal equilibrium if and only if function

$$J = s(1 - f)(2h - 1)x + fhs + s - hs = 0$$

has a solution in $(0, 1)$, or equivalently

$$0 \leq \frac{fh + 1 - h}{(1 - f)(1 - 2h)} \leq 1,$$

which is again equivalent to $h + f - hf \leq 0$. Thus a population with overdominance has an internal equilibrium if and only if

$$h \leq -\frac{f}{1 - f}, \tag{3}$$

where f is the inbreeding coefficient. Hence an inbreeding population $(1, 1 - hs, 1 - s)$ is of pseudo-overdominance if and only if (3) holds, For a population of self-fertilization plants, where $f = 0.5$, it is pseudo-overdominant if the degree of dominance $h \geq -1$.

Next we discuss relationship with quantitative trait loci. Suppose that initially we have two homozygous, inbred lines which are used as parents, P_1 and P_2. These parents are identical except for one gene, denoted by A. Assume that A has two alleles A^+ and A^-, which is responsible for the higher and lower phenotypic scores, respectively, of the character being studied. Let $P_1 > P_2$. The genotypes of P_1 and P_2 fort this gene will be A^+A^+ and A^-A^-; while their hybrid will be A^+A^-. Let the average phenotype of the two parents, namely P_1

and P_2, be m and the additive and dominance genetic component of means (the average phenotypic value) be a_A ($=\bar{P}_1 - m$) and $d_A (= \bar{F}_1 - m)$, where $\bar{P}$ is the mean of P. Thus we have that

- if $d_A/a_a = 1$, allele A^+ is complete dominant,
- if $d_A/a_a = -1$, allele A^+ is complete recessive;
- if $0 \le d_A/a_a < 1$, allele A^+ is partially dominant;
- if $-1 \le d_A/a_a < 0$, allele A^+ is partially recessive;
- if $d_A/a_a > 1$, allele A^+ is over-dominant; and
- if $d_A/a_a < -1$, allele A^+ is under-dominant.

Then we consider the selection type $(1, 1 - hs, 1 - s)$ of the two alleles A^+ and A^-. Let $a = s/2$, $d = s(1 - 2h)/2$ and $m = 1 - s/2$. Thus the above conclusions hold true from the point of view of population genetics. For example, the inequality $d/a > 1$ holds if and only if $0 > h$. In addition, for self-fertilization population ($f = 0.5$), it is of pseudo-overdominance if and only if $0 \ge h \ge -1$. Although it is not clear how this quantity changes when considering multiple loci, there exists a subclass of pseudo-overdominant populations satisfying

$$1 \le \frac{d}{a}(= 1 - 2h) \le 3.$$

2.4   Purging Rate

The following result gives a quantitative calculation which shows that inbreeding mating purges deleterious mutants much efficiently than random mating. We will also show that purging deleterious alleles may occur in pseudo-overdominant populations. The reason of this observation is that inbreeding mating population with overdominnce may not have internal equilibrium ($0 < x < 1$). Note that a well-known property of a random-mating population with overdominance is that there is always an internal equilibrium ($0 < x < 1$). For an inbreeding population with the relative genotype fitnesses $(1, 1 - hs$ and $1 - s)$ and the inbreeding coefficient f satisfying $-\frac{f}{1-f} \le h \le 1$, the average time $t(x, x_t)$ required for allele frequency x to x_t is equal to

$$(fhs + s - hs) \ln \frac{x_t}{x} - (hs + fhs) \ln \frac{1 - x_t}{1 - x}$$

$$- \frac{(1 - f)(2hs - s)}{(hs + fhs)(fhs + s - hs)} \ln \frac{(1 - f)(2hs - s)x_t + fhs + s - hs}{(1 - f)(2hs - s)x + fhs + s - hs}. \quad (4)$$

From this formula, the strength of purging increases when h decreases. When $h = 0$ and $f = 0.25$, inbreeding mating takes only less than 10% to purge the deleterious allele from 0.1 to 0.01, comparing with random mating. It is interesting to note that inbreeding mating only take two third to purge the deleterious allele from 0.1 to 0.01, comparing random mating. This is another advantage of selfing fertilisation plants. This result explains why there are so many selfing fertilisation plants although they have various degree of inbreeding depression. We also see that purring is much efficient for over-dominance than dominance-recessive genotypes.

3 Theory of Inbreeding at Multiple Loci

Genes can interact with each other in many possible ways. But often the fitness-reducing effects of homozygosity for deleterious alleles (mutant alleles or alleles at loci with overdominance) act roughly multiplicatively [3], [12]. Thus the relative viability of an individual over all loci is the product of the individual locus viability values, given by

$$w = w_A w_B w_C \ldots,$$

where w is the overall viability of an individual, and, for example, w_A is the relative viability of the individual at the A locus. This multiplicative model is often assumed for viability loci [5]. In addition, all the loci are assumed to assort independently, that is they are unlinked. It is also assumed that there is no mutation at these loci over the period of inbreeding. The selective disadvantage s is assumed be to the same for all loci and a similar assumption is applied to the initial frequency. Then each w_A corresponds to an $h \leq 1/s$ such that

$$w_A = 1 - 2shxy - sy^2 + fxy(2hs - s).$$

The overall inbreeding mean fitness is

$$w_s = \Pi_h(1 - 2shxy - sy^2 + fxy(2hs - s))$$

and the overall outcross (random-mating) mean fitness is

$$w_o = \Pi_h(1 - 2shxy - sy^2).$$

The following example is supportive to this multiplication hypothesis. Willis [23] defined the average dominance, given by

$$\bar{h} = \frac{\sum_i p_i q_i s_i^2 h_i}{\sum_i p_i q_i s_i^2 h_i}. \tag{5}$$

Since

$$\min_i\{h_i\} = \min_i \frac{p_i q_i s_i^2 h_i}{p_i q_i s_i^2} \leq \frac{\sum_i p_i q_i s_i^2 h_i}{\sum_i p_i q_i s_i^2} \leq \max_i\{h_i\},$$

$\bar{h}$ may not be significant from h_i. From their measurement, Carr et al. [4] obtained the data of imbreeding depression that are shown in Table 1. These values are now calculated by our proposed formula, given by

$$\delta = 1 - \frac{w_s}{w_o},$$

where

$$w_s = (1 - 2shxy - sy^2 + fxy(2hs - s))^{20},$$
$$w_o = (1 - 2shxy - sy^2)^{20},$$

$x = 0.97$, $y = 0.03$ and $f = 0.5 \sim 0.67$ (due to uncertainty). The calculated values of inbreeding depression is also presented in Table 1.

Table 1. Inbreeding depression δ^* calculated from our formula comparing with the value of δ in literature [4] (SD: standard deviation and $\bar{h}$ in Eq. (5)).

Trait	δ	SD	h	δ^*	$|\delta - \delta^*|$
Proportion of germination	0.179	0.033	0.213	0.225	0.036
Number of flowers	0.272	0.044	0.057	0.323	0.051
Viable pollen per flower	0.488	0.022	0.087	0.305	0.18
Fraction of viable pollen	0.324	0.02	0.063	0.32	0.006
Pollen grains per flower	0.293	0.02	0.122	0.284	0.009
Ovules per flower	0.441	0.026	0.179	0.248	0.192

For inbreeding mating, we have that

$$w_s(h) = 1 - 2shxy - sy^2 + fxy(2hs - s)$$

and for random mating we have

$$w_0(h) = 1 - 2shxy - sy^2.$$

Thus we have that

$$\ln w_s = \sum_h \ln w_s(h)$$

$$\ln w_o = \sum_h \ln w_o(h).$$

To estimate the numbers of recessive deleterious alleles and dominant deleterious alleles, Carr et al. [4] carries out three survey experiments for ten populations of Amsinckia gloriosa, Amsinckia spectabilis and Mimulus guttatus and M. micranthus. Among these ten populations, there are five populations that are low inbreeding depression, while the other five populations are mixed-mating. These ten populations contain 36 loci. And these 36 loci include ∼6 additive or dominant deleterious and 2 overdominant. The remaining ∼28 loci are partial recessive deleterious alleles. Thus it is reasonable to assume the occurrence of partial recessive genes is more than 3-fold of partial dominant genes. It is also assumed that the deleterious recessive genotype are lethal ($s = 1$). Based on these assumptions, we have the following result regarding the mean fitness of a population with multiple loci. The proof will be provided in the expanded version of this manuscript.

Theorem 3.1 Assume that all genes are of partial dominance or partial recessive. It is also assumed that the number of genes with partial recessive is 3-fold more than that with partial dominance. Then inbreeding mating reduces the mean fitness of the population.

Theorem 3.1 in fact shows that, if there are many loci which exhibit overdominance, then inbreeding mating could severely reduce the mean fitness of the population. In particular, in populations of mammals, there are many overdominance genotypes (i.e., sickle cell anaemia). Thus inbreeding mating reduces the mean fitness of such populations.

Recent research works have showed that there were many overdominant loci in rice [19] [20]. Dong (2007) showed that overdominance and epistasis might play an important role as the genetic basis of heterosis in Brassica rapa. Suppose that there are many overdominant loci whose genotype fitnesses are $(1, 1 - hs, 1 - s)$ and $-N_0$ is the least value of h. Then we see

$$
\ln w_s = \int_{-N_0}^{1} \ln(1 - 2shxy - sy^2 + fxy(2hs - s))dh
$$

$$
= \int_{-N_0}^{1} \ln w_0(h)dh,
$$

$$
\ln w_0 = \int_{-N_0}^{1} \ln(1 - 2shxy - sy^2)dh
$$

$$
= \ln w_s(h)dh.
$$

To show whether inbreeding reduces the mean fitness of a population, it is sufficient to show whether $\ln w_s \leq \ln w_0$. We have the following result regarding the mean fitness. The proof will be provided in the expanded version of this paper.

Proposition 3.2 If the number of genes N is large enough, then inbreeding mating reduces the mean fitness of the population, that is, $w_s \leq w_o$.

4   Conclusions and Discussions

For a large population with a single locus, we have shown that for any inbreeding coefficient $0 \leq f \leq 1$, the mean fitness increases monotonically for directional selection. This result substantially extends the existing result for the case of random mating [13]. In addition, considering the average mean fitness (AMF) which is the integral of mean fitness of frequency x going through over the interval $[0, 1]$, Theorem 2.2 shows that the AMF of a population always increases monotonically for any inbreeding population. These results have substantial application to the case with one-gene traits (i.e. the traits determined by a single gene) and explain why inbreeding mating do not produce serious impact on the mean fitness of the population.

It is known that Fisher's fundamental theorem of natural selection (FTNS) [12] is not the same as the so-called "mean fitness increase theorem (MFIT)". FTNS in its full generality is deeper, more general and more complex than MFIT, since MFIT is simpler and only holds under limited conditions, e.g., under the condition of non-random mating and when fitness depends on a single gene [13]. Our introduced notion of average mean fitness (TMF) may serve as an intermediate between FTNS and MFIT, which shows that inbreeding mating, which contains the random mating as a special case, do not affect the increase of TMF if fitness depends on only one gene.

We also presented a new application of multiplication hypothesis to link the results about each single gene and the whole genomic impact of multi-genes in Table 1. It was widely accepted that the results of population genetics for the case of a single gene is well-developed, but it is very hard to obtain fruitful results for the case of multiple genes. The multiplication hypothesis allows us to produce an reliable estimation for genomic measurement from the results of single genes (see Table 1).

However, for a small population, because of genetic drift and other reasons, our theoretical analysis suggested that inbreeding mating might produce severe consequence on reducing the mean fitness of the population and even causes population extinction due to the non-constant inbreeding coefficients.

Acknowledgements. F.K. acknowledges support from the Australian Research Council (ARC) through Discovery Project (DP120102728). T.T. acknowledges support from the Australian Research Council (ARC) through Discovery Project (DP120104460) and is also a recipient of the Australian Research Council Future Fellowship (FT100100748).

References

1. Barrett, S.C.H., Charlesworth, D.: Effect of a change in the level of inbreeding on the genetic load. Nature 352, 522–524 (1991)
2. Bodmer, W.H., Cavalli-Storza, L.L.: Genetics, Evolution, and Man, p. 446. W. H. Freeman, San Francisco (1976)
3. Carr, D.E., Dudash, M.R.: Inbreeding depression in two species of Mimulus (Scrophulariaceae) with contrasting mating systems. Am. J. Bot. 83, 586–593 (1996)
4. Carr, D.E., Dudash, M.R.: Recent approaches into the genetic basis of inbreeding depression in plants. Phil. Trans. R. Soc. Lond. B 358, 1071–1084 (2003)
5. Charlesworth, D., Charlesworth, B.: Inbreeding Depression and its Evolutionary Consequences. Annu. Rev. Ecol. Syst. 18, 237–268 (1987)
6. Charlesworth, B., Charlesworth, D.: The genetic basis of inbreeding depression. Genet. Res. 74, 329–340 (1999)
7. Charlesworth, D., Willis, J.H.: The genetics of inbreeding depression. Nat. Rev. Genet. 10, 783–796 (2009)
8. Dolgin, E.S., Charlesworth, B., Baird, S., Cutter, A.D.: Inbreeding and Outbreeding Depression in Caenorhabditis Nematodes. Evolution 61, 1339–1352 (2007)
9. Dong, D., Cao, J., Kai, S., Liu, L.: Overdominance and Epistasis Are Important for the Genetic Basis of Heterosis in Brassica rapa. Hortscience 42, 1207–1211 (2007)
10. Dudash, M.R., Carr, D.E.: Genetics underlying inbreeding depression in Mimulus with contrasting mating systems. Nature 393, 682–684 (1998)
11. Fenster, C.B., Galloway, L.F.: Inbreeding and ourbreeding Depression in Nature Populations of Chamaecrista fasciculata(Fabaceae). Conserv. Biol. 14, 1406–1412 (2000)
12. Fisher, R.A.: The Genetical Theory of Natural Selection. Clarendon Press, Oxford (1930)
13. Ewens, W.J.: Mathematical Population Genetics I, Theoretical Introduction, 2nd edn. Springer (2004)

14. Goodwillie, C., Knight, M.-C.: Inbreeding Depression and Mixed Mating in Leptosiphon jepsonii: A Comparison of Three Populations. Ann. Bot. 98, 351–360 (2006)
15. Hedrick, P.W.: Purging inbreeding depression. Heredity 73, 363–372 (1994)
16. Hedrick, P.W.: Genetics of Populations, 4th edn. Jones and Bartlet Publication (2010)
17. Johnston, M.O., Schoen, D.J.: Mutation rates and dominance levels of genes affecting fitness in two angiosperm species. Science 267, 226–229 (1995)
18. Li, Y., Zhang, X., Ma, C., Shen, J., Chen, Q., Wang, Q., Fu, T., Tu, J.: QTL and epistatic analyses of heterosis for seed yield and three yield component traits using molecular markers in rapeseed (Brassica napus L.). Russ. J. Genet. 48, 1001–1008 (2012)
19. Li, Z.K., Luo, L.J., Mei, H.W., Wang, D.L., Shu, Q.Y., Tabien, R., Zhong, D.B., Ying, C.S., Stansel, J.W., Khush, G.S., Paterson, A.H.: Overdominant epistatic loci are the primary genetic basis of inbreeding depression and heterosis in rice. I. Biomass and grain yield. Genetics 158, 1737–1753 (2001)
20. Luo, L.J., Li, Z.-K., Mei, H.W., Shu, Q.Y., Tabien, R., Zhong, D.B., Ying, C.S., Stansel, J.W., Khush, G.S., Patterson, A.H.: Overdominant epistatic loci are the primary genetic basis of inbreeding depression and heterosis in rice. II. Grain yield components. Genetics 158, 1755–1771 (2001)
21. Roff, D.A.: Life History Evolution. Sinauer Associates, Sunderland (2002)
22. Swindell, W.R., Bouzat, J.L.: Gene flow and adaptive potential in Drosophila melanogaster. Conserv. Genet. 7, 79–89 (2006)
23. Willis, J.H.: Inbreeding Load, Average Dominance and the Mutation Rate for Mildly Deleterious Alleles in Mimulus guttatus. Genetics 153, 1885–1898 (1999)
24. Winn, A.A., Elle, E., Kalisz, S., Cheptou, P., Eckert, C.G., Goodwillie, C., Johnston, M.O., Moeller, D.A., Ree, R.H., Sargent, R.D., Vallejo-Marin, M.: Analysis of inbreeding depression in mixed-mating plants provides evidence for selective interference and stable mixed mating. Evolution 65, 3339–3359 (2011)
25. Wright, L.I., Tregenza, T., Hosken, D.J.: Inbreeding, inbreeding depression and extinction. Conserv. Genet. 9, 833–843 (2008)
26. Wright, S.: Coefficients of inbreeding and relationship. Am. Nat. 56, 330–338 (1922)

dSpliceType: A Multivariate Model for Detecting Various Types of Differential Splicing Events Using RNA-Seq

Nan Deng and Dongxiao Zhu*

Department of Computer Science
Wayne State University
Detroit, MI 48202, USA
{ndeng,dzhu}@wayne.edu

Abstract. Alternative splicing plays a key role in regulating gene expression. Dysregulated alternative splicing events have been linked to a number of human diseases. Recently, the high-throughput RNA-Seq technology provides unprecedented opportunities and holds a strong promise for better characterizing and dissecting alternative splicing events on a whole transcriptome scale. Therefore, efficient and effective computational methods and tools for detecting differentially spliced genes and events in human disease are urgently needed. We present a novel and efficient computational method, dSpliceType, to detect five most common types of differential splicing events between two conditions using RNA-Seq. dSpliceType is among the first to utilize sequential dependency of normalized base-wise read coverage signals and capture biological variability among replicates using a multivariate statistical model. dSpliceType substantially reduces sequencing biases by taking ratio of normalized RNA-Seq splicing indexes at each nucleotide between disease and control conditions. Our method employs a change-point analysis followed by a parametric statistical test using Schwarz Information Criterion (SIC) on each candidate splicing event for differential splicing event detection. We evaluated and compared the performance of dSpliceType with the other two existing methods, MATS and Cuffdiff. The result demonstrates that dSpliceType is a fast, effective and accurate approach, which can detect various types of differential splicing events from a wide range of expressed genes, including genes with lower abundances. dSpliceType is freely available at http://orleans.cs.wayne.edu/dSpliceType/.

Keywords: Differential Splicing Detection, Next-Generation Sequencing, RNA-Seq, Multivariate Conditional Gaussian, Schwarz Information Criterion, Change Point Analysis, Hypothesis Testing.

1 Introduction

Alternative splicing plays a key role in regulating process during gene expression in higher eukaryotes [12]. It occurs in more than 90% of human genes in different

---

* To whom correspondence should be addressed.

M. Basu, Y. Pan, and J. Wang (Eds.): ISBRA 2014, LNBI 8492, pp. 322–333, 2014.
© Springer International Publishing Switzerland 2014

types [20], including skipped exon (SE), retained intron (RI), alternative 3' or 5' splice sites (A3SS or A5SS), and mutually exclusive exons (MXE) [12]. Studies have shown that dysregulation of alternative splicing events may lead to various human diseases [12,20]. Recently, the RNA-Seq technology shows promise to interrogate transcriptomes more accurately [4,21]. Therefore, efficient and effective computational methods for detecting differentially spliced genes and events associated with diseased-specific conditions using RNA-Seq are urgently needed.

In general, differential splicing detection is to detect whether the difference in the relative transcript abundances in a gene across samples between conditions is significant [10]. A natural idea is to estimate relative transcript abundances followed by a statistical test of relative abundances within a gene between conditions to quantify the differences. This type of methods, such as Cufflinks/Cuffdiff [19] and others [5,8], is powerful. However, they rely on accurate estimation of transcript relative abundances, which is a non-trivial problem. Other methods detect differentially spliced genes by comparing read counts either on all exons within a gene, such as SplicingCompass [2] and FDM [18], or on a single exon, e.g. DEXSeq [1]. These methods can potentially detect differentially spliced genes, but can not specify the regions or associated types of differential splicing events.

Another type of methods is event-based, directly detecting differential splicing events. Some methods, such as MISO [11] and SpliceTrap [22], focus on the detection of SE events. More recently, MATS [17] and DiffSplice [10] are capable of detecting multiple types of events. However, both Markov Chain Monte Carlo (MCMC) based MATS and permutation test based DiffSplice suffer from excessive computational loads.

In this paper, we propose a novel and efficient computational method, dSpliceType, to detect various types of differential splicing events using RNA-Seq. Compared with the existing methods, dSpliceType enjoys the following novel features. First and most importantly, instead of using read counts, dSpliceType is among the first to detect five types of differential splicing events using read coverage signals from both exonic and junction reads. It utilizes sequential dependency of normalized base-wise read coverage signals and captures biological variability among the biological replicates using a multivariate conditional normal model. Second, since we observed that sequencing and alignment biases are likely to affect read coverage signals the same way at each exonic nucleotide in both conditions, dSpliceType is expected to significantly reduce biases by taking the ratio of normalized RNA-Seq splicing indexes at each nucleotide between two conditions. Third, according to the results of simulation studies and real-world RNA-Seq data analysis, dSpliceType has demonstrated itself as an efficient yet accurate method to detect various types of differential splicing events from a wide range of expressed genes, including relatively low abundance genes. Our previous work on detecting differential splicing events between two individual samples (without replicate) has been published in [6].

2   Methods

2.1   Overview

dSpliceType is a parametric statistical framework for detecting five types of differential splicing events using RNA-Seq data. Figure 1 shows the workflow of dSpliceType.

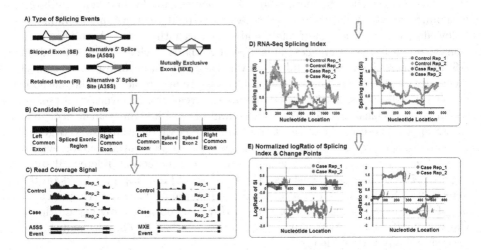

Fig. 1. The workflow of dSpliceType for detecting various types of differential splicing events. A) Five most common types of splicing events. Left panel represents SE, RI, A3SS and A5SS events, and right panel represents MXE event. B) Candidate splicing events are compiled by removing introns and concatenating left common exon, spliced exon(s) or exonic region and right common exon. C) For each candidate splicing event (illustrated by A5SS and MXE events), read coverage signals are calculated on nucleotides for each replicate in both conditions. D) and E) RNA-Seq splicing indexes and normalized logRatio of splicing indexes are calculated based on read coverage signals. dSpliceType detects the differential splicing events by identifying change points on the ending locations of exon(s) or exonic region.

2.2   Extracting Candidate Splicing Events

dSpliceType extracts candidate splicing events for the five most common types of alternative splicing from gene annotation database along with supported junction reads as shown in Figure 1A and Figure 1B. With intron removal, candidate splicing events consist of concatenating left common exon, spliced exon(s) (for SE and MXE events) or exonic region (for RI, A3SS and A5SS events) and right common exon. Two spliced exons are for MXE event. The detailed strategies for extracting different types of candidate splicing events are described in [6]. Novel candidate splicing events can be extracted by incorporating novel junction reads.

2.3   Calculating Normalized logRatio of RNA-Seq Splicing Indexes

After extracting candidate splicing events, the read coverage signal (Figure 1C) and the RNA-Seq splicing index (Figure 1D) at each nucleotide location are calculated in terms of differential splicing for each replicate in both conditions. The RNA-Seq splicing index at the ith nucleotide location is denoted as SI_i. The formula of calculation of SI_i has been given in [6]. After that, the logRatio of normalized RNA-Seq splicing indexes (Figure 1E) on each nucleotide of a candidate splicing event between two conditions is calculated. We denote it as $\log(SI_{\text{caseSample}_{im}}/\overline{SI}_{\text{controlSample}_i})$, where m is the index of replicates in case condition and $\overline{SI}_{\text{controlSample}_i}$ is the average of RNA-Seq splicing indexes at ith nucleotide location of replicates in control condition.

Since the sequencing and alignment biases are more likely to affect read coverage signals at the same nucleotide locations on all samples in the same way, the effect of biases from RNA-Seq is substantially reduced by taking ratio of normalized RNA-Seq splicing indexes at each nucleotide location of replicates in two conditions.

2.4   Detecting Differential Splicing Events

The Multivariate Conditional Normal Distribution Model for the Normalized Logratio of RNA-Seq Splicing Indexes. As shown in Figure 1E, we denote the normalized $\log(SI_{\text{caseSample}_{im}}/\overline{SI}_{\text{controlSample}_i})$ at the ith nucleotide along the candidate splicing event as $\mathbf{X}_i$, which is a m-dimensional normal random vector from $N_m(\boldsymbol{\mu}_i, \boldsymbol{\Sigma}_i)$, for $i = 1,...,n$. For computational simplicity, we capture the sequential dependency between $\mathbf{X}_i$ and $\mathbf{X}_{i-1}$, which follows [7]:

$$\mathbf{X}_i|\mathbf{X}_{i-1} = \mathbf{x}_{i-1} \sim N_m(\tilde{\boldsymbol{\mu}}, \tilde{\boldsymbol{\Sigma}}),$$

where

$$\tilde{\boldsymbol{\mu}} = \boldsymbol{\mu} + \boldsymbol{\Sigma}_{i,i-1}\boldsymbol{\Sigma}_{i-1,i-1}^{-1}(\mathbf{x}_{i-1} - \boldsymbol{\mu}),$$

$$\tilde{\boldsymbol{\Sigma}} = \boldsymbol{\Sigma}_{i,i} - \boldsymbol{\Sigma}_{i,i-1}\boldsymbol{\Sigma}_{i-1,i-1}^{-1}\boldsymbol{\Sigma}_{i-1,i}.$$

The sequence of $\{\mathbf{X}_i|\mathbf{X}_{i-1}\}$ can be considered as a series of multivariate conditional normal random variables from $N_m(\tilde{\boldsymbol{\mu}}_i, \tilde{\boldsymbol{\Sigma}}_i)$, for $i = 1, ..., n$, where n is the total exonic length of the candidate splicing event. If no differential splicing happens, $\tilde{\boldsymbol{\mu}}_i$ and $\tilde{\boldsymbol{\Sigma}}_i$ are assumed to be constant mean vector of $\tilde{\boldsymbol{\mu}}$ and covariance matrix of $\tilde{\boldsymbol{\Sigma}}$; while deviations from the constant mean vector and covariance matrix in the spliced region may indicate a differential splicing event.

The Hypothesis Testing. The identification of differential splicing event among multiple samples can be transformed to identify multiple change points at exon boundaries according to different types of candidate splicing events, and can be further defined as testing the null hypothesis for both mean and covariance parameters in the series of $\{\mathbf{X}_i|\mathbf{X}_{i-1}\}$ [3]:

$$H_0 : \tilde{\boldsymbol{\mu}}_1 = \tilde{\boldsymbol{\mu}}_2 = ... = \tilde{\boldsymbol{\mu}}_n = \tilde{\boldsymbol{\mu}} \text{ and}$$

$$\tilde{\Sigma}_1 = \tilde{\Sigma}_2 = \dots = \tilde{\Sigma}_n = \tilde{\Sigma}. \tag{1}$$

For SE, A3SS, A5SS and RI, the alternative hypothesis is:

$$H_1 : \tilde{\mu}_1 = \dots = \tilde{\mu}_i \neq \tilde{\mu}_{i+1} = \dots = \tilde{\mu}_j \neq \tilde{\mu}_{j+1} = \dots = \tilde{\mu}_n \text{ and}$$

$$\tilde{\Sigma}_1 = \dots = \tilde{\Sigma}_i \neq \tilde{\Sigma}_{i+1} = \dots = \tilde{\Sigma}_j \neq \tilde{\Sigma}_{j+1} = \dots = \tilde{\Sigma}_n, \tag{2}$$

where i and j, $1 < i < j < n$, are the ending locations of the left common exon (in black) and the spliced exon/exonic region (in purple), respectively, as shown on the left panel of Figure 1B. For each candidate splicing event of the four types, a significant differential splicing event is detected when the null hypothesis (1) is rejected at a given significance level α.

For MXE, the alternative hypothesis is:

$$H_1 : \tilde{\mu}_1 \dots = \tilde{\mu}_i \neq \tilde{\mu}_{i+1} \dots = \tilde{\mu}_j \neq \tilde{\mu}_{j+1} \dots = \tilde{\mu}_k \neq \tilde{\mu}_{k+1} \dots = \tilde{\mu}_n \text{ and}$$

$$\tilde{\Sigma}_1 \dots = \tilde{\Sigma}_i \neq \tilde{\Sigma}_{i+1} \dots = \tilde{\Sigma}_j \neq \tilde{\Sigma}_{j+1} \dots = \tilde{\Sigma}_k \neq \tilde{\Sigma}_{k+1} \dots = \tilde{\Sigma}_n, \tag{3}$$

where i, j and k, $1 < i < j < k < n$, are the ending locations of the left common exon (in black) and the two spliced exons (in purple and green), respectively, as shown on the right panel of Figure 1B. A significant differential splicing MXE event is detected when the null hypothesis (1) is rejected at a given significance level α.

The Schwarz Information Criterion.   In order to test the null hypothesis (1) against the alternative hypothesis (2) or (3), the Schwarz information criterion (SIC)-based method [16] is employed. The smaller SIC score indicates the better data fitting of a model. Thus, the hypothesis testing can be converted into selecting a model such that the null hypothesis (1) represents a model without change of mean and covariance parameters, while the alternative hypothesis (2) or (3) represents models with different means and covariances specified by two or three change points. Since, on average, more than 100 nucleotides are in the common and spliced exons/exonic regions, number of X_i's are considered to be sufficient for estimating model parameters and calculating SIC scores.

We denote SIC(n) as the SIC corresponding to the null hypothesis (1), which is derived as :

$$SIC(n) = -2 \log L_0(\hat{\tilde{\mu}}, \hat{\tilde{\Sigma}}) + \frac{m(m+3)}{2} \log n,$$

where the log likelihood is

$$\log L_0(\hat{\tilde{\mu}}, \hat{\tilde{\Sigma}}) = -\frac{1}{2} mn \log 2\pi - \frac{n}{2} \log |\hat{\tilde{\Sigma}}| - \frac{n}{2}.$$

So, we have

$$SIC(n) = mn \log 2\pi + n \log |\hat{\tilde{\Sigma}}| + n + \frac{m(m+3)}{2} \log n.$$

Corresponding to $H_1(2)$ with two change points i and j, the SIC for differential splicing events (SE, RI, A3SS and A5SS), denoted by $SIC(i, j)$ for fixed i and j, $m \leqslant i, j \leqslant n - m$, is derived as:

$$
SIC(i,j) = -2\log L_1(\hat{\hat{\boldsymbol{\mu}}}_1, \hat{\hat{\boldsymbol{\mu}}}_2, \hat{\hat{\boldsymbol{\mu}}}_3, \hat{\hat{\boldsymbol{\Sigma}}}_1, \hat{\hat{\boldsymbol{\Sigma}}}_2, \hat{\hat{\boldsymbol{\Sigma}}}_3) + \frac{3m(m+3)}{2}\log n
$$
$$
= mn\log 2\pi + i\log|\hat{\hat{\boldsymbol{\Sigma}}}_1| + (j-i)\log|\hat{\hat{\boldsymbol{\Sigma}}}_2| + (n-j)\log|\hat{\hat{\boldsymbol{\Sigma}}}_3|
$$
$$
+ n + \frac{3m(m+3)}{2}\log n.
$$

Similarly, corresponding to $H_1(3)$ with three change points i, j and k, the SIC for differential splicing events of MXE , denoted by $SIC(i, j, k)$ for fixed i, j and k, $m \leqslant i, j, k \leqslant n - m$, is derived as:

$$
SIC(i,j,k) = -2\log L_2(\hat{\hat{\boldsymbol{\mu}}}_1, \hat{\hat{\boldsymbol{\mu}}}_2, \hat{\hat{\boldsymbol{\mu}}}_3, \hat{\hat{\boldsymbol{\mu}}}_4, \hat{\hat{\boldsymbol{\Sigma}}}_1, \hat{\hat{\boldsymbol{\Sigma}}}_2, \hat{\hat{\boldsymbol{\Sigma}}}_3, \hat{\hat{\boldsymbol{\Sigma}}}_4) + 2m(m+3)\log n
$$
$$
= mn\log 2\pi + i\log|\hat{\hat{\boldsymbol{\Sigma}}}_1| + (j-i)\log|\hat{\hat{\boldsymbol{\Sigma}}}_2| + (k-j)\log|\hat{\hat{\boldsymbol{\Sigma}}}_3|
$$
$$
+ (n-k)\log|\hat{\hat{\boldsymbol{\Sigma}}}_4| + n + 2m(m+3)\log n.
$$

According to the principle of information criterion [3], the null model fits the data in the sequence of $\{\mathbf{X}_i|\mathbf{X}_{i-1}\}$ better if

$$
SIC(n) < SIC(i,j) \text{ or } SIC(n) < SIC(i,j,k).
$$

Otherwise, the model with two change points better fits the data in the sequence of $\{\mathbf{X}_i|\mathbf{X}_{i-1}\}$ for differential splicing events SE, A3SS, A5SS and RI, and the change points i and j are at the ending locations of the left common exon and the spliced exon or exonic region.

Similarly, the model with three change points better fits the data in the sequence of $\{\mathbf{X}_i|\mathbf{X}_{i-1}\}$ for differential splicing event MXE, and the change points i, j and k are the ending locations of the left common exon and the two spliced exons.

The Test Statistic. According to [3], the difference between the SIC scores of the models with and without change points,

$$
\Delta_n = SIC(i,j) - SIC(n) \text{ or } \Delta_n = SIC(i,j,k) - SIC(n)
$$

can be used as a statistic, and we use the asymptotic null distribution of Δ_n to calculate the approximate p-value for the test of the null hypothesis (1) against the alternative hypothesis (2) or (3). The raw p-values of the multiple tests are adjusted using the stringent Bonferroni's procedure.

3   Results

3.1   Simulation Studies

Simulation Datasets. We evaluated the accuracy of dSpliceType and compared the performance with two existing methods, MATS [17] and Cufflinks/

Cuffdiff [19], using simulation studies. FluxSimulator [9] was used to simulate various splicing ratios of events and generate 4 groups of RNA-Seq datasets on the entire human transcriptome. Each group includes 3 replicates in control and case conditions, respectively; and each replicate consists of 30 million, 50 million, 100 million and 200 million paired-end reads with 100bp in length in each group.

We mapped the simulated RNA-Seq datasets uniquely to the human reference genome (hg19/GRCh37) using Tophat2 [13] and Bowtie2 [14]. To evaluate and compare the three methods, the alignment results in BAM format were served as inputs for the latest version of MATS (3.0.8) and Cufflinks/Cuffdiff (2.1.1) using default parameters. Read coverage signals (.bedgraph files) converted from alignment results (.bam files) using BEDtools [15] and read junctions (.bed files) were used as inputs for dSpliceType. The complete Ensembl annotation database and the significance level of 0.05 for adjusted p-values were used to detect differentially spliced genes for Cufflinks/Cuffdiff and differential splicing events for dSpliceType and MATS. To control false positives and biological significance of events, we further set parameters of dSpliceType such that the average read coverage on the spliced exonic region is more than 5, the average ratio of normalized RNA-Seq splicing indexes on the spliced exonic region is greater than 1.2 or smaller than 0.8. Please note that the detected differentially spliced genes in Table 1 and Figure 2 are all true positives, and no false positive is detected by all three methods with their parameter settings.

Table 1. Comparison of the differentially spliced genes detected by dSpliceType, MATS and Cuffdiff in 4 groups of simulation datasets. For each method, the highest detection rate is in bold face.

# of Reads	# of Spliced Genes[1]	Methods					
		dSpliceType		MATS		Cuffdiff	
30M		8,054	78%	7,148	70%	2,086	20%
50M	10,275	8,701	85%	7,977	78%	2,626	26%
100M		9,170	89%	8,704	85%	3,425	33%
200M		9,467	**92%**	9,154	**89%**	4,527	**44%**

[1]The total number of differentially spliced genes in the simulation datasets.

Simulation Results of Detecting Differentially Spliced Genes. We compared the overall performances of the three computational methods on detecting differentially spliced genes in 4 groups of simulation datasets. We collected the detected differentially spliced genes directly from the result file of Cuffdiff (splicing.diff). For dSpliceType and MATS, a gene is considered to be differentially spliced if any type of differential splicing event was detected by the method for that gene. Table 1 shows that dSpliceType outperforms the other two methods by achieving the highest numbers and detection rates in all simulation datasets. One possible reason for the lowest detection rate of Cuffdiff is the inaccurate estimation of transcript relative abundances of genes when a gene has many annotated transcripts.

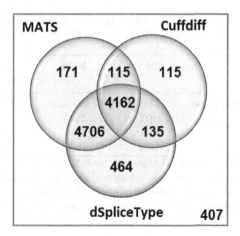

Fig. 2. The comparison of the detected differentially spliced genes by dSpliceType, MATS and Cuffdiff (200M simulation dataset)

To better evaluate the performance of dSpliceType, we compared the differentially spliced genes detected by the three methods in the 200M simulation dataset. As shown in Figure 2, there are 4,162 true differentially spliced genes detected by all the methods, and 464, 171 and 115 genes were exclusively detected by dSpliceType, MATS and Cuffdiff, respectively. We further examine the 464 genes detected exclusively by dSpliceType, and found that our method is able to better detect differentially spliced genes of low abundances ($1<\text{FPKM}<5$) than the competing methods.

In addition to differentially spliced genes of relatively low abundances, a large number of differentially spliced genes detected by dSpliceType is overlapped with that detected by the other two methods as shown in Figure 2. Therefore, dSpliceType is demonstrated to be able to detect differentially spliced genes in a large dynamic range of expressed genes.

Runtime Comparison. Table 2 shows the runtime comparison of the three methods among the 4 groups of simulation datasets on the same Linux Ubuntu Server with 4 x Twelve-Core AMD Opteron 2.6GHz and 256GB RAM. For each dataset, the runtime of dSpliceType is faster than the other two. The runtime of dSpliceType on each dataset can be separated into two parts, the time of converting alignment results to read coverage signals using BEDtools [15] and the time of detecting differential splicing events by dSpliceType. The increase in number of reads reflects more of the increase in conversion time, not detection time.

Table 2. Runtime comparison of dSpliceType, MATS and Cuffdiff in 4 groups of simulation datasets. The shortest runtimes are highlighted in bold.

Methods	30M	50M	100M	200M
		(Hours : Minutes)		
BEDTools + dSpliceType	0:36+0:25	0:48+0:29	1:30+0:31	2:30+0:32
Total	**1:01**	**1:17**	**2:01**	**3:02**
Cuffdiff	2:52	3:01	3:31	4:38
MATS	17:02	19:13	30:35	40:41

dSpliceType (1 thread), Cuffdiff (6 threads), and MATS (1 thread). The runtime of Cuffdiff includes gene and transcript relative abundance estimation, differential expression analysis and differential splicing analysis. The runtime of MATS includes the conversion time from .bam to .sam, and differential splicing analysis.

3.2   Real-World Data Analysis

RNA-Seq Data and Pre-processing. We applied dSpliceType to a public paired-end Illumina RNA-Seq dataset of human H1 and H1 derived neuronal progenitor cell lines (shorted as H1 and H1-npc). The dataset can be accessed from NIH Roadmap Epigenomics Project (http://www.roadmapepigenomics.org/) with NCBI SRA number SRR488684, SRR488685, SRR486241 and SRR486242 as two replicates of H1 and H1-npc cell lines, respectively. For each replicate, about 200 million reads (100bp × 2) were sequenced. For real-world RNA-Seq data analysis, the alignment procedure, the input files and parameters for dSpliceType are as same as simulation studies.

Detection of Differential Splicing Events. For the five types of splicing events, the numbers of corresponding genes of the differential splicing events detected by dSpliceType are 1,915 of SE, 247 of A5SS, 282 of A3SS, 621 of RI, and 19 of MXE. For illustration purpose, we present five differential splicing events detected by dSpliceType with different types of alternative splicing in Figure 3, in which MATS can detect four except the A3SS differential splicing event of gene DNAJC10, and Cuffdiff only detected gene CLK4 as a differentially spliced gene.

4   Discussion and Conclusion

As studies increasingly shift from DNA microarrays, RNA-Seq holds the promise to better interrogate transcriptomes, particularly splicing mechanisms. The method, dSpliceType, is designed specifically to utilize read coverage signals and work with multiple biological replicates.

Compared with read-count based methods, dSpliceType has the following major advantages. dSpliceType detects accurately differential splicing events. Instead of complex model for bias correction, we believe that taking ratio of

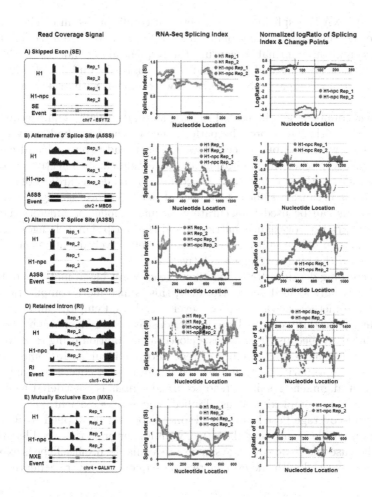

Fig. 3. The five case studies of detected differential splicing events from multiple samples by dSpliceType. Each row indicates a case study, and three columns show the plots of read coverage signal, RNA-Seq splicing index, and logRatio of splicing index and detected change points, respectively. (A) A skipped exon (SE) differential splicing event is detected for the gene chr7 - ESYT2 with two change points i and j at the ending locations of the left common exon and the spliced exon. (B) An alternative 5' splice site (A5SS) differential splicing event is detected for the gene chr2 + MBD5 with two change points i and j at the ending locations of the left common exon and the spliced exonic region. (C) An alternative 3' splice site (A3SS) differential splicing event is detected for the gene chr2 + DNAJC10 with two change points i and j at the ending locations of the left common exon and the spliced exonic region. (D) A retained intron (RI) differential splicing event is detected for the gene chr5 - CLK4 with two change points i and j at the ending locations of the left common exon and the spliced exonic region. (E) A mutually exclusive exon (MXE) differential splicing event is detected for the gene chr4 + GALNT7 with three change points i, j and k at the ending locations of the left common exon and the two spliced exons.

normalized RNA-Seq splicing indexes between conditions is an efficient way to eliminate the effect of sampling biases from RNA-Seq. We use a model of multivariate conditional normal to capture the sequential dependency of the read coverage signals after normalization and taking logRatio, and detect differential splicing events by comparing SIC scores between models with or without change points.

The read counts based differential splicing methods usually require sequencing at a certain depth. Therefore, the performance of these methods may be limited for genes with low abundances. However, as a read-coverage based method, dSpliceType overcomes this limitation; the detection can be effective as long as the nucleotides of the splicing event are covered by reads, regardless of coverage depth. This is because a sufficient number of per-base coverage signal values in exonic regions can be used to accurately estimate model parameters (i.e., the means and variance covariance matrices), and the sharp signal changes on exon-exon boundaries of the splicing events can be effectively identified as change points, even if the read coverage is relatively low.

The increasing depth of RNA-Seq allows read-count based methods to detect more differentially spliced genes. However, this is also more computational intensive if the method needs to process every reads and employs an iterative or re-sampling procedure. dSpliceType is not sensitive to the increase in the number of short reads because it provides a closed-form solution using read coverage signals, and the increasing number of reads do not incur extra computational load since they primarily result in elevated values of read coverage signals. Converting read alignment result to read coverage signals is considered to linearly increase the time complexity. Thus, dSpliceType is time efficient.

As a read-coverage based method, dSpliceType can be applied to RNA-Seq data from multiple sequencing platforms, with longer or shorter read lengths, as long as the base-wise read coverage signals are available. dSpliceType is expected to be more powerful with the ever-increasing sequencing coverage depth.

Acknowledgements. We would like to thank Dr. Jiangsheng Yu for helpful discussion of the method, and the reviewers for their constructive feedbacks.

References

1. Anders, S., Reyes, A., Huber, W.: Detecting differential usage of exons from RNA-seq data. Genome Research 22(10), 2008–2017 (2012)
2. Aschoff, M., Hotz-Wagenblatt, A., Glatting, K.-H., Fischer, M., Eils, R., König, R.: Splicingcompass: differential splicing detection using RNA-Seq data. Bioinformatics 29(9), 1141–1148 (2013)
3. Chen, J.: Parametric statistical change point analysis. Birkhauser, Boston (2012)
4. Deng, N., Puetter, A., Zhang, K., Johnson, K., Zhao, Z., Taylor, C., Flemington, E.K., Zhu, D.: Isoform-level microRNA-155 target prediction using RNA-seq. Nucleic Acids Research 39(9), e61 (2011)
5. Deng, N., Sanchez, C.G., Lasky, J.A., Zhu, D.: Detecting splicing variants in idiopathic pulmonary fibrosis from non-differentially expressed genes. PloS One 8(7), e68352 (2013)

6. Deng, N., Zhu, D.: Detecting various types of differential splicing events using RNA-Seq data. In: Proceedings of the International Conference on Bioinformatics, Computational Biology and Biomedical Informatics, p. 124. ACM (2013)

7. Eaton, M.L.: Multivariate statistics: a vector space approach. Wiley, New York (1983)

8. Gonzàlez-Porta, M., Calvo, M., Sammeth, M., Guigó, R.: Estimation of alternative splicing variability in human populations. Genome Research 22(3), 528–538 (2012)

9. Griebel, T., Zacher, B., Ribeca, P., Raineri, E., Lacroix, V., Guigó, R., Sammeth, M.: Modelling and simulating generic RNA-Seq experiments with the flux simulator. Nucleic Acids Research 40(20), 10073–10083 (2012)

10. Hu, Y., Huang, Y., Du, Y., Orellana, C.F., Singh, D., Johnson, A.R., Monroy, A., Kuan, P.F., Hammond, S.M., Makowski, L., et al.: DiffSplice: the genome-wide detection of differential splicing events with RNA-seq. Nucleic Acids Research 41(2), e39 (2013)

11. Katz, Y., Wang, E.T., Airoldi, E.M., Burge, C.B.: Analysis and design of RNA sequencing experiments for identifying isoform regulation. Nature Methods 7(12), 1009–1015 (2010)

12. Keren, H., Lev-Maor, G., Ast, G.: Alternative splicing and evolution: diversification, exon definition and function. Nature Reviews Genetics 11(5), 345–355 (2010)

13. Kim, D., Pertea, G., Trapnell, C., Pimentel, H., Kelley, R., Salzberg, S.L.: TopHat2: accurate alignment of transcriptomes in the presence of insertions, deletions and gene fusions. Genome Biology 14(4), R36 (2013)

14. Langmead, B., Salzberg, S.L.: Fast gapped-read alignment with Bowtie 2. Nature Methods 9(4), 357–359 (2012)

15. Quinlan, A.R., Hall, I.M.: BEDTools: a flexible suite of utilities for comparing genomic features. Bioinformatics 26(6), 841–842 (2010)

16. Schwarz, G.: Estimating the dimension of a model. The Annals of Statistics 6(2), 461–464 (1978)

17. Shen, S., Park, J.W., Huang, J., Dittmar, K.A., Lu, Z.X., Zhou, Q., Carstens, R.P., Xing, Y.: MATS: a Bayesian framework for flexible detection of differential alternative splicing from RNA-Seq data. Nucleic Acids Research 40(8), e61 (2012)

18. Singh, D., Orellana, C.F., Hu, Y., Jones, C.D., Liu, Y., Chiang, D.Y., Liu, J., Prins, J.F.: FDM: a graph-based statistical method to detect differential transcription using RNA-seq data. Bioinformatics 27(19), 2633–2640 (2011)

19. Trapnell, C., Roberts, A., Goff, L., Pertea, G., Kim, D., Kelley, D.R., Pimentel, H., Salzberg, S.L., Rinn, J.L., Pachter, L.: Differential gene and transcript expression analysis of RNA-seq experiments with TopHat and Cufflinks. Nature Protocols 7(3), 562–578 (2012)

20. Wang, E.T., Sandberg, R., Luo, S., Khrebtukova, I., Zhang, L., Mayr, C., Kingsmore, S.F., Schroth, G.P., Burge, C.B.: Alternative isoform regulation in human tissue transcriptomes. Nature 456(7221), 470–476 (2008)

21. Wang, Z., Gerstein, M., Snyder, M.: RNA-Seq: a revolutionary tool for transcriptomics. Nature Reviews Genetics 10(1), 57–63 (2009)

22. Wu, J., Akerman, M., Sun, S., McCombie, W.R., Krainer, A.R., Zhang, M.Q.: SpliceTrap: a method to quantify alternative splicing under single cellular conditions. Bioinformatics 27(21), 3010–3016 (2011)

Conformational Transitions and Principal Geodesic Analysis on the Positive Semidefinite Matrix Manifold

Xiao-Bo Li and Forbes J. Burkowski

University of Waterloo, Canada
{x22li,fjburkowski}@uwaterloo.ca

Abstract. Given an initial and final protein conformation, generating the intermediate conformations provides important insight into the protein's dynamics. We represent a protein conformation by its Gram matrix, which is a *point* on the rank 3 positive semidefinite matrix manifold, and show matrices along the geodesic linking an initial and final Gram matrix can be used to generate a feasible pathway for the protein's structural change. This geodesic is based on a particular quotient geometry. If a protein is known to contain domains or groups of atoms that act as rigid clusters, facial reduction can be used to decrease the size of the Gram matrices before calculating the geodesic. The geodesic between two conformations is only one path a protein's Gram matrix can follow; principal geodesic analysis (PGA) is one possible strategy to find other geodesics.

Keywords: cliques, rigid clusters, elastic network interpolation, Euclidean distance matrix, Gram matrix, positive semidefinite, facial reduction, geodesic, Riemannian manifold, principal geodesic analysis.

1 Introduction

Modelling conformational transitions of proteins is part of the effort to gain insight into the mechanics of biological processes. To efficiently generate transition conformations, numerous interpolation approaches have been suggested. Examples include [29] [19], and [11]. Recently, [16] and [17] proposed elastic network interpolation (ENI). This methodology interpolates the inter-atomic distances between an initial and final protein conformation subject to optimizing a simple potential function. Rather than distances, we consider the matrix of *distance-squared* entries, also called the *Euclidean distance matrix* (EDM). Proteins are in 3D space, so their EDMs have an embedding dimension of 3. These EDMs are in 1-to-1 correspondence with rank 3 positive semidefinite (PSD) centred Gram matrices, which are commonly used in semidefinite optimization to represent a protein conformation [5] [3] [4] [20]. The set of PSD matrices of fixed rank p is a Riemannian manifold; the points on this manifold being the matrices themselves. The geometry is not unique. In section 2 we discuss the geometry we chose and its associated geodesic formulation.

M. Basu, Y. Pan, and J. Wang (Eds.): ISBRA 2014, LNBI 8492, pp. 334–345, 2014.

Rigid clusters are parts of a protein that are known to act as rigid bodies during large conformational changes. *Rigid cluster ENI* is the extension of ENI to model only the rigid clusters [18]. Given rigid clusters, *facial reduction* [20] [21] can be used to find a smaller Gram matrix from which we can calculate the geodesic. If no information on rigid clusters is available, we can use the peptide planes, tetrahedral carbons, and phenol rings present within the protein molecular structure to perform facial reduction [3]. In section 3 we define face and facial reduction for semidefinite matrices.

The geodesic between two conformations is only one path a Gram matrix can change along. The extension of principal component analysis (PCA) to Riemannian manifolds, called *principal geodesic analysis* [9][10][12][13], is one way to find other geodesics. In section 5 we explain our formulation of PGA.

We use lactoferrin to demonstrate moving along the geodesic between two conformations, and use HIV-1 protease as an example to illustrate PGA. Due to space limitations, we will not discuss matrix manifolds in-depth and have provided a detailed reference section.

2   From Distance Geometry to Riemannian Geometry

Let Y be the $n \times 3$ matrix holding the *centred* atomic coordiantes of the protein's n atoms. The centred Gram matrix X is given by $X = YY^T$. The rank of a Gram matrix X is the rank of Y. Since proteins are 3D objects, both X and Y are rank 3. The centred Gram matrix is related to the Euclidean distance matrix (EDM) by a linear bijective mapping [20][2].

The space of PSD matrices of fixed rank can be viewed as a Riemannian manifold. The geometry is not unique, and some are discussed in [28][23][15][6]. The quotient structure presented in [6][22] is the choice for this paper because it explicitly accounts for the "facial factorization" explained later.

This quotient structure is given as follows: Let $St(p, n)$ denote the set of $n \times p$ orthonormal matrices, also called the *Stiefel manifold*. Let $O(n)$ denote the set of orthogonal $n \times n$ matrices. Let $\mathbf{S}_{++}^n$ denote the set of $n \times n$, full rank, positive definite matrices and $\mathbf{S}_+^{n,p}$ denote the set of $n \times n$ rank p positive semidefinite matrices. The set $\mathbf{S}_+^n$ is the union of $\mathbf{S}_{++}^n$ with $\mathbf{S}_+^{n,p}$ for $p = 1, \ldots, n - 1$. The *polar factorization* of the matrix Y containing atomic coordinates is given by $Y = UR$ with $U \in St(p, n)$ and $R \in \mathbf{S}_{++}^p$; consequently, the Gram matrix can be written

$$X = YY^T = UR^2U^T = USU^T \text{ with } R^2 = S \in \mathbf{S}_{++}^p . \tag{1}$$

For any orthogonal matrix $O \in O(p)$, UO and OSO gives the same Gram matrix. Matrices that are the result of this action by O are referred to as belonging to the same *equivalence class*. Thus, $\mathbf{S}_+^{n,p}$ admits the quotient representation: $St(p, n) \times \mathbf{S}_{++}^p / O(p)$. The space $St(p, n) \times \mathbf{S}_{++}^p$ is called the *total space*.

2.1   Approximate Geodesics

Horizontal geodesics in the total space move orthogonal to the equivalence classes. These curves can be used to approximate curves in the quotient space. For $A, B \in \mathbf{S}_+^{n,p}$, their construction was provided in [6]. Let $V_A, V_B \in St(p, n)$ be two matrices in the span of range(A) and range(B) respectively. Performing singular value decomposition (SVD) on $V_A^T V_B$, we have:

$$V_A^T V_B = O_A \mathrm{diag}(\sigma_1, \ldots, \sigma_p) O_B^T . \tag{2}$$

The *principal angles* θ_is are determined by the singular values, $\sigma_i = \cos(\theta_i)$ and the *principal vectors* are given by $U_A = V_A O_A$ and $U_B = V_B O_B$ [14]. The horizontal curve connecting range(A) and range(B) is:

$$U(t) = U_A \cos(\Theta t) + (I - U_A U_A^T) U_B G \sin(\Theta t) , \tag{3}$$

where $\Theta = \mathrm{diag}(\theta_1, \ldots, \theta_p)$ and G is the pseudoinverse of $\mathrm{diag}(\sin(\theta_1), \ldots, \sin(\theta_p))$. At each t, $U(t)$ needs to be rotated onto V_A by solving the Procrustes problem to avoid arbitrary rotations along the geodesic, see [20] and Algorithm 12.4.1 of [14]; [7] suggests the rotation matrix should have determinant 1.

The geodesic $S(t)$ in $\mathbf{S}_{++}^p$ connecting S_A and S_B is:

$$S(t) = S_A^{\frac{1}{2}} \exp(t \log S_A^{-\frac{1}{2}} S_B S_A^{-\frac{1}{2}}) S_A^{\frac{1}{2}} . \tag{4}$$

$U(t)S(t)U^T(t)$ is the desired geodesic in $\mathbf{S}_+^{n,p}$. These calculations are efficient to perform. Note that $V_A^T V_B$ is 3×3, $U(t)$ is $n \times 3$ and $S(t)$ is 3×3.

3   Face

For the nonnegative orthant $\mathbb{R}_+^n = \{x \in \mathbb{R}^n : x \geq 0\}$, a particular face, F, is given by restricting some coordinates of $x = (x_1, \ldots, x_n)$ to zero [20]. For an abstract convex cone K, the definition of a face is as follows:

Definition 1. *[20] Let K be a convex cone, then a face F of this cone, is a subset $F \subseteq K$ with $x, y \in K$ and $x + y \in F \to x, y \in F$.*

For $K = \mathbf{S}_+^n$, a face F is determined by a fixed $U \in St(p, n)$:

$$F := U \mathbf{S}_+^p U^T = \left\{ Q \begin{bmatrix} B & 0 \\ 0 & 0 \end{bmatrix} Q^T : B \in \mathbf{S}_+^p \right\} . \tag{5}$$

For any $p \leq n$. $Q = [U \quad V]$ is an orthogonal matrix. Restricting B to be $B \in \mathbf{S}_{++}^p$ and $p < n$ gives the relative interior of a face F containing all rank deficient matrices of fixed rank p. The quotient manifold structure applies to this relative interior. The polar decomposition can be used to find the matrix $U \in St(p, n)$.

A group of atoms whose mutual distances are fixed is called a *clique* or *rigid cluster*. Cliques restrict the Gram matrix to be on a *particular* face of the semidefinite cone $\mathbf{S}_+^n$ [21]. A protein contains many cliques; tetrahedral carbons, and peptide planes are all examples given in [3]. In addition, the rigid clusters mentioned in [16][18] are large cliques.

3.1   Facial Reduction

If the Gram matrix $X = USU^T$ is on the face defined by V containing all the cliques, then we can get a smaller Gram matrix representing the flexible part of the protein by noting that $U = VW$. From:

$$X = USU^T = VWSW^TV^T \, , \tag{6}$$

we have WSW^T is the smaller Gram matrix from which the geodesic is calculated.

Facial reduction is the process of finding the matrix V. It starts by constructing faces for each single clique independently, assuming no other cliques exist; then, these faces are intersected to obtain the face containing all cliques. Finally, the face is centred.

For a single clique, [20] [21] provides the algorithm to determine its face. Rigid clusters are groups of atoms that do not overlap. Thus, rigid clusters are disjoint cliques. The face containing all these disjoint cliques is given by a block diagonal matrix where the diagonals are the faces of the individual cliques, also shown in [20] [21].

If we were to use the tetrahedral carbons and peptide planes as cliques as done in [3], then these cliques are not disjoint. To determine the face of all these cliques, we can intersect them using Theorem 12.4.2 of [14], as demonstrated in [3]. Disjoint cliques can also be intersected using [14], see [3] for an example showing this.

Intersecting non-disjoint cliques using [14] requires the SVD of larger and larger matrices as the face grows to its final size. This is the main bottleneck of facial reduction. For disjoint cliques, the complexity of facial intersection is trivial since we only need to form a diagonal block matrix.

4   Geodesic Interpolation

4.1   Methodology

We will now use the geodesic formulation in section 2.1 to interpolate lactoferrin's intermediate structures. Using UCSF Chimera [26], we fetched two models with PDB identifiers 1LFG and 1LFH. To ensure both structures have the same atoms, we deleted all side chain atoms, including β-carbons, and ligand atoms [1], leaving behind two structures with the same backbone atoms. There are 2765 atoms in total after deletion. We also overlapped the two proteins using the methodology described in [7] to ensure they have the same frame of reference.

The three disjoint rigid clusters identified in [16] were used as the cliques for facial reduction. After facial reduction, the protein's Gram matrix size decreased from 2765×2765 to 11×11. We only need to calculate the horizontal geodesic for this small 11×11 Gram matrix. Starting from 1LFG's Gram matrix, we moved forward in time from $t = 0$ to $t = 1$ for 1LFG at intervals of $\Delta t = 0.01$. Unlike the procedure described in [16], no further optimization was required at each time step.

---

[1] Also called Het atoms by Chimera.

4.2   Results

Figure 1 shows that by time $t = 1.0$, the two conformations lined up closely, while for intermediate values of t, the conformation smoothly advanced. The black structure is the end conformation.

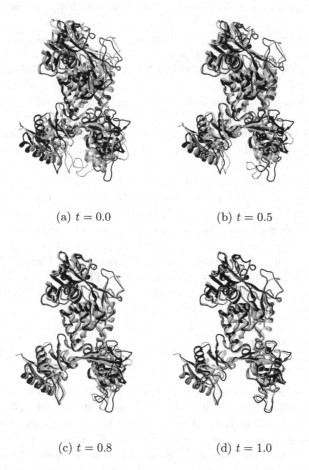

(a) $t = 0.0$ (b) $t = 0.5$

(c) $t = 0.8$ (d) $t = 1.0$

Fig. 1. Conformations along different times on the geodesic from 1LFG to 1LFH

The main observation to note is the α-helices have moved to align with the ending conformation. However, some of the loops seems to have not aligned by $t = 1$. Throughout the interpolation, the same loops also moved as a rigid body.

We also computed the geodesic without facial reduction. At the end when $t = 1$, all the loops did line up as shown in Fig. 2. This indicates that facial reduction allowed the rigid clusters to move together rigidly, whereas without facial reduction the same atoms showed more extensive range of motion.

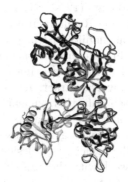

Fig. 2. 1LFG's geodesic at time $t = 1$ with no facial reduction; the loops align with 1LFH more closely.

Structure Quality. Since without facial reduction, the protein showed more degrees of freedom, we decided to check the quality of the structures in this case.

Similar to [3], we checked the chirality of the α-carbons since this is not encoded in the Gram matrix. We used the CORN rule by calculating the dihedral angle of CA, N, C, CB [3] [25]; therefore, we did not delete the β carbons in the two ending structures, although side chains were deleted as described in section 4.1. The CORN rule saw no violations for $0 \leq t \leq 1$.

We also compared the bond and geminal distances of the interpolated structures with that of the initial conformation. Let d_{ab}^{I} denote the bond or geminal distances between atoms a and b of the initial structure and let d_{ab}^{t} be those distances in the interpolated structure at time t. For all these distances we calculated $|d_{ab}^{I} - d_{ab}^{t}|$, and observed how the mean and standard deviation of these absolute values changed from $0 \leq t \leq 1$. We observed the average and standard deviation were both under $0.1\,\text{Å}$ all along the geodesic. This suggests bond and geminal distances were well-preserved while the points on the geodesic are in between the two ending conformations.

5   Principal Geodesic Analysis

5.1   Introduction

The geodesic connecting two conformations is only one path the Gram matrix can move. One way to find other geodesics is by using principal geodesic analysis (PGA).

PGA is a generalization of principal component analysis (PCA) to data on a Riemannian manifold [9] [10] [12] [13]. PCA is an established tool for understanding protein flexibility. In [27], the authors applied this methodology to HIV-1 protease data. They found the principal components by column-wise concatenating the atomic coordinates of selected sample conformations into a matrix, then the left singular vectors of this matrix give different, orthogonal, directions of motion called principal components (PCs). PCA assumes the data lie in Euclidean space, PGA is suited for any Riemannian manifold and can be adopted to $St(p,n)$, $\mathbf{S}^p_{++}$ and the quotient manifold $St(p,n) \times \mathbf{S}^p_{++}/O(p)$.

The idea behind PGA is to first determine the *intrinsic mean* of the data on the Riemannian manifold. Then the data is mapped onto the tangent space at the intrinsic mean using an "inverse retraction", also known as a *logarithmic map*. Once all the data is mapped onto this tangent space, they are vectorized (stacking the columns of the matrix to form a vector), PCA is performed on these vectors. The resulting PCs are tangent vectors with their origin at the intrinsic mean, and determine the principal geodesics in those directions.

For proteins, PGA does not work with the atomic coordinates directly. Since we have a quotient manifold, $St(p,n) \times \mathbf{S}^p_{++}/O(p)$, where $p = 3$, we first need to find PCs on the tangent space of $St(p,n)$ and $\mathbf{S}^p_{++}$ separately, then combine these vectors to form vectors pointing in the horizontal direction of the total space $St(p,n) \times \mathbf{S}^p_{++}$.

We now describe the PGA methodology and use it to find HIV-1 protease's principal geodesics.

5.2   Methodology

PGA Data. The various crystallizations of HIV-1 protease found in the PDB all have sequence differences due to mutations. To ensure the data is consistent, we again removed all non-backbone atoms, except for the β-carbons, using UCSF Chimera [26]. In addition, only structures with the same number of atoms after deletion were included in our analysis. 193 structures were selected.

We performed facial reduction using the peptide planes and tetrahedral α-carbons for the cliques as done in [3]. The Gram matrix size decreased from 966×966 to 571×571. Then, 1HIV was used as a reference structure, and all proteins were rotated onto it to ensure the coordinates have the same frame of reference [7].

Next, for each protein, its coordinates were placed in an $n \times 3$ matrix Y; then, Y is polar factorized to give $Y = UR$. Let $U = VW$, where V is the face from facial reduction, PGA is performed on this sample of W and $S = R^2$ matrices. The full Gram matrix is given by $VWSW^TV^T$. However, for conciseness, we ignore V and show what is to be done on WSW^T. In the following, we refer to W as the W component and S as the S component (of the Gram matrix X). This is not to be confused with *principal components*, which refer specifically to the vectors resulting from PCA.

We now introduce the Logarithmic (Log) and Exponential (Exp) maps. These are needed to calculate the intrinsic mean and for PGA. In the following discussion, $Sym(A) = (A + A^T)/2$ and $Skew(A) = (A - A^T)/2$.

Logarithmic (Log) and Exponential (Exp) Maps. The Exp map maps points on the tangent space onto the manifold. For any $\eta_S \in T_S\mathbf{S}^p_{++}$ we map it onto $\mathbf{S}^p_{++}$ using the formulae from [24] :

$$Exp_S(\eta_S) = S^{\frac{1}{2}}exp(S^{-\frac{1}{2}}\eta_S S^{-\frac{1}{2}})S^{\frac{1}{2}} . \tag{7}$$

For $\eta_W \in T_W St(p, n)$, we have the mapping from [1][24]:

$$Exp_W(\eta_W) = \mathrm{uf}(W + \eta_W) , \tag{8}$$

where $\mathrm{uf}(\cdot)$ extracts the orthogonal factor of the polar factorization.

The Log map maps points on the manifold onto the tangent space. For mapping any point $W' \in St(p, n)$ onto the tangent space at W, we proceed as suggested in [24] by first obtaining a direction in Euclidean space:

$$Z_W = W' - W , \tag{9}$$

then projecting Z_W onto $T_W St(p, n)$ using [24]:

$$\Psi(Z_W) = Z_W - W\mathrm{Sym}(W^T Z_W) . \tag{10}$$

$\mathbf{S}^3_{++}$ has a well known Logarithmic map, see for example [9][24]. For any $S' \in \mathbf{S}^3_{++}$, we have :

$$\mathrm{Log}_S(S') = S^{\frac{1}{2}} log(S^{-\frac{1}{2}} S' S^{-\frac{1}{2}}) S^{\frac{1}{2}} . \tag{11}$$

Intrinsic Mean. The algorithm for computing the intrinsic mean is given in [9] [10] [12] [13]. Let W_M and S_M denote the intrinsic mean for the sample of W matrices and S matrices respectively. Fig. 3 shows the conformation of 1HIV after setting its coordinates using W_M and S_M.

Fig. 3. 1HIV at W_M and S_M

As for the case of the geodesic connecting two points, if we calculate $|d^I_{ab} - d^M_{ab}|$ where d^I_{ab} represent bond and geminal distances in the original 1HIV conformation, and d^M_{ab} those distances in the intrinsic mean, the average of these absolute values are under 0.1 Å. This shows the intrinsic mean preserved bond and geminal distances well.

Horizontal Geodesics. For all proteins, we map their W and S components onto their respective tangent spaces at W_M and S_M using the Log map and performed PCA on these two tangent spaces, giving two sets of principal components.

At $X_M = W_M S_M W_M^T$, the tangent space of the total space is the product of the two tangent spaces: $T_{X_M}\mathcal{M} = T_{W_M}St(p,n) \times T_{S_M}\mathbf{S}_{++}^p$. Let $\eta_W \in T_{W_M}St(p,n)$ and $\eta_S \in T_{S_M}\mathbf{S}_{++}^p$. Any vector in the total tangent space, $\eta_X \in T_{X_M}\overline{\mathcal{M}}$ [2] can be projected onto the horizontal direction via:

$$\Pi_X(\eta_X) = (\eta_W - W\Omega, \eta_S - (S\Omega - \Omega S)) . \tag{12}$$

Here, Ω is a skew symmetric matrix found by solving the Lyapunov equation [22] [24].

$$\Omega S^2 + S^2\Omega + S\Omega S = SSkew(W^T\eta_W) + SSym(\eta_S) - Sym(\eta_S)S . \tag{13}$$

This equation is derived from the metric on the total tangent space at X_M, see [24] [22]. Therefore, the horizontal direction is uniquely determined by a pair of vectors $\eta_W \in T_{W_M}St(p,n)$ and $\eta_S \in T_{S_M}\mathbf{S}_{++}^p$.

Given the horizontal directions, we calculated the geodesic using formulae from [8] and [6]. A geodesic curve emanating from W_M in the horizontal direction Δ is given by:

$$W(t) = W_M Q^T \cos(\Gamma t)Q + P\sin(\Gamma t)Q , \tag{14}$$

where $P\Gamma Q = \Delta$ is the compact SVD of Δ. The horizontal geodesic emanating from S_M in the direction D is [6]:

$$S(t) = S_M^{\frac{1}{2}}\exp(tS_M^{-\frac{1}{2}}DS_M^{-\frac{1}{2}})S_M^{\frac{1}{2}} . \tag{15}$$

5.3   Results

We examined the geodesic for various U and S PCs. Figure 4 gives some examples. We observed changing S modes had little influence on the protein's dynamics. Most of dominant U modes showed the binding pocket or the flap deform in various ways. For example, figure 4 shows at the first U and S PC the flaps twist. The fourth PC for U was more interesting in that it showed the flaps open and close. These structures were taken at $t = 0.25$.

Structure Quality. The geodesic for the above structures all maintained the correct chirality at time $t = 0.25$. The bond length and geminals show slightly more stretching than in the case where the geodesic was between two structures. Using the 4th PC for U and 1st PC for S, both in the positive direction at time $t = 0.25$ as an example, we calculated $|d_{ab}^I - d_{ab}^t|$ for all bonds and geminals in the conformation. We found the average of these absolute values for bond and geminal distance to be 0.113Å and 0.186Å respectively, with a standard deviation of 0.2086Å and 0.3292Å. The stretching increased with t. We will address this issue more in future papers.

---

[2] Vectors in the total tangent space are often denoted with a bar on top, e.g. $\overline{\eta}_X$. We drop the bar for simplicity

(a) 1st PC of U in negative direction, 1st PC of S in positive direction

(b) 1st PC of U in positive direction, 1st PC of S in positive direction

(c) 4th PC of U in negative direction, 1st PC of S in positive direction

(d) 4th PC of U in positive direction, 1st PC of S in positive direction

Fig. 4. 1HIV's various U and S PC combinations on the geodesic at time $t = 0.25$

6   Conclusion

In this paper we viewed the protein as a point on the positive semidefinite matrix manifold using a quotient geometry. We used the manifold's geodesic to generate intermediate conformations and performed PGA. In both cases, we used facial reduction to work with a smaller Gram matrix with the cost being performing SVD. We feel these tools are in the early stages and hope this Riemannian perspective can lead to more efficient algorithms for studying structural changes.

Acknowledgment. We like to thank B. Mishra for sharing his insights on matrix manifold optimization algorithms.

References

1. Absil, P.A., Mahony, R., Sepulchre, R.: Optimization Algorithms on Matrix Manifolds. Princeton University Press, Princeton (2008)
2. Alfakih, A., Khandani, A., Wolkowicz, H.: Solving euclidean distance matrix completion problems via semidefinite programming. Computational Optimization and Applications 12(1-3), 13–30 (1999)
3. Alipanahi, B.: New Approaches to Protein NMR Automation. Ph.D. thesis, University of Waterloo (2011)
4. Alipanahi, B., Krislock, N., Ghodsi, A., Wolkowicz, H., Donaldson, L., Li, M.: Determining protein structures from noesy distance constraints by semidefinite programming. Journal of Computational Biology 20(4), 296–310 (2013)
5. Biswas, P., Toh, K.C., Ye, Y.: A distributed SDP approach for large-scale noisy anchor-free graph realization with applications to molecular conformation. SIAM Journal on Scientific Computing 30(3), 1251–1277 (2008)
6. Bonnabel, S., Sepulchre, R.: Riemannian metric and geometric mean for positive semidefinite matrices of fixed rank. SIAM J. Matrix Anal. Appl. 31(3), 1055–1070 (2009), http://dx.doi.org/10.1137/080731347
7. Burkowski, F.J.: Structural Bioinformatics: An Algorithmic Approach. Chapman & Hall/CRC (2009)
8. Edelman, A., Arias, T.A., Smith, S.T.: The geometry of algorithms with orthogonality constraints. SIAM J. Matrix Anal. Appl. 20(2), 303–353 (1998)
9. Fletcher, P.T., Joshi, S.: Principal geodesic analysis on symmetric spaces: Statistics of diffusion tensors. In: Sonka, M., Kakadiaris, I.A., Kybic, J. (eds.) CVAMIA-MMBIA 2004. LNCS, vol. 3117, pp. 87–98. Springer, Heidelberg (2004)
10. Fletcher, P.T., Joshi, S.: Riemannian geometry for the statistical analysis of diffusion tensor data. Signal Processing 87(2), 250–262 (2007)
11. Gerstein, M., Krebs, W.: A database of macromolecular motions. Nucleic Acids Res. 26, 4280–4290 (1998)
12. Goh, A.: Riemannian manifold clustering and dimensionality reduction for vision-based analysis. In: Machine Learning for Vision-Based Motion Analysis, pp. 27–53. Springer (2011)
13. Goh, A., Vidal, R.: Clustering and dimensionality reduction on Riemannian manifolds. In: IEEE Conference on Computer Vision and Pattern Recognition, CVPR 2008, pp. 1–7. IEEE (2008)
14. Golub, G.H., Van Loan, C.F.: Matrix Computations, 3rd edn. Johns Hopkins University Press, Baltimore (1996)
15. Journée, M., Bach, F., Absil, P.A., Sepulchre, R.: Low-rank optimization on the cone of positive semidefinite matrices. SIAM J. Optim. 20(5), 2327–2351 (2010)
16. Kim, M.K.: Elastic Network Models of Biomolecular Structure and Dynamics. Ph.D. thesis, The Johns Hopkins University (2004)
17. Kim, M.K., Jernigan, R.L., Chirikjian, G.S.: Efficient generation of feasible pathways for protein conformational transitions. Biophysical Journal 83(3), 1620–1630 (2002)
18. Kim, M.K., Jernigan, R.L., Chirikjian, G.S.: Rigid-cluster models of conformational transitions in macromolecular machines and assemblies. Biophysical Journal 89(1), 43–55 (2005)
19. Kleywegt, G.J., Jones, T.A.: Phi/psi-chology: Ramachandran revisited. Structure 4, 1395–1400 (1996)

20. Krislock, N.: Semidefinite facial reduction for Low-Rank Euclidean Distance Matrix Completion. Ph.D. thesis, School of Computer Science, University of Waterloo (2010)
21. Krislock, N., Wolkowicz, H.: Explicit sensor network localization using semidefinite representations and facial reductions. SIAM Journal on Optimization 20(5), 2679–2708 (2010)
22. Meyer, G.: Geometric optimization algorithms for linear regression on fixed-rank matrices. Ph.D. thesis, University of Liège (2011)
23. Mishra, B., Meyer, G., Sepulchre, R.: Low-rank optimization for distance matrix completion. In: 2011 50th IEEE Conference on Decision and Control and European Control Conference (CDC-ECC), Orlando, FL, USA, December 12-15, pp. 4455–4460 (2011)
24. Mishra, B., Meyer, G., Bach, F., Sepulchre, R.: Low-rank optimization with trace norm penalty. arXiv preprint arXiv:1112.2318 (2011)
25. Morris, A.L., MacArthur, M.W., Hutchinson, E.G., Thornton, J.M.: Stereochemical quality of protein structure coordinates. Proteins: Structure, Function, and Bioinformatics 12(4), 345–364 (1992)
26. Pettersen, E.F., Goddard, T.D., Huang, C.C., Couch, G.S., Greenblatt, D.M., Meng, E.C., Ferrin, T.E.: UCSF Chimera–a visualization system for exploratory research and analysis. J. Comp. Chem. 25(13), 1605–1612 (2004)
27. Teodoro, M.L., Phillips Jr., G.N., Kavraki, L.E.: Understanding protein flexibility through dimensionality reduction. Journal of Computational Biology 10(3-4), 617–634 (2003)
28. Vandereycken, B.: Riemannian and multilevel optimization for rank-constrained matrix problems. Ph.D. thesis, Department of Computer Science, KU Leuven (2010)
29. Vonrhein, C., Schlauderer, G.J., Schulz, G.E.: Movie of the structural changes during a catalytic cycle of nucleoside monophosphate kinases. Structure 3, 483–490 (1995)

Joint Analysis of Functional and Phylogenetic Composition for Human Microbiome Data

Xingpeng Jiang[1], Xiaohua Hu[1], and Weiwei Xu[2]

[1] College of Computing and Informatics, Drexel University, Philadelphia, 19104 PA
{xpjiang,xh29}@drexel.edu
[2] International School of Software, Wuhan University, Wuhan, Hubei, China
jennifer84830@gmail.com

Abstract. With the advance of high-throughput sequencing technology, it is possible to investigate many complex biological and ecological systems. The objective of Human Microbiome Project (HMP) is to explore the microbial diversity in our human body and to provide experimental and computational standards for subsequent similar studies. The first-stage HMP generated a lot of data for computational analysis and provided a challenge for integration and interpretation of various microbiome data. In this paper, we introduce a data integration method –Laplacian-regularized **J**oint **N**on-negative **M**atrix **F**actorization (LJ-NMF) for analyzing functional and phylogenetic profiles from HMP jointly. The experimental results indicate that the proposed method offers an efficient framework for microbiome data analysis.

Keywords: Metagenomics, Non-negative matrix factorization, Laplacian regularization, Human Microbiome Project.

1 Introduction

It was very hard to study the functional and phylogenetic diversity of a microbial ecosystem until Handelsman introduced the technology of metagenomics [Handelsman et al., 1998]. In recent years, the cost of high-throughput sequencing technology including metagenomic and rRNA (16S or 18S) target sequencing has decreased dramatically. Consequently the investigation of microbial community in large-scale is possible and a lot of big projects have been launched. The Global Ocean Sampling Expedition (GOS)[Venter et al., 2004, Rusch et al., 2007] was the first large-scale metagenomic project which used Sanger sequencing to explore hundreds of marine microbial communities. The first stage of Human Microbiome Poject (HMP) was successively completed in 2012 and the second stage merits attention [Cho and Blaser, 2012, Consortium, 2012]. HMP generated a lot of DNA sequences data which can be computationally summarized to represent the functional or phylogenetic compositions of the microbiome community. These composition profiles are usually analyzed separately to explore sample variety. Another large-scale project is MetaHIT whose goal is to explain the associations between the genes of the human intestinal microbiota and health

M. Basu, Y. Pan, and J. Wang (Eds.): ISBRA 2014, LNBI 8492, pp. 346–356, 2014.

and disease [Qin et al., 2010]. Using MetaHIT data and two other datasets, Arumugam et al.'s analysis indicated that gut microbiome could be classified into enterotypes [Arumugam et al., 2011]. However, several other studies suggested that more dataset should be added and rigid methods should be proposed to draw a reasonable conclusion about gut microbiome classification. For example, Koren et al. studied how various factors influenced the detection of enterotypes including clustering methodology and distance metrics [Koren et al., 2013]. We should mention that these methods are all based on clustering methods although Jeffery et al. [Jeffery et al., 2012] argued that gut microbiome may be described better by gradients instead of enterotype classifications. There is few method to integrate functional and phylogenetic information together to study microbial communities although the combination of different measurements could provide a comprehensive view.

Non-negative matrix factorization (NMF) has also been applied to analyze microbiome datasets [Jiang et al., 2012b, Jiang et al., 2012a]. NMF can be viewed as a data representation method–samples or variables could be represented by a linear combination of a few number of NMF basis vectors or a clustering method–samples or variables can be assigned a membership label by the basis vectors of NMF. NMF has an advantage that it is a soft-clustering approach by which a sample can be classified into several clusters if it tend to play multiple roles. NMF has many successful applications in different fields including but not limited to gene expression [Brunet et al., 2004] analysis, fMRI [Lohmann et al., 2007] and computer vision [Guillamet and Vitria, 2002]. A joint NMF (JNMF) approach has been proposed for integrating multimodality data for multiview clustering in image analysis [Akata et al., 2011]. JNMF simultaneously determine the latent dimensions (basis) in different feature spaces which is faithful to the different characteristics of various measurements. In this study, we develope a Laplacian-regularized JNMF (LJ-NMF) to integrate functional and phylogenetic information for analyzing microbiome data from HMP. LJ-NMF is an extension of JNMF which adopts the manifold structure of data by using a Laplacian regularization [Cai et al., 2011]. The experimental results of LJ-NMF on HMP data show that it achieves superior results over state-of-the-art in terms of computational efficiency and interpretation.

The main contribution in this paper presents an effective approach for data integration of various microbiome data. LJ-NMF is a common framework that can be applied in other fields if there are multiple measurements for samples. The rest of the paper is organized as follows: Section 2 reviews some basic knowledge of NMF, and JNMF extension in [Akata et al., 2011] and our proposed extension LJ-NMF based on Laplacian regularization. Section 3 presents our evaluation on HMP data and compare our method with other NMF-based approaches.

2   Methods

2.1   Data Decomposition Using NMF

Given a $p \times s$ matrix X, NMF is seeking to find two matrices W and H with lower dimensions (W is a $p \times k$ and H is a $k \times s$ respectively, where $k \ll \{p, s\}$ is the *degree* of a factorization) so that $WH \approx X$, by minimizing an objective function under the prerequisite that W and H must be non-negative. The objective function we use is the least squares

$$J := \min_{s.t. W, H \geq 0} \|X - WH\|^2 \ . \tag{1}$$

NMF can be used as a data representation approach in two dually view. Firstly, let's consider the columns of the matrix X, and denote them by $X_1, X_2, ..., X_s$. If we regard $W_1, W_2, ..., W_k$ (the columns of W) as the basis vector and H_{ij} as the coefficients, then X_j can be represented by a linear additive sum of $W_1, W_2, ..., W_k$:

$$X_j \approx \sum_i^k H_{ij} W_i$$

Dually, We denote the transpose of X's rows as $L_1, L_2, ..., L_m$ and consider $H_1, H_2, ..., H_k$ (H_j is the transpose of the jth row of H) as the basis and W_{ij} as the coefficients, then L_j can be represented by a linear additive sum of $H_1, H_2, ..., H_k$:

$$L_j \approx \sum_{i=1}^k W_{ji} H_i$$

NMF can also be used for classification. Naturally, the columns of the coefficient matrix H can be used to assign membership labels for each sample, and dually the rows of W can be used to classify variables or features.

2.2   Multiview Clustering Using JNMF

In microbiome studies, it is common to investigate functional or phylogenetic composition profiles for microbial community studies. However, current methods are often used to analyze the two profiles separately without taking advantage of data fusion. Our main motivation in this study is to integrate the two profiles. We adopt a recent expansions of NMF to multiview learning–Joint NMF. Given a $p \times s$ matrix X and $t \times s$ matrix Y, the basic idea of JNMF is to find k basis vectors $W \in R^{p \times k}$ and $V \in R^{t \times k}$ for both profiles that share a common coefficient matrix $H \in R^{k \times s}$. That is:

$$X \approx WH$$

and

$$Y \approx VH$$

JNMF is formulated as a convex combination of two single NMF on X and Y respectively. The objective function of JNMF is based on constrained least square construction:

$$J := \min_{s.t.W,V,H\geq 0} (1 - \alpha) \|X - WH\|^2 + (\alpha) \|Y - VH\|^2 . \tag{2}$$

where $\alpha \in [0,1]$ can be viewed as the weight of each factorization. Optimization of the objective function can be obtained by a similar multiplicative update rules as presented in [Akata et al., 2011, Eweiwi et al., 2013], with a small adjustment for updating H. The updating rules for W and V are

$$W_{au} \leftarrow W_{au} \frac{(XH^T)_{au}}{(WHH^T)_{au}} \tag{3}$$

and

$$V_{au} \leftarrow V_{au} \frac{(YH^T)_{au}}{(VHH^T)_{au}} \tag{4}$$

while the updating rule for the common coefficient matrix H is

$$H_{au} \leftarrow H_{au} \frac{((1 - \alpha)W^T X + \alpha V^T Y)_{au}}{(((1 - \alpha)W^T W + \alpha V^T V)H)_{au}} \tag{5}$$

2.3   LJ-NMF for Data Integration

Many studies of NMF approach and its extensions have used regularization to penalize the coefficient matrix W or H. Among these, Cai et al. proposed a Graph constrained NMF (GNMF) by using Laplacian regularization for single-view NMF by keeping manifold structure of data [Cai et al., 2011]. In this study, we introduce Laplacian regularization on JNMF and we derive the updating formula for this new method.

Graph Construction for Laplacian Regularization. Before introducing LJ-NMF, we first construct a weight graph G from X. A number of techniques for defining a weight graph G can be used but we only use one method– heat kernel weighting method, because it is a widely used method in most manifold learning algorithms. The testing of the graph construction methods is interesting but not the principle scope of this study. This technique is to define the weight matrix A for graph G using distance matrix d among data points:

$$A_{ij} = e^{(-d_{ij}^2)/\sigma^2}$$

After getting A, the graph Laplacian is defined as $L = D - A$, where D is a degree matrix. The off-diagonal elements of D are all zeros and along the diagonal is the sum of all edge weights to the data points. The graph Laplacian keeps the information of the geometric structure of the data [Cai et al., 2011].

LJ-NMF Objective Function and Optimization. The objective function for LJ-NMF is defined by adding a Laplacian penalty terms to JNMF:

$$J := \min_{s.t.W,V,H \geq 0} (1 - \alpha) \|X - WH\|^2 + (\alpha) \|Y - VH\|^2 + \lambda Tr(H^T LH) . \quad (6)$$

where L is the Laplacian matrix of the constructed weight graph from data. Next we try to solve the optimization problem by a Lagrange multiplier method. Considering the constraint that W,V and H are nonnegative and let ϕ, ψ and χ be the Lagarange multiplier matrices for the respective constraints, the Lagrange L of the objective function is

$$L = (1 - \alpha) \|X - WH\|^2 + (\alpha) \|Y - VH\|^2 + \lambda Tr(H^T LH)$$
$$+ Tr(\phi W^T) + Tr(\psi V^T) + Tr(\chi H^T). \quad (7)$$

The partial derivatives of L, with respect to W,V and H are

$$\frac{\partial L}{\partial W} = -2XW^T + 2HWW^T + \phi \quad (8)$$

$$\frac{\partial L}{\partial V} = -2YV^T + 2HVV^T + \psi \quad (9)$$

$$\frac{\partial L}{\partial H} = -2(1-\alpha)W^T X - 2\alpha(V^T Y) + 2H((1-\alpha)W^T W + \alpha V^T V) + 2\lambda HL + \chi \quad (10)$$

By using Karush-Kuhn-Tucker conditions $(\phi W)_{ij} = 0, (\psi V)_{ij} = 0$ and $(\chi H)_{ij} = 0$, we can see that the updating rules for W and H derived from the equations 8 and 9 are the same to the equations 3 and 4. For H, we can get its updating formula from the equation 10:

$$H_{au} \leftarrow H_{au} \frac{((1 - \alpha)W^T X + \alpha V^T Y + \lambda HA)_{au}}{(((1 - \alpha)W^T W + \alpha V^T V)H + \lambda HD)_{au}} \quad (11)$$

Combing equations 3, 4 and 11 we can get the optimization solution for LJ-NMF objective function in equation 6.

3   Data Integration of HMP Data by LJ-NMF

3.1   Data from HMP

We use the functional and phylogentic profiles from Human Microbiome Project. By filtering body sites with less than 50 samples, the dataset contains 613 samples drawn from six body sites including one vagina(posterior fornix), one gut(stool), one nasal(anterior nares), and three oral sites (supragingival plaque, tongue dorsum and buccal mucosa). The phylogenetic profile which contains the microorganism relative abundances was estimate by software MetaPhlAn at species level (X has size 696×613) [Abubucke et al., 2012]. For functional profile, we investigate the transporter profile (Y has size 366×613) by filtering out those with low variances. The data were all downloaded from HMP data site: http://hmpdacc.org/. The computational analysis is run on R environment and all codes are available upon request.

3.2 Results

Model Selection and Method Comparison. From the data, we can see that six is a reasonable k because there are six distinct body sites, so that we investigate the influence of α and λ on the model. After clustering samples into six groups by using the coefficient matrix H, we compare the ratio of intra-class similarity to inter-class similarity to qualify the clustering performance. This ratio is analogue to the objective function of Fisher's linear discriminant which is defined as the ratio of the variance between the classes to the variance within the classes. From a number of experiments, we find that the similarity ratio decrease dramatically when $\alpha > 0.5$ or $\lambda > 50$, thus we provide the following candidate sets for model selection: $\alpha \in \{0, 3^{-(5:1)}, 0.5\}$ and $\lambda \in \{0, 0.001, 0.01, 10, 20, 30, 50\}$. Figure 1 shows the ratios for model selection. The largest ratio indicate that the model has the best performance at $\alpha = 0.012$ and $\lambda = 10$, thus they will be used for detail investigations of HMP data.

It is necessary to remind that LJ-NMF is a general framework and NMF, GNMF and JNMF are its special cases. It will be reduced to the original JNMF if $\lambda = 0$. Furthermore, it will become a single-view NMF on a phylogenetic profile X when $\alpha = 0$ and $\lambda = 0$. When $\alpha = 0$ and $\lambda \neq 0$, LJ-NMF will be reduced to graph-constrained NMF [Cai et al., 2011]. We also duplicate the experiment of a single-view NMF on a functional profile Y, with this case $\alpha = 1$ and the performane is not shown in Figure 1. We find that its performance is not better than the selected model. All our results suggest that LJ-NMF has better performance than NMF, GNMF and JNMF for distinguishing sample clusters. The proposed data integration method could be a promising framework for microbiome analysis.

Habitat Clustering by JL-NMF. Clustering microbiome is an important method to interpret the relationship and difference among microbial communities. Recent analysis of human-associated microbiome has categorized individuals into enterotypes or clusters based on the abundances of microorganisms in the gut microbiota [Wu et al., 2011, Arumugam et al., 2011]. Jeffery et al. argued that the grouping of individual subjects into categories may have oversimplified a complex situation [Jeffery et al., 2012]. Recently Koren et al. redefined the "enterotype" as microbiota types across different body sites and they tested many computational and experimental factors that may influence the detection of enterotypes[Koren et al., 2013]. They recommended that multiple approaches should be used and compared when testing enterotypes. Contrarily, NMF and its extensions are not hard clustering technology. NMF based approach don't

Model Selection of LJ-NMF

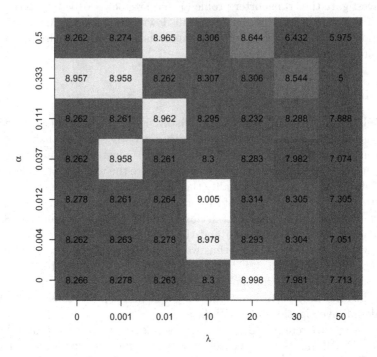

Fig. 1. The model selection of α and λ for LJ-NMF. The number (thus the color) in each box indicate the ratio of intra-class similarity to inter-class similarity after the LJ-NMF clustering under the corresponding parameters combination.

classify a sample into one certain group. Based on the coefficient matrix H, a sample may belongs to several groups if its weights in different column of H are high. This suggest that NMF based methods may be a good candidates to describe and interpret microbiomes.

We investigate the clustering structure of HMP data using LJ-NMF. As Figure 2 shows, LJ-NMF clearly identifies clusters which correspond to microbiome samples of six body sites. The most distinct clusters are from nasal(black), gut(blue) and vagina(green). Although three oral sites also have clustering structure, samples in one oral site may have overlapping with another oral site. This is not surprising that these three sites are all from mouth and they may have similar bacterial composition and diversity.

Visualization of data is at the central of exploratory data analysis. The first stage of data computational analysis is to make sense of the data intuitively before going ahead with more objective-directed modeling and analysis. NMF is not designed for visualization naturally because its objective function is not projecting high-dimensional space into two or three spaces. Thus we used a Multidimensional Scaling (MDS) [Torgerson, 1952] on H to visualize relationships among samples in two or three dimensional space. In two dimensional space,

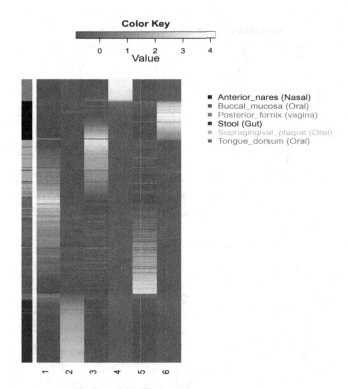

Fig. 2. Habitat clustering by H from LJ-NMF. The 613×6 matrix H visualized in a heatmap with colors (see Color Key above) indicate the coefficient strength. The left of the figure indicates the habitat of microbiome samples with colors representing different body sites. The right legend provides the names of these body sites with corresponding coloring code.

the result is similar to the Koren's paper [Koren et al., 2013]. But the three dimensional visualization provide more surprising patterns (see Figure 3). This support Jeffery et al. 's argument that the variation at the species level of human microbiome is continuous and irrelevant to the existence of discontinuous clusters (enterotype).

4   Discussion

In this report, we combine functional and phylogenetic composition profiles into one framework to analyze microbiome data. We develop a novel algorithm which can be used in any data integration tasks where multiple measurements are available. The Laplacian penalty of LJ-NMF captures the geometric structure (manifold) form both phylogenetic and functional composition data. Microbial community is complex not only because of the delicacy interactions among species but also the complicate impacts from host environment. Thus the nonlinear geometric structure is more suitable for modeling microbial community than

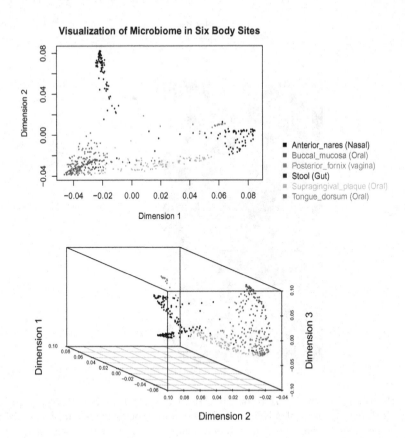

Fig. 3. The visualization of samples in two dimension and three dimension Euclidean space from LJ-NMF

linear or Euclidean space-based methods such as (Principle Component Analysis) PCA [Abdi et al., 2010] and MDS [Torgerson, 1952]. We compares the results with several previous state-of-art extension of NMF including GNMF and JNMF. The result suggests that LJ-NMF is a more versatile framework than these methods for data integration and dimension reduction.

On application, we find that the proposed method is very useful for analyzing HMP data. By analyzing and visualizing coefficient matrix H we can get clear patterns among microbiome from different body sites. Because the method is not a hard-clustering method and it do not assign sampels to a single membership, LJ-NMF provide more flexible analysis than hard-clustering based methods. In addition, LJ-NMF can be easily extended to multiple data types when necessary although we only combine two data types inspired by HMP for this study.

References

[Abdi et al., 2010] Abdi, H., Williams, L.J., Abdi, H., Williams, L.J.: Principal component analysis. Wiley Interdisciplinary Reviews: Computational Statistics 2(4), 433–459 (2010)

[Abubucke et al., 2012] Abubucker, S., Segata, N., Goll, J., et al.: Metabolic Reconstruction for Metagenomic Data and Its Application to the Human Microbiome. PLoS Comput. Biol. 8(6), e1002358 (2012)

[Akata et al., 2011] Akata, Z., Thurau, C., Bauckhage, C.: Non-negative matrix factorization in multimodality data for segmentation and label prediction (2011)

[Arumugam et al., 2011] Arumugam, M., Raes, J., Pelletier, E., et al.: Enterotypes of the human gut microbiome. Nature 473(7346), 174–180 (2011), PMID: 21508958

[Brunet et al., 2004] Brunet, J.-P., Tamayo, P., Golub, T.R., Mesirov, J.P.: Metagenes and molecular pattern discovery using matrix factorization. PNAS 101(12), 4164–4169 (2004), PMID: 15016911

[Cai et al., 2011] Cai, D., He, X., Han, J., Huang, T.: Graph regularized nonnegative matrix factorization for data representation. IEEE Transactions on Pattern Analysis and Machine Intelligence 33(8), 1548–1560 (2011)

[Cho and Blaser, 2012] Cho, I., Blaser, M.J.: The human microbiome: at the interface of health and disease. Nature Reviews Genetics 13(4), 260–270 (2012)

[Consortium, 2012] The Human Microbiome Project Consortium: A framework for human microbiome research. Nature 486(7402), 215–221 (2012)

[Eweiwi et al., 2013] Eweiwi, A., Cheema, M.S., Bauckhage, C.: Discriminative joint non-negative matrix factorization for human action classification. In: Weickert, J., Hein, M., Schiele, B. (eds.) GCPR 2013. LNCS, vol. 8142, pp. 61–70. Springer, Heidelberg (2013)

[Guillamet and Vitria, 2002] Guillamet, D., Vitria, J.: Classifying Faces with Nonnegative Matrix Factorization (2002)

[Handelsman et al., 1998] Handelsman, J., Rondon, M.R., Brady, S.F., Clardy, J., Goodman, R.M.: Molecular biological access to the chemistry of unknown soil microbes: a new frontier for natural products. Chem. Biol. 5(10), R245–R249 (1998), PMID: 9818143

[Jeffery et al., 2012] Jeffery, I.B., Claesson, M.J., O'Toole, P.W., Shanahan, F.: Categorization of the gut microbiota: enterotypes or gradients? Nat. Rev. Micro. 10(9), 591–592 (2012)

[Jiang et al., 2012a] Jiang, X., Langille, M.G.I., Neches, R.Y., Elliot, M., Levin, S.A., Eisen, J.A., Weitz, J.S., Dushoff, J.: Functional biogeography of ocean microbes revealed through non-negative matrix factorization. PLoS ONE 7(9), e43866 (2012a)

[Jiang et al., 2012b] Jiang, X., Weitz, J., Dushoff, J.: A non-negative matrix factorization framework for identifying modular patterns in metagenomic profile data. Journal of Mathematical Biology 64(4), 697–711 (2012b)

[Koren et al., 2013] Koren, O., Knights, D., Gonzalez, A., Waldron, L., Segata, N., Knight, R., Huttenhower, C., Ley, R.E.: A guide to enterotypes across the human body: Meta-analysis of microbial community structures in human microbiome datasets. PLoS Comput. Biol. 9(1), e1002863 (2013)

[Lohmann et al., 2007] Lohmann, G., Volz, K.G., Ullsperger, M.: Using non-negative matrix factorization for single-trial analysis of fMRI data. NeuroImage 37(4), 1148–1160 (2007)

[Qin et al., 2010] Qin, J., Li, R., Raes, J.: A human gut microbial gene catalogue established by metagenomic sequencing. Nature 464(7285), 59–65 (2010)

[Rusch et al., 2007] Rusch, D.B., Halpern, A.L., Sutton, G.: The sorcerer II global ocean sampling expedition: Northwest atlantic through eastern tropical pacific. PLoS Biology 5(3), e77 (2007)

[Torgerson, 1952] Torgerson, W.S.: Multidimensional scaling: I. Theory and method. Psychometrika 17(4), 401–419 (1952)

[Venter et al., 2004] Venter, J.C., Remington, K., Heidelberg, J.F., Halpern, A.L., Rusch, D., Eisen, J.A., Wu, D., Paulsen, I., Nelson, K.E., Nelson, W., et al.: Environmental genome shotgun sequencing of the sargasso sea. Science 304(5667), 66–74 (2004)

[Wu et al., 2011] Wu, G.D., Chen, J., Hoffmann, C., et al.: Linking long-term dietary patterns with gut microbial enterotypes. Science 334(6052), 105–108 (2011)

[Yooseph et al., 2007] Yooseph, S., Sutton, G., Rusch, D.B., et al.: The sorcerer II global ocean sampling expedition: Expanding the universe of protein families. PLoS Biology 5(3), e16 (2007)

schematikon: Detailed Sequence-Structure Relationships from Mining a Non-redundant Protein Structure Database

(Extended Abstract)

Boris Steipe[1,2] and Bhooma Thiruv[3]

[1] Department of Biochemistry, University of Toronto,
1 Kings College Circle, Toronto, Ontario, M5S 1A8, Canada
[2] Department of Molecular Genetics, University of Toronto,
1 Kings College Circle, Toronto, Ontario, M5S 1A8, Canada
boris.steipe@utoronto.ca
[3] The Center for Applied Genomics (TCAG), The Hospital for Sick Children,
MaRS Centre, East Tower, 101 College Street, Room 14-701,
Toronto, Ontario, M5G 1L7, Canada
bthiruv@sickkids.ca

Abstract. If a "protein folding code" exists, it ought to give rise to detectable sequence propensities that are associated with low energy conformations, *i.e.* native structure. To the degree that the frequency of structure patterns in folded proteins has a Boltzmann-like behaviour, such conformations should be detectable by their excess occurrence over random. We have mined a database of non-homologous, well resolved protein structure domains – Nh3D – and have discovered an abundance of such sets of overrepresented structurally similar patterns. We designate the best representatives of a set a *motif*. Our motif dictionary *schematikon* shows significant and interesting sequence propensities and is predictive regarding the experimentally determined consequences of sequence change on stability.

Keywords: structure motif, 3D motif, structure space, data mining, protein structure, protein engineering, protein stability.

1 Introduction

All known life forms utilize linear sequential information storage systems (DNA) to produce complex, self-organizing systems in space (proteins). It is well known that the linear information encoded in the sequence of the 20 proteinogenic amino acids in general suffices to uniquely specify a protein's folded 3D structure. However defining the precise rules of self-organization, and exploiting them in applications of structure prediction and protein engineering is not a generally solved problem, despite recent accomplishments (see e.g. [6], [4], [9]). Protein sequences arise from a stochastic

M. Basu, Y. Pan, and J. Wang (Eds.): ISBRA 2014, LNBI 8492, pp. 357–366, 2014.

evolutionary process, subject to selection of function over long timescales. We can reason that such a process leads to Boltzmann-like frequency distributions of features with regard to their role in determining the free energy of structures [11] and such features include sequences and 3D patterns. Since low-energy patterns should be observed with high frequency, this motivates an interest in defining overrepresented patterns both in sequence and in structure databases. A significant number of algorithms and approaches towards finding recurring patterns in a structure dataset have been proposed. However any interpretation of such recurring patterns in terms of their statistical free-energy properties must take sampling bias into account. It is well known that homologous proteins (i.e. proteins that have descended in evolution from a common ancestor) have similar structures – even in the absence of detectable sequence similarity. Recurrence of structural patterns in homologous proteins thus gives no useful information regarding their underlying properties. Therefore we have previously defined Nh3D – a database of non-homologous protein domains [12]. Here we present results on exhaustively mining this database for recurring, short linear patterns with an unsupervised procedure to discover structural *motifs* from the peptide backbones of patterns that are 3 to 15 amino acids in length. We have constructed a database that retains only such motifs that recur significantly more frequently than expected by random chance, we report motifs of maximal length, not their shorter fragments, we filter motifs to be robust against random deletions of the source data and we ensure that backbone atoms in longer motifs are spatially coherent. With this motif dictionary – *schematikon* – we have generated an unbiased, comprehensive resource of the structural building blocks of proteins. Finally, we show that the amino acid propensities derived from the motifs are predictive for protein stability changes.

Recurrent structural patterns in proteins are hierarchically organized. At the level of protein folds or topologies the number of folds in nature is thought to be finite [5], on the order of 1000 to 5000. Databases like SCOP and CATH have provided hierarchical descriptions of protein structure where folds/topologies comprise one level. In general, proteins of different topologies do not share detectable common ancestors. This provides a measure on which to base a selection of non-homologous proteins. At the level of local 3D structure or super-secondary structure two or more secondary structural elements interact. At the level of secondary structure, stretches of amino acids with recurring backbone torsion angles form helices or extended strands, on average accounting for over 50% of protein structure. The rest of the structure can be viewed as connecting loops. We focus here on small, local patterns that represent a limited set of possibilities to achieve energetically favorable conformations in each context. These short local patternsinclude secondary structural elements as well as organizing principles in super-secondary structure.

Previous approaches have often focused on sequence patterns in proteins that correlate with structure. Such approaches suffer from being able to detect only motifs with significant information at the sequence level, or an *ad hoc* selection of patterns according to preconceived notions of motifs of interest, many of which go back to a classic review by Jane Richardson [8]. Nevertheless the general utility of such patterns has been amply demonstrated e.g. with the Isites library [2] used in the Rosetta server for *ab initio* structure prediction.

2 Methods

2.1 The Non-redundant Structure Dataset

Our non redundant structure subset has been previously described [12]. Briefly, we extract co-ordinate subsets from the Protein Data Bank driven by CATH domain topology information [7]. The experimentally best-defined representative for each distinct topology is selected and validated to exclude domains that display sequence/structure similarity that might be indicative of homology. Finally homologous internal repeats are purged. This gives us a database of well defined structures which can be assumed to have no residual recurrences that are due to common ancestry; this constitutes an unbiased ensemble for pattern discovery.

2.2 A Definition of a Structural Motif

In order to define *motifs* as abstractions over a set of 3D structure patterns that capture the characteristic features of the set, we adopt the following definitions.

Since we want to be able to *discover* structure/sequence relationships, we do not filter by sequence but consider patterns defined only by the backbone atoms of amino acids. Specifically we use only the set {N, C^α, and C} since the position of the O atom is computable with small error from the position of the subsequent N-atom; for the first and the last residues of a pattern the C atom co-ordinates of the residue preceding the first residue and the N atom co-ordinate of the residue after the last residue are also used. Thus we include information on the ϕ angle of the first residue and the ψ angle of the last residue consistently with all other residues.

Definition 1. A structural pattern p_j^i of length i beginning at position j of a structure thus consists of the following atoms taken from sequential residues :

$$p_j^i = \left\{ C_{j-1}, \left(N_j, C_j^\alpha, C_j \right) ..., \left(N_{j+i-1}, C_{j+i-1}^\alpha, C_{j+i-1} \right) N_{j+i} \right\} \quad (1)$$

A simple unique code is used to identify each pattern. This code is a string such as *1f0j_A_329_0_3* made up of five parts: PDB ID of the structure from which the domain was taken, chain identifier, start residue number of the pattern in the PDB file, the insert code of the residue (0 if none), and the length i of the pattern.

Definition 2. A **structural similarity space** S^i is the pair $\left(P^i, d \right)$ where P^i is the set of all structural patterns of length i from a set of protein structure domains and $d : P^i \times P^i \rightarrow$ is the root-mean square deviation after optimal superposition (RMSD) of equivalent atoms, which is a metric [10].

Let a probability density in S^i be the probability of observing a pattern $p^i \in P^i$ in some volume of S^i, then:

Definition 3. A **motif** is the pattern with the highest local probability density in a structural similarity space.

Note that this definition ties the *motif* to its interpretation as an energetically most favorable (i.e. observationally most probable) paradigm of a set of similar structures, while suggesting a strategy to identify motifs by computing and identifying local maxima of probability density. As well, a motif is both an abstraction over patterns and an instance – analogous to a cluster medoid –, it is uniquely defined with reference to public data (PDB), and the set of similar patterns it represents is unambiguously defined and straightforward to compute.

2.3 Finding Local Density Maxima

We estimate the local probability density function (PDF) at each pattern p_x^i of length i by calculating a density

$$\rho(p_x^i) = \sum_{p_y^i \in P^i} \frac{1}{\sigma\sqrt{2\pi}} e^{-\frac{1}{2}\left(\frac{d(p_x^i, p_y^i)}{\sigma}\right)^2} \tag{2}$$

with $\sigma = 0.1\text{Å}$ and $d\left(p_x^i, p_y^i\right)$ the RMSD of patterns x and y, while applying a 5σ cutoff. Note that this kernel has the form of a Gaussian between pairs, but the structural space we apply this to is high-dimensional, i.e. this is the 1D projection of the true PDF. The heuristic choice of $\sigma = 0.1\text{Å}$ determines the *roughness* of our PDF and allows us to express the bandwidth of the kernel in relation to a measure that is familiar to structural biologists. We can then find local maxima – i.e. *motifs* – with a hill-climbing procedure that identifies the pattern with the highest local density. The radius of the sphere in which our hill-climbing algorithm searches for local maxima determines the *resolution* of the search process. The algorithm to find the local density maxima is a process of comparing the PDF for patterns within a RMSD threshold and choosing the pattern with the highest PDF. This RSMD threshold is referred to as the *neighborhood* of a pattern. The neighborhood is defined as the minimum distance that can distinguish two local density maxima. The effect of a large neighborhood is similar to that of over-smoothing where very few peaks are observed while the choice of a small radius leads to large numbers of patterns with poor support. The RMSD threshold was varied from 0.2Å to 1.7Å for different pattern lengths, and lists of recurrent patterns at each threshold were compiled.

Each motif thus represents a class of patterns that lie within a fixed distance (and satisfy a coherence requirement, see below), which we call the *neighborhood* of the motif.

Geometric Search Tree Pruning. In principle, finding the densities of all observations is an $O(n^2)$ problem where n is the number of patterns. Since we use a 5σ cutoff for density evaluations, the use of a geometric data structure can reduce the complexity of the operation. We implemented a GNAT, a Geometric Near-neighbor Access Tree, which is a data structure designed to search for approximate matches in large databases given a distance metric [1]. The tree is constructed as a hierarchical Dirichlet-domain based structure. The Dirichlet domain of a branch, consists of all the patterns that are structurally most similar to its root pattern. This allows to prune branches that can't contain target patterns.

2.4 Further Constraints

In order to obtain a meaningful list of motifs, we impose additional constraints: the motifs should represent **significant** deviations of the PDF from random, the list should be **robust** against removal of source data, and the motifs should display structural **coherence**. Finally, we wish to retain only motifs, of **maximal length**.

Significance. To estimate the significance of motifs as an excess of support over random expectation, we compared the number of patterns found in our source data set , Nh3D to the number of patterns found in a pseudo-randomly generated dataset of decoy structures. These were generated by randomly sampling the source data for a peptide unit, with replacement, and attaching it to a growing peptide chain, until the length of the source domain was reached. Steric overlap during construction was avoided by resampling and backtracking if necessary. This procedure preserved detailed geometry at the residue level and has nearly the same amino acid and backbone angle distribution as the source domain, while removing all propensities arising from residue-residue interactions.

Motifs were deemed significant at each length if they were detected with a neighborhood cutoff at which 99% or greater of patterns in the pseudo-random dataset are unique i.e. they do not recur. The significant neighborhood radius is very well approximated by an exponential: RMSDcutoff = $0.0974 \times \exp(0.1616 \times \text{motif length})$ ($R^2 = 0.978$).

Robustness. We require a minimum absolute level of recurrence to ensure its independence of random inclusions or deletions of source data. If 10% of the dataset were replaced with proteins that would not contribute support for to a motif, in effect the support for all motifs would be reduced by 10%. We define a robustness cut-off as the minimum support that ensures that 99% of the motifs would still have significant support after deleting 10% of their support.

Coherence. Since the neighborhood of longer motifs can be quite large (>1.0Å for motifs of length > 14), and individual amino acid configurations contribute relatively less to the overall RMSD, we require that none of the atoms of a neighbor deviate more than one van der Waals radius from the corresponding atom of the motif after superposition. Imposing the coherence cut-off resulted in fewer, better-defined clusters and reduced the noise.

Maximal Motifs. We attempt to describe motifs at the longest length at which they are significantly supported. Many motifs of length i are components of one or more motifs of length $i+1$. We call a neighbor of a motif of length i that cannot be extended by one residue to become neighbor of a motif of length $i+1$ a *unique neighbor* of the motif. Motifs that are completely included in a motif of longer length – *i.e.* the number of unique neighbors does not exceed our significance threshold – are removed from our list.

3 Results

Overall, our procedures define 859 maximal motifs of lengths between 3 and 15 amino acids. We call this dictionary of motifs *schematikon*.

3.1 Length Distribution of Motifs

Among the top ten motifs ranked by unique support, we find five α-helix motifs of different lengths but only one, short, β-strand motif. In general, we find β-strands to have a lot of structural variability leading to them to be poorly represented in the *schematikon* database. A log(rank)/log(support) plot of motifs shows a power-law distribution, typical of many biological phenomena. However it is not the case that the lower-ranked motifs are dominated by long motifs (Fig. 1).

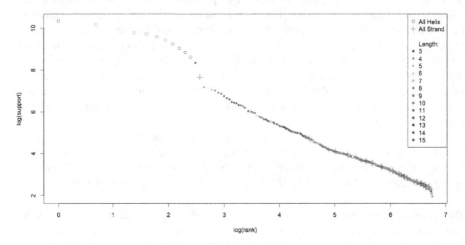

Fig. 1. log/log plot of *schematikon* motifs. All the highest-ranked motifs are helical, and the support decreases strictly ordered by length. They are followed by a group of motifs in which the first residue is not helical. As the color coded symbols indicate, frequencies of these motifs follow an approximate power-law behaviour as a function of length. Strand motifs are more distributed and cluster in the less highly supported region. Motif lengths are color coded and . Data analysis and plotting was done with **R** [14].

3.2 Sequence Structure Correlations

Significance of Observed Sequence Propensities. Small sample sizes have to be taken into account when comparing observed residue distributions against database background. To assess significance, we generated residue sets with the same length and support as the database at random from the background frequencies and compared the observed frequency deviation to the random set deviations. A motif was deemed to have significant sequence propensities if it deviated with a p value < 0.01 in at least one position, Bonferroni corrected for the length of the motif. Both single amino acids and groups of amino acids with shared biophysical properties were considered, using nine predefined residue groups: {I, L, V, M}, {F, W, Y}, {D, E}, {H, K, R}, {N, Q}, {S, T}, {G, A}, {P}, and {C}. We find that over 90% of motifs have a significant sequence propensity in at least one of their positions. This indicates that our dataset contains motifs that are associated with backbone determinants, as well as one or more side chain–backbone or side chain–side chain interactions. These side chain interactions give rise to sequence propensities which in turn express the sequence-structure relationship represented by the motif.

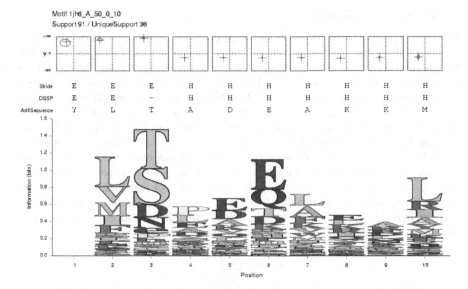

Fig. 2. *1jh6_A_50_0_10*: A complex motif of length 10, the top ranked strand-cap-helix motif. Residue-wise (ϕ,ψ) angles are shown in the top row, followed by structural annotations with Stride [15] and DSSP [16], and the motif sequence. Residue propensities are illustrated with a sequence logo. The core of the motif, from position 3 to 6, corresponds to a well-known helix capping motif, but we find additional hydrophobic propensities in positions 2, 7 and 10. Note that the actual sequence of the motif (YLTADEAKKM) corresponds to the most frequent residue in the three most informative positions. The residue propensities in position 1 are not significantly different from background, thus no characters are shown.

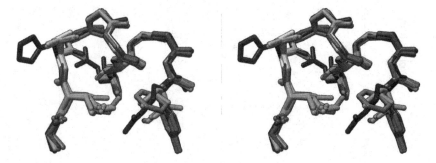

Fig. 3. Stereo view of 1jh6_A_50_0_10 – the strand-cap-helix motif shown in Figure 2 – and its nine closest neighbors. The N-terminus is to the left, side chains are shown for positions 2, 3, 7 and 10 of the motif. Colors and positions correspond to Figure 2. The helix-capping serine or threonine residue is seen at the left-top corner and the hydrophobic side-chain cluster is at the bottom of the scene. The hydrophobic propensities in the helix portion of the motif (positions 7 and 10) are clearly co-determined by accomodating packing of the the strand residue (position 2). In this way an approximate 90° angle between a β–strand and an α–helix is encoded. Figure created with UCSF Chimera [13].

3.3 Biological Significance

The most important question for the observed positional amino acid propensities however is whether they carry predictive information about sequence changes. To address
this question we accessed the protherm database [3] and retrieved the stability change
data for 243 experimentally determined, reversible, two-state, single point mutations in
the well-studied protein barnase (PDB ID: 1BNI). We then assessed positional propensities for the protein structure by retreiving from *schematikon* all motifs that matched a
sequential pattern of the crystal structure with an RMSD of 0.67Å or better. This provides an anotation of the structure with motifs. To compile positional propensities for
the 1BNI residues, we chose for each position the residues of the matching motifs and
its nineteen closest neighbors, if the information for the position in a motif was greater
than 1.2 bit relative to the amino acid distributions for all matching motifs. This ensures that we consider only residues for comparison that actually have significant propensities in their respective motif neighborhood. Using these residue sets we calculated
an empirical prediction of a free energy change for the mutation as

$$\Delta\Delta G \propto \log\left(\frac{f_{WT}}{f_{MUT}}\right) \qquad (3)$$

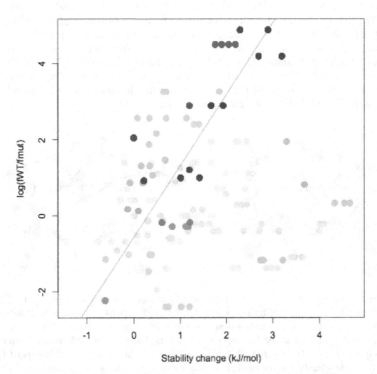

Fig. 4. Experimentally determined stability changes vs. motif-derived predictions for barnase.
All predictions are shown but a color scale has been applied to emphasize predictions for which
the sum of observed native and mutant residues was larger than a cutoff. The 22 mutations for
which more than 60 observations were compiled achieve a correlation coefficient of > 0.8. The
regression line for this set of predictions has been drawn. Data analysis was done with **R** [14].

where f_{WT} is the frequency of observing the native "wild type" residue in the motif-derived residue set and f_{MUT} is the frequency of the mutated residue – this provides a computational prediction of the stability change [11]. We expect predictions to be strongest for positions in which larger numbers of native and mutated residues are observed. This is indeed the case (Fig. 4).

We see that not all mutations can be well predicted, subject to the availability of data. But for those positions where predictions are based on significant numbers of mutations, the predictions are quite good, achieving quantitative correlation coefficients of better than 0.8, and qualitatively correct (stabilizing vs. destabilizing) predictions of 80% and better.

4 Conclusion

With *schematikon* we have constructed a motif database, based on unsupervised pattern mining in a non-homologous dataset, that allows us to put the description of structural motifs on a quantitative basis. The motifs are highly predictive for experimentally determined stability changes. We thus note that the assumption of Boltzmann like frequency distributions of patterns in protein structures is substantially correct and can be exploited for applications in modeling and biotechnology.

Acknowledgements. The authors would like to thank Philip Tong for valuable discussions and him and Zuolin Bai, Andy Gao, Qing Gao, Kristina Grabowski, Haerin Kim, Gerald Quon, Adrian Saldanha, Jennifer Tsai, and Carol Yong for contributions to earlier stages of this project.

References

1. Brin, S.: Near Neighbor Search in Large Metric Spaces. In: VLDB 1995: Proceedings of the 21st International Conference on Very Large Data Bases, Zurich, Switzerland, pp. 574–584. Morgan Kaufmann Publishers, Burlington (1995)
2. Bystroff, C., Baker, D.: Prediction of Local Structure in Proteins Using a Library of Sequence-Structure Motifs. J. Mol. Biol. 281, 565–577 (1998)
3. Kumar, M.D., Bava, K.A., Gromiha, M.M., Parabakaran, P., Kitajima, K., Uedaira, H., Sarai, A.: ProTherm and ProNIT: Thermodynamic Databases for Proteins and Protein-Nucleic Acid Interactions. Nuleic Acids Res. 34, D204–D206 (2006)
4. Lane, T.J., Shukla, D., Beauchamp, K.A., Pande, V.S.: To Milliseconds and Beyond: Challenges in the Simulation of Protein Folding. Curr. Opin. Struct. Biol. 23, 58–65 (2013)
5. Liu, X., Fan, K., Wang, W.: The Number of Protein Folds and their Distribution over Families in Nature. Proteins 54, 491–499 (2004)
6. Marks, D.S., Hopf, T.A., Sander, C.: Protein Structure Prediction from Sequence Variation. Nat. Biotechnol. 30, 1072–1080 (2012)
7. Orengo, C.A., Michie, A.D., Jones, S., Jones, D.T., Swindells, M.B., Thornton, J.M.: CATH–a Hierarchic Classification of Protein Domain Structures. Structure 5, 1093–1108 (1997)
8. Richardson, J.S.: The Anatomy and Taxonomy of Protein Structure. Adv. Protein Chem. 34, 167–339 (1981)

9. Richter, F., Leaver-Fay, A., Khare, S.D., Bjelic, S., Baker, D.: *De Novo* Enzyme Design Using Rosetta3. PLoS One 6, e19230 (2011)
10. Steipe, B.: A Revised Proof of the Metric Properties of Optimally Superimposed Vector Sets. Acta Crystallogr. A 58, 506 (2002)
11. Steipe, B.: Consensus-Based Engineering of Protein Stability: From Intrabodies to Thermostable Enzymes. Methods Enzymol. 388, 176–186 (2004)
12. Thiruv, B., Quon, G., Saldanha, S.A., Steipe, B.: Nh3D: A Reference Dataset of Non-Homologous Protein Structures. BMC Struct. Biol. 5, 12 (2005)
13. Pettersen, E.F., Goddard, T.D., Huang, C.C., Couch, G.S., Greenblatt, D.M., Meng, E.C., Ferrin, T.E.: UCSF Chimera–A Visualization System for Exploratory Research and Analysis. J. Comput. Chem. 25, 1605–1612 (2004)
14. R Core Team: R: A Language and Environment for Statistical Computing. R Foundation for Statistical Computing, Vienna, Austria (2013), http://www.R-project.org/
15. Frishman, D., Argos, P.: Knowledge-Based Protein Secondary Structure Assignment. Proteins 23, 566–579 (1995)
16. Kabsch, W., Sander, C.: Dictionary of Protein Secondary Structure: Pattern Recognition of Hydrogen-Bonded and Geometrical Features. Biopolymers 22, 2577–2637 (1983)

PNImodeler: Web Server for Inferring Protein Binding Nucleotides from Sequence Data

Jinyong Im, Narankhuu Tuvshinjargal, Byungkyu Park,
Wook Lee, and Kyungsook Han[*]

Department of Computer Science and Engineering
Inha University, Incheon, South Korea
khan@inha.ac.kr

Interactions between DNA and proteins are essential to many biological processes such as transcriptional regulation and DNA replication. With the increased availability of structures of protein-DNA complexes, several computational studies have been conducted to predict DNA binding sites in proteins. However, little attempt has been made to predict protein binding sites in DNA. From an extensive analysis of protein-DNA complexes obtained from the Protein Data Bank (PDB), we identified powerful features of DNA and protein sequences which can be used in predicting protein binding sites in DNA sequences. The features can be classified into three types: original DNA sequence, DNA sequence fragments from the original DNA sequence, and protein sequence interacting with the DNA. The original DNA sequence is represented by its nucleotide composition. DNA sequence fragments are represented by nucleotide triplet composition, normalized position, molecular mass, molecular pKa and interaction propensity of nucleotide triplets. For protein, which is an interaction partner of DNA, the sum of normalized position of twenty amino acids in the sequence and dipeptide composition are represented.

We have developed two SVM models to predict protein binding nucleotides in DNA. One model uses DNA sequence data alone and predicts all potential binding sites with unknown protein partners. The other model uses both DNA and protein sequences to predict protein binding nucleotides with the specific protein. One SVM model that used DNA sequence data alone achieved a sensitivity of 73.4%, a specificity of 64.8%, an accuracy of 68.9% and a Matthews correlation coefficient (MCC) of 0.382 with a test dataset that was not used in training. Another SVM model that used both DNA and protein sequences achieved a sensitivity of 67.6%, a specificity of 74.3%, an accuracy of 71.4% and MCC of 0.418. The SVM model that used both DNA and protein sequences yielded better overall performance than the model that used DNA sequence alone.

The SVM models have been implemented as a web server called PNImodeler (Protein-Nucleic acid Interaction modeler), and the web server is available at http://bclab.inha.ac.kr/pnimodeler. PNImodeler will be useful to find protein-binding sites in DNA with unknown structure. To the best of our knowledge, this is the first attempt to predict protein-binding DNA nucleotides with sequence data alone.

---

[*] Corresponding author.

M. Basu, Y. Pan, and J. Wang (Eds.): ISBRA 2014, LNBI 8492, p. 367, 2014.
© Springer International Publishing Switzerland 2014

A MCI Decision Support System Based on Ontology

Xiaowei Zhang[1,*], Yang Zhou[1], Bin Hu[2,1], Jing Chen[1], and Xu Ma[1]

[1] School of Information Science and Engineering, Lanzhou University, Lanzhou, China
[2] College of Electronic Information and Control Engineering,
Beijing University of Technology, Beijing, China
{zhangxw,yzhou11,bh,max2012,jchen10}@lzu.edu.cn

Mid Cognitive Impairment (MCI) [1] threatens the health of the elderly around the world and could progress to Alzheimer's Disease (AD) [2] with a high risk. It is necessary to detect MCI earlier to reduce the occurrence rate of AD.

So we develop a Decision Support System (DSS) for detecting subjects with MCI. This system is based on fMRI-Bayesian ontology (FB-Ontology) combined with Bayesian networks algorithm. The DSS employs Functional Magnetic Resonance Imaging (fMRI) techniques to distinguish MCI patients from normal controls (NC). We preprocess fMRI data and calculate path length, global efficiency and hub node features based on the automated anatomical labeling (AAL) template for DSS. By using Bayesian networks algorithm, our DSS could provide uncertain reasoning results for clinicians. Meanwhile, the FB-Ontology acts like a bridge between reasoning engine and low-level database in system, it could provide transparent, unified, normalized, shareable knowledge to users. Finally, we select 22 subjects with MCI and 18 normal controls from Alzheimer's Disease Neuroimaging Initiative (ADNI) [3]. Using a 5-fold cross validation method for training and testing, the system could reach an average classification rate of 90%.

References

1. Gauthier, S., et al.: Mild cognitive impairment. The Lancet 367(9518), 1262–1270 (2006)
2. Petersen, R.C.: Mild Cognitive Impairment: Aging to Alzheimer's Disease. Brain 127(1), 231–233 (2004)
3. http://www.loni.ucla.edu/adni/

---

* Corresponding author.

M. Basu, Y. Pan, and J. Wang (Eds.): ISBRA 2014, LNBI 8492, p. 368, 2014.
© Springer International Publishing Switzerland 2014

Context Similarity Based Feature Selection Methods for Protein Interaction Article Classification*

Yifei Chen[1], Yuxing Sun[1], and Ping Hou[2]

[1] School of Information Science, Nanjing Audit University, 86 Yushan Rd(W),
Nanjing, P.R. China
[2] Fondazione Bruno Kessler (FBK-irst), Trento, Italy

An overwhelming amount of biological articles are published daily online as a result of growing interest in biological research, especially the study of protein-protein interactions. It is essential to classify which articles describe the protein interactions. Therefore study on automatic protein interaction articles classification (IAC) has become a task with practical significance to the text classification in biological domain.

Since the feature space in text classification is high dimensional, feature selection techniques are widely used for reducing the dimensionality of features to speed up the computation of the classifier. However, the existing feature selection methods are mostly based on the term frequency or document frequency. These approaches are context independent, that is, they do not utilize the context information in a document when judging the importance of features, such as word order, multi-word phrases and semantic relationships, which are important for the IAC tasks. Hence, based on the study of four well-known frequency based feature selection methods, Gini Index (GI), Document Frequency (DF), Class Discriminating Measure (CDM) and Accuracy Balanced (Acc2), we propose four context similarity based feature selection methods, GI_{cs}, DF_{cs}, CDM_{cs} and $Acc2_{cs}$, to introduce the similarity measure of context multi-word phrases.

In order to evaluate the performances of the proposed context similarity based feature selection methods, two data sets ($Data_{BCII}$ and $Data_{BCIII}$) are used in our experiments, which are both extracted from the BioCreAtIvE challenges. The experimental results reveal that all the context similarity based methods outperform the corresponding frequency based methods in terms of the micro-F1 measure. On the $Data_{BCII}$, when top 4.3% features are selected, GI_{cs} acquires the highest $F1$ measure value, which effectively improves the performance when all the features are used by 2.54. And on the $Data_{BCIII}$ when the top 7.4% are selected, CDM_{cs} acquires the highest $F1$ measure value, which improves the performance when all the features are used by 1.12. Moreover, through the analysis on the comparison of selected features and the dimension reduction rate, the proposed methods provide better performances by bring more distinguishing information with the fewer selected features for the text classifier.

---

* This work is supported by the National Natural Science Foundation of China (No.61202135), the Natural Science Foundation of Jiangsu Province (No.BK2012472) and the Qing Lan Project.

M. Basu, Y. Pan, and J. Wang (Eds.): ISBRA 2014, LNBI 8492, p. 369, 2014.

Genome-Wide Analysis of Transcription Factor Binding Sites and Their Characteristic DNA Structures

Zhiming Dai[1], Dongliang Guo[1], Xianhua Dai[1], and Yuanyan Xiong[2,3]

[1] Department of Electronics and Communication Engineering,
School of Information Science and Technology,
Sun Yat-Sen University, Guangzhou 510006, China
[2] State Key Laboratory for Biocontrol, Sun Yat-Sen University, Guangzhou 510275, China
[3] SYSU-CMU Shunde International Joint Research Institute, Shunde, China
mody0911@gmail.com

Transcription factors (TF) regulate gene expression by binding DNA regulatory regions. Transcription factor binding sites (TFBSs) are conserved not only in primary DNA sequences but also in DNA structures [1,2]. However, the global relationship between TFs and their preferred DNA structures remains to be elucidated. In this paper, we have developed a computational method to generate a genome-wide landscape of TFs and their characteristic binding DNA structures in *Saccharomyces cerevisiae*. TFBSs are conserved in different DNA structures, independent of sequence conservation. We revealed DNA structural features for different TFs. The structural conservation shows positional preference in TFBSs. Structural levels of DNA sequences are correlated with TF-DNA binding affinities. Our findings will have implications in understanding TF regulatory mechanisms.

References

[1] Parker, S.C., Hansen, L., Abaan, H.O., Tullius, T.D., Margulies, E.H.: Local DNA topography correlates with functional noncoding regions of the human genome. Science 324, 389–392 (2009)
[2] Broos, S., Soete, A., Hooghe, B., Moran, R., van Roy, F., De Bleser, P.: PhysBinder: improving the prediction of transcription factor binding sites by flexible inclusion of biophysical properties. Nucleic Acids Res., W531–W534 (2013)

M. Basu, Y. Pan, and J. Wang (Eds.): ISBRA 2014, LNBI 8492, p. 370, 2014.
© Springer International Publishing Switzerland 2014

A Comparative Study of Disease Genes and Drug Targets in the Human Protein Interactome

Jingchun Sun, Kevin Zhu, W. Jim Zheng, and Hua Xu

School of Biomedical Informatics, The University of Texas Health Science Center at Houston, Houston, TX 77030, USA

Most complex diseases are caused by variation in many genes, which are defined as disease genes [1]. Medicines (drugs) are major choices to treat the diseases or reduce their symptoms as they act through interacting with some proteins [2]. These proteins are defined as drug targets. Thus, disease genes contribute to the pathology of one disease while drug targets are critical for the efficacy of disease treatment. However, the interrelationship between the disease genes and drug targets is not clear.

In this study, we collected disease genes from GWAS catalog database and drug targets from DrugBank and TTD databases. We compared them and found that, though their intersections were small, disease genes were significantly enriched in targets compared to their enrichment in the human protein-coding genes. We further compared network properties of the proteins encoded by disease genes and drug targets in the human interactome. The results showed that the drug targets tended to have a higher degree, a higher betweenness, and a lower clustering coefficient. Additionally, we observed a clear fraction increase of disease proteins or drug targets in the near neighborhood compared with the randomized genes, which is consistent with previous results [3].

The study first comprehensively compared the disease genes and drug targets. The results provide network characteristics for designing computational strategies to predict novel drug targets and drug repurposing.

References

[1] Lander, E.S., Schork, N.J.: Genetic dissection of complex traits. Science 265, 2037–2048 (1994)
[2] Schreiber, S.L.: Target-oriented and diversity-oriented organic synthesis in drug discovery. Science 287, 1964–1969 (2000)
[3] Yildirim, M.A., Goh, K.I., Cusick, M.E., Barabasi, A.L., Vidal, M.: Drug-target network. Nat. Biotechnol. 25, 1119–1126 (2007)

M. Basu, Y. Pan, and J. Wang (Eds.): ISBRA 2014, LNBI 8492, p. 371, 2014.
© Springer International Publishing Switzerland 2014

Efficient Identification of Endogenous Mammalian Biochemical Structures

Mai A. Hamdalla, Reda A. Ammar, and Sanguthevar Rajasekaran*

Computer Science and Engineering Department, University of Connecticut,
Connecticut, USA
rajasek@engr.uconn.edu

Metabolomics is the comprehensive, qualitative, and quantitative study of all the small molecules, called metabolites, in an organism [1]. A major challenge in metabolomics is the interpretation of the vast amount of data produced by the high-throughput techniques used for information extraction and data interpretation [2]. The existence of several on-line chemical structure databases has provided a vital support for molecular identification by allowing the search for candidate compounds using experimentally determined features with computationally simulated features. Such searches often result in a large number of false positives, making identification of the compound under investigation extremely difficult. Hence, cheminformatics methods are needed to efficiently search such large chemical databases and potentially identify unknown endogenous biochemical compounds. Several methods [3–6] have been developed with the objective of discriminating between candidate structures that are synthetic and those that are biochemical. In the talk, we will present an efficient cheminformatics tool that uses known endogenous mammalian biochemicals and graph matching methods to identify endogenous mammalian biochemical structures in chemical structure space.

References

1. Villas-Bôas, S.G., Bruheim, P.: The potential of metabolomics tools in bioremediation studies. Omics: A Journal of Integrative Biology 11, 305–313 (2007)
2. Kertesz, T., Hill, D.W., Albaugh, D., Hall, L., Hall, L., Grant, D.F.: Database searching for structural identification of metabolites in complex biofluids for mass spectrometry-based metabonomics. Bioanalysis 1, 1627–1643 (2009)
3. Nobeli, I., Ponstingl, H., Krissinel, E.B., Thornton, J.M.: A structure-based anatomy of the E.coli metabolome. Journal of Molecular Biology 334, 697–719 (2003)
4. Gupta, S., Aires-de Sousa, J.: Comparing the chemical spaces of metabolites and available chemicals: models of metabolite-likeness. Molecular Diversity 11, 23–36 (2007)
5. Peironcely, J.E., Reijmers, T., Coulier, L., Bender, A., Hankemeier, T.: Understanding and classifying metabolite space and metabolite-likeness. PloS One 6 (2011)
6. Hamdalla, M.A., Mandoiu, I.I., Hill, D.W., Rajasekaran, S., Grant, D.F.: BioSM: A chemoinformatics tool for identifying biochemical structures in chemical structure space. Journal of Chemical Information and Modeling (2012)

---

* Corresponding author.

M. Basu, Y. Pan, and J. Wang (Eds.): ISBRA 2014, LNBI 8492, p. 372, 2014.
© Springer International Publishing Switzerland 2014

LncRNA2Function: A Comprehensive Resource for Functional Investigation of Human lncRNAs Based on RNA-seq Data

Qinghua Jiang[1,*], Rui Ma[2,*], Jixuan Wang[3], Xiaoliang Wu[3], Shuilin Jin[4], Jiajie Peng[2], Renjie Tan[2], Tianjiao Zhang[2], Yu Li[1], and Yadong Wang[2,**]

[1] School of Life Science and Technology,
Harbin Institute of Technology, Harbin, Heilongjiang 150001, China
[2] School of Computer Science and Technology,
Harbin Institute of Technology, Harbin, Heilongjiang 150001, China
ydwang@hit.edu.cn
[3] School of Software, Harbin Institute of Technology,
Harbin, Heilongjiang 150001, China
[4] Department of Mathematics, Harbin Institute of Technology,
Harbin, Heilongjiang, 150001, China

The GENCODE project has collected over 10,000 human long non-coding RNA (lncRNA) genes. However, the vast majority of them remain to be functionally characterized. Computational investigation of potential functions of human lncRNA genes is helpful to guide further experimental studies on lncRNAs. In this study, based on expression correlation between lncRNAs and protein-coding genes across 19 human normal tissues, we used the hypergeometric test to functionally annotate a single lncRNA or a set of lncRNAs with significantly enriched functional terms among the protein-coding genes that are co-expressed with the lncRNA(s). The functional terms include all nodes in the Gene Ontology (GO) and 4,380 human biological pathways collected from 12 pathway databases. We mapped 9,625 human lncRNA genes to GO terms and biological pathways. Finally, we developed the first ontology-driven tool named lncRNA2Function, which enables researchers to browse the lncRNAs associated with a specific functional term, the functional terms associated with a specific lncRNA, or to assign functional terms to a set of human lncRNA genes such as a cluster of co-expressed lncRNAs. The lncRNA2Function is freely available at http://mlg.hit.edu.cn/lncrna2function.

---

[*] Contributed equally.
[**] Corresponding author.

M. Basu, Y. Pan, and J. Wang (Eds.): ISBRA 2014, LNBI 8492, pp. 373–374, 2014.
© Springer International Publishing Switzerland 2014

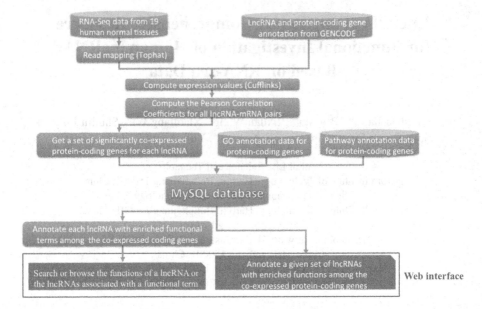

Fig. 1. Overview of the lncRNA2Function

Network Propagation Reveals Novel Genetic Features Predicting Drug Response of Cancer Cell Lines

Jiguang Wang[*], Judith Kribelbauer[*], and Raul Rabadan[**]

Department of Biomedical Informatics and Center for Computational Biology
and Bioinformatics, Columbia University, New York, NY 10032 USA
rabadan@dbmi.columbia.edu

Translating data derived from cancer genomes into personalized cancer therapy is a holy grail of computational biology. An important, yet challenging, question in this undertaking is to relate features of tumor cells to clinical outcomes of anticancer drugs. Recent progress in large pharmacogenomic studies has provided a wealth of data about cancer cell lines, indicating that many genetic and gene expression candidates might predict the drug response of cancer cells [1-3]. Unfortunately, most of the predicted features lack underlying mechanisms and are not consistent with our prior knowledge [4].

To address this question, we have developed a new method, named dNetFS, to prioritize gene expression features, as well as genetic features of cancer cell lines, that predict drug response by integrating genomic/pharmaceutical data, protein-protein interaction networks, and prior knowledge of drug-targets interaction with the techniques of network propagation. Compared with previous methods, dNetFS is better than other simple network-based methods and dramatically improves the accuracy of prediction of traditional correlation-based methods by means of cross-validation analysis. Our dNetFS software will be available upon request.

By applying dNetFS in the study of an inhibitor of Insulin-like Growth-Factor-Receptor (IGF1R), BMS-754807 [5], we were able to show that the sensitivity of BMS-754807 could be accurately predicted by the expression levels of some important genes, including proto-oncogene tyrosine-protein kinase Src, and neuroblastoma RAS viral (v-ras) oncogene homolog.

References

1. Barretina, J., et al.: The Cancer Cell Line Encyclopedia enables predictive modelling of anticancer drug sensitivity. Nature 483, 603–607 (2012)
2. Basu, A., et al.: An interactive resource to identify cancer genetic and lineage dependencies targeted by small molecules. Cell 154, 1151–1161 (2013)
3. Garnett, M.J., et al.: Systematic identification of genomic markers of drug sensitivity in cancer cells. Nature 483, 570–575 (2012)
4. Haibe-Kains, B., El-Hachem, N., Birkbak, N.J., Jin, A.C., Beck, A.H., Aerts, H.J., Quackenbush, J.: Inconsistency in large pharmacogenomic studies. Nature 504, 389–393 (2013)

---

[*] Contributed equally.
[**] Corresponding author.

M. Basu, Y. Pan, and J. Wang (Eds.): ISBRA 2014, LNBI 8492, p. 375, 2014.
© Springer International Publishing Switzerland 2014

Splice Site Prediction Using Support Vector Machine with Markov Model and Codon Information

Dan Wei[1,2], Yin Peng[3,4], Yanjie Wei[2,*], and Qingshan Jiang[2,*]

[1] Institute of Graphics and Image, Hangzhou Dianzi University, Hangzhou 310018, China
[2] Shenzhen Key Lab. for High Performance Data Mining,
Shenzhen Institutes of Advanced Technology,
Chinese Academy of Sciences, Shenzhen 518055, China
[3] Department of Pharmacology, Sun Yat-Sen University, Guangzhou, 510275, China
[4] Department of Center Laboratory, The First Affiliated Hospital to Shenzhen University,
Shenzhen 518035, China

Prediction of donor and acceptor splice sites plays a central role for detecting the gene structure for the eukaryotes. In this paper, we combine the sequence conservativeness and codon usage bias to predict splice sites. Our method is based on SVM with Markov model and codon usage information (MC-SVM). The method first extracts two features of the candidate sequences, including the conserved features described by the probabilistic parameters of the first Markov model (MM1) and the codon usage bias information. Then an F-score based feature selection is used to select the most discriminative features. Finally, MC-SVM applies SVM on the training sequences with sequence-based vectors as input to obtain the SVM prediction model, and uses the model to predict the splice sites of testing sequences.

The proposed method is tested using 10-fold cross-validation on two 1:1 and 1:10 datasets, with all the true splice sites taken from Homo Sapiens Splice Sites Data set (HS3D) and equal/decuple number of false sites randomly selected from the same data set. The evaluation shows that MC-SVM is highly accurate compared to MM1-SVM[1], Reduced MM1-SVM [2] and some other methods [3] in terms of sensitivity, specificity and global accuracy Q^9. Furthermore, ROC curves show that MC-SVM exhibits better overall prediction performance than MM1-SVM, Reduced MM1-SVM and MEM [4] methods for predicting both acceptor and donor sites.

References

1. Baten, A.K.M.A., Chang, B.C.H., Halgamuge, S.K., Li, J.: Splice site identification using probabilistic parameters and SVM classification. BMC Bioinformatics 7(suppl. 5), S15 (2006)
2. Baten, A.K.M.A., Halgamuge, S.K., Chang, B.C.H.: Fast splice site detection using information content and feature reduction. BMC Bioinformatics 9(suppl. 12), S8 (2008)
3. Zhang, Q., Peng, Q., Zhang, Q., Yan, Y., Li, K., Li, J.: Splice sites prediction of human genome using length-variable Markov model and feature selection. Expert Syst. Appl. 37, 2771–2782 (2010)
4. Yeo, G., Burge, C.: Maximum entropy modeling of short sequence motifs with application to RNA splicing signals. J. Comput. Biol. 11(2-3), 377–394 (2004)

---

* Corresponding author.

M. Basu, Y. Pan, and J. Wang (Eds.): ISBRA 2014, LNBI 8492, p. 376, 2014.
© Springer International Publishing Switzerland 2014

Similarity Analysis of DNA Sequences Based on Frequent Patterns and Entropy*

Xiaojing Xie[1], Jihong Guan[2], and Shuigeng Zhou[1,**]

[1] School of Computer Science, and Shanghai Key Lab. of Intelligent Information Processing, Fudan University, China
{xiexiaojing,sgzhou}@fudan.edu.cn
[2] Department of Computer Science and Technology, Tongji University, China
jhguan@tongji.edu.cn

Abstract. DNA sequence analysis has been an important research topic in Bioinformatics. Evaluating the similarity between sequences, which is crucial for sequence analysis, has attracted much research effort, and dozens of algorithms and tools have been developed. These methods are based on either alignment or word frequency and geometric representation etc., each of which has its advantage and disadvantage. In this paper, for effectively computing the similarity between DNA sequences, we introduce a novel method based on frequency patterns and entropy to construct representative vectors of DNA sequences. Concretely, each sequence is first divided into blocks of the same length. Then, a modified PrefixSpan [1] algorithm is used to discover the maximal frequent patterns in each block. Finally, with the probabilities of these patterns, the entropy of each block is calculated. The resulting entropies of the blocks constitute the components of the sequence vector. Our method is able to capture fine-granularity information (location and ordering) of DNA sequences, via sequence blocking. As only the maximal frequent patterns are considered, our method is insensitive to noise and sequence rearrangement. Experiments are conducted to evaluate the proposed method, which is compared with two existing methods [2,3]. When testing on the β-globin genes of 11 species and using the results from MEGA as the baseline, our method achieves higher correlation coefficients than the two existing methods.

References

1. Pei, J., et al.: Mining sequential patterns by pattern-growth: The prefixspan approach. IEEE Transactions on Knowledge and Data Engineering 16, 1424–1440 (2004)
2. Yu, H.J., Huang, D.S.: Graphical Representation for DNA Sequences via Joint Diagonalization of Matrix Pencil. IEEE Journal of Biomedical and Health Informatics 17, 503–511 (2013)
3. Li, C., et al.: Similarity analysis of DNA sequences based on the weighted pseudo-entropy. Journal of Computational Chemistry 32, 675–680 (2011)

---

* This work was supported by National Natural Science Foundation of China (NSFC) under grants No. 61173118 and No. 61272380.
** Corresponding author.

M. Basu, Y. Pan, and J. Wang (Eds.): ISBRA 2014, LNBI 8492, p. 377, 2014.

Exploiting Topic Modeling
to Boost Metagenomic Sequences Binning*

Ruichang Zhang[1], Zhanzhan Cheng[1], Jihong Guan[2], and Shuigeng Zhou[1,**]

[1] Shanghai Key Lab. of Intelligent Information Processing, Fudan University, China
{rczhang,chengzhanzhan,sgzhou}@fudan.edu.cn
[2] Department of Computer Science and Technology, Tongji University, China
jhguan@tongji.edu.cn

With the rapid development of high-throughput technologies, researchers can sequence the whole metagenome of a microbial community sampled directly from the environment. The assignment of these sequence reads into different species or taxonomical classes is a vital step for metagenomic analysis, which is referred to as *banning* of metagenomic data.

In this paper, we propose a new method *TM-Cluster* for banning metagenomic reads. First, we represent each metagenomic read as a set of "k-meres" with their frequencies appearing in the read. Then, we employ a probabilistic topic model — the Latent Dirichlet Allocation (LDA) model [1] to the reads, which generates a number of hidden "topics" such that each read can be represented by a distribution vector of the generated topics. Finally, as in the Cluster method TCluster [3], we apply SKWIC [2] — a variant of the classical K-means algorithm with automatic feature weighting mechanism to clustering these reads.

Our method can achieve stable and better overall performance on datasets with from several thousands to millions of reads of a number of species and various relative abundance ratios, compared to existing banning methods including AbundanceBin, MetaCluster 3.0 and MCluster [3]. Analysis on the topic number of LDA model in our method also implies that the topic number hidden in the metagenomic data is related to the species number to some extent. In summary, our experiments indicate that the incorporation of topic modeling can effectively improve the banning performance of metagenomic reads.

References

1. Blei, D.M., Ng, A.Y., Jordan, M.I.: Latent dirichlet allocation. Journal of Machine Learning Research 3, 993–1022 (2003)
2. Frigui, H., Nasraoui, O.: Simultaneous clustering and dynamic keyword weighting for text documents. In: Survey of Text Mining, pp. 45–72. Springer (2004)
3. Liao, R., Zhang, R., Guan, J., Zhou, S.: A new unsupervised binning approach for metagenomic sequences based on n-grams and automatic feature weighting. IEEE/ACM Transactions on Computational Biology and Bioinformatics (2013), http://doi.ieeecomputersociety.org/10.1109/TCBB.2013.137

---

* This work was supported by National Natural Science Foundation of China (NSFC) under grants No. 61173118 and No. 61272380.
** Corresponding author.

Network-Based Method for Identifying Overlapping Mutated Driver Pathways in Cancer

Hao Wu, Lin Gao[*], Feng Li, Fei Song, and Xiaofei Yang

School of Computer Science and Technology, Xidian University,
Xi'an, Shaanxi 710071, China
lgao@mail.xidian.edu.cn, haowu@nwsuaf.edu.cn,
{lifeng_10_28,ronasong,yangxiaofeihe}@163.com

Abstract. Large-scale cancer genomics projects are providing lots of data on genomic, epigenomic and gene expression aberrations in many cancer types [1]. One key challenge is to detect functional driver pathways and to filter out nonfunctional passenger genes in cancer genomics. In this study, we present a network-based method (Net-Dendrix) to detect overlapping driver pathways automatically. This algorithm can directly find driver pathways or gene sets de novo from somatic mutation data utilizing two combinatorial properties, high coverage and high exclusivity [2,3,4], without any prior information. Vandin et al. introduce the Maximum Weight Submatrix Problem to find driver pathways and show that it is an NP-hard problem [2]. To solve it better and reduce the complexity of the problem, we firstly construct gene network based on the approximate exclusivity between each pair of genes using somatic mutation data from lots of cancer patients. Secondly, we present a new greedy strategy to add or remove genes for getting overlapping gene sets with driver mutations according to the properties of high exclusivity and high coverage. To assess the efficiency of Net-Dendrix, we apply it onto simulated data and compare it with Iterative versions of MCMC [2] and RME [4]. Net-Dendrix can obtain the optimal results in less than eight seconds, while Iter-IME can get them in more than 20s and Iter-MCMC can get them in more than 600s. To further verify the performance of Net-Dendrix, we apply it to analyze somatic mutation data from five real biological data sets such as the mutation profiles of 90 glioblastoma tumor samples and 163 lung carcinoma samples. Net-Dendrix detects groups of genes which overlap with known pathways, including P53, RB and PI(3)K signaling pathways. Many gene sets with p-value<1e-3 are found from the somatic mutation data. So Net-Dendrix can detect more biological relevant gene sets. Results show that Net-Dendrix outperforms other algorithms for detecting driver pathways or gene sets.

Keywords: Driver pathway, Network-based method, Somatic mutation, Mutually exclusivity, High coverage.

---

[*] Corresponding author.

M. Basu, Y. Pan, and J. Wang (Eds.): ISBRA 2014, LNBI 8492, pp. 379–380, 2014.
© Springer International Publishing Switzerland 2014

References

1. Zhao, J., Zhang, S., Wu, L.-Y., Zhang, X.-S.: Efficient methods for identifying mutated driver pathways in cancer. Bioinformatics 28(22), 2940–2947 (2012)
2. Vandin, F., Upfal, E., Raphael, B.J.: De novo discovery of mutated driver pathways in cancer. Genome Research 22(2), 375–385 (2012)
3. Leiserson, M.D., Blokh, D., Sharan, R., Raphael, B.J.: Simultaneous identification of multiple driver pathways in cancer. PLoS Computational Biology 9(5), e1003054 (2013)
4. Miller, C.A., Settle, S.H., Sulman, E.P., Aldape, K.D., Milosavljevic, A.: Discovering functional modules by identifying recurrent and mutually exclusive mutational patterns in tumors. BMC Medical Genomics 4(1), 34 (2011)

Completing a Bacterial Genome
with *in silico* and Wet Lab Approaches

Rutika Puranik[1], Jacob Werner[1], Guangri Quan[2],
Rong Zhou[3], and Zhaohui Xu[1,*]

[1] Department of Biological Sciences,
Bowling Green State University, Bowling Green, OH 43403, USA
zxu@bgsu.edu
[2] School of Software, Harbin Institite of Technology,
Weihai, Shandong, 264209, China
[3] Department of Mathematics, Yuncheng University, Shanxi, 044000, China

The existence of gaps in draft genome assemblies compromises our ability to take full advantage of genome data. In this study, a pipeline is developed to assemble complete genomes primarily from the next generation sequencing (NGS) data. The input of the pipeline are paired-end Illumina sequence reads, and the output is a high quality complete genome sequence. The pipeline alternates the employment of computational and biological methods in seven steps. It combines the strengths of *de novo* assembly, reference based assembly, customized programming, public databases utilization, and wet lab experimentation.

The application of the pipeline is demonstrated by the completion of a bacterial genome, *Thermotoga* sp. strain RQ7, a potential biohydrogen production strain. Illumina sequencing produced 400 Mb of clean data. Initial assembling with SOAPdenovo [1] and SOAPaligner [2] generated a scaffold of 1,822,593 bp that contained 27 mini gaps, ranging from 1 bp to 3.2 kb, and one big gap of ∼ 38 kb. After running through the pipeline, the genome was closed at 1,851,618 bp, with a GC content of 46.13%. The annotation of 63 ORFs were updated, affecting the prediction of many essential cellular processes.

This work distinguishes itself from similar studies [3, 4] due to its multi-phase interactions between computational and biological approaches. The constituting principles and methods are applicable to similar studies of both prokaryotic and eukaryotic genomes.

References

[1] Li, R., Zhu, H., Ruan, J., Qian, W., Fang, X., Shi, Z., Li, Y., Li, S., Shan, G., Kristiansen, K., et al.: De novo assembly of human genomes with massively parallel short read sequencing. Genome Research 20(2), 265–272 (2010)
[2] SOAPaligner, http://soap.genomics.org.cn/soapaligner.html
[3] Nadalin, F., Vezzi, F., Policriti, A.: GapFiller: a de novo assembly approach to fill the gap within paired reads. BMC Bioinformatics 13(suppl. 14), S8 (2012)
[4] Xing, Y., Medvin, D., Narasimhan, G., Yoder-Himes, D., Lory, S.: CloG: A pipeline for closing gaps in a draft assembly using short reads. In: 2011 IEEE 1st International Conference on Computational Advances in Bio and Medical Sciences (ICCABS), February 3-5, pp. 202–207 (2011)

---

* Corresponding author.

M. Basu, Y. Pan, and J. Wang (Eds.): ISBRA 2014, LNBI 8492, p. 381, 2014.
© Springer International Publishing Switzerland 2014

Screening Ingredients from Herbs against Pregnane X Receptor in the Study of Inductive Herb-Drug Interactions: Combining Pharmacophore and Docking-Based Rank Aggregation

Zhijie Cui[1], Hong Kang[1], Kailin Tang[1], Qi Liu[1], Zhiwei Cao[1,2,*], and Ruixin Zhu[1,3,*]

[1] Department of Bioinformatics, Tongji University, Shanghai, P.R. China
[2] Shanghai Center for Bioinformation Technology, Shanghai, P.R. China
[3] School of Pharmacy, Liaoning University of Traditional Chinese Medicine, Dalian, Liaoning, P.R. China

An issue of integrative medicine about herb-drug interactions has been increasing concerned [1]. Herbal ingredients can activate nuclear receptors to induce drug-metabolizing enzyme and/or transporter expression, which result in altering efficacy and toxicity of co-administered drugs. This process is called inductive herb-drug interactions [2]. Pregnane X Receptor (PXR) and drug-metabolizing target genes are involved in most of inductive herb-drug interactions. To predict this kind of herb-drug interaction, identifying ligands of nuclear receptors and drug-metabolizing enzyme/transporter could be done respectively. In addition, because drugs and their metabolizing enzymes are well known, the prediction would be simplified to only screen agonists of nuclear receptors. Finally, 421 herbs were collected to build a curated herb-drug interaction database, which records 380 herb-drug interactions including 90 herbs and 230 drugs. This database was used to validate our computational results.

A combinational *in silico* strategy of pharmacophore and docking-based rank aggregation (DRA) was employed to identify PXR's agonists. Firstly, 305 ingredients were screened out from 820 ingredients as candidate agonists of PXR with our pharmacophore model. Secondly, DRA was used to re-rank the result of pharmacophore filtering. Finally, the top 10 ingredients were mapped to 14 herbs, and 5 of these herbs were involved to the reported herb-drug interactions. This study demonstrated that the computational strategy was a promising way to investigate inductive herb-drug interactions.

References

1. Lopez-Picazo, J.J., Ruiz, J.C., Sanchez, J.F., Ariza, A., Aguilera, B., Lazaro, D., Sanz, G.R.: Prevalence and typology of potential drug interactions occurring in primary care patients. The European Journal of General Practice 16(2), 92–99 (2010)
2. Reitman, M.L., Chu, X., Cai, X., Yabut, J., Venkatasubramanian, R., Zajic, S., Stone, J.A., Ding, Y., Witter, R., Gibson, C., et al.: Rifampin's acute inhibitory and chronic inductive drug interactions: experimental and model-based approaches to drug-drug interaction trial design. Clinical Pharmacology and Therapeutics 89(2), 234–242 (2011)

---

* Corresponding authors.

M. Basu, Y. Pan, and J. Wang (Eds.): ISBRA 2014, LNBI 8492, p. 382, 2014.
© Springer International Publishing Switzerland 2014

Improving Multiple Sequence Alignment by Using Better Guide Trees

Qing Zhan[1,*], Yongtao Ye[2,*], Tak-Wah Lam[2],
Siu-Ming Yiu[2], Hing-Fung Ting[2,**], and Yadong Wang[1,**]

[1] School of Computer Science and Technology,
Harbin Institute of Technology, Harbin 150001, China
[2] HKU-BGI Bioinformatics Algorithms & Core Technology Research Lab.,
Department of Computer Science, University of Hong Kong

A commonly used approach for multiple sequence alignment (MSA) is the progressive alignment approach, which first constructs a guide tree that is supposed to capture the phylogenetic relationship of the input sequences, and then aligns the sequences progressively according to the topology of the tree. Previous studies have verified that guide trees are very important to the quality of the resulting alignments. In this work, we investigated how to construct better guide trees for better MSAs. In particular, we study an adaptive guide tree construction method, which was introduced by Ye *et al.*[1] for their MSA tool GLProbs. This method first computes the average percent identity $\overline{\text{PID}}$ of the input sequences, and if $\overline{\text{PID}}$ is small, it explores local information to construct the guide trees, and if $\overline{\text{PID}}$ is large, it focuses on global information. We study whether this adaptive method constructs the best guide trees for GLProbs. We also study whether it can improve the output quality of other MSA tools.

First, we have modified GLProbs to GLProbs-Random and GLProbs-Reference in which the adaptively constructed guide tree used by GLProbs is replaced by a randomly generated tree and an estimated phylogenetic tree (from the reference MSA) respectively. The three columns labeled GLProbs in Table 1 compares their performances with the sum of pairs score (SP): comparing the columns for GLProbs and GLProbs-Random confirmed that the guide trees constructed by the adaptive method do better, and comparing the columns for GL-Probs and GLProbs-Reference suggested that the adaptively constructed guide trees are among the best. Next, we have modified five leading tools by replacing their original guide tree construction steps with the adaptive one, and keeping other steps intact. The result in Table 1 shows that all the five modified tools achieved significant improvements, especially when the sequences have low similarity, e.g. ClustalW-Adaptive outperformed its original by 12.3% for sequences of 0-20% similarity.

---

* Joint first authors.
** Corresponding authors.

M. Basu, Y. Pan, and J. Wang (Eds.): ISBRA 2014, LNBI 8492, pp. 383–384, 2014.
© Springer International Publishing Switzerland 2014

Table 1. Mean SP scores on Benchmark OXBench

similarity	GLProbs			ClustalW		MSAProbs		Probalign		ProbCons		T-Coffee	
	Ref	Ran	Ori	Ada	Ori	Ada	Ori	Ada	Ori	Ada	Ori	Ada	Ori
0%-100%	90.30	90.06	**90.38**	**89.83**	89.44	**90.09**	90.06	**89.99**	89.96	**89.72**	89.68	**89.53**	89.51
0%-20%	**47.63**	46.03	47.33	**48.22**	42.94	**45.14**	44.84	**44.31**	43.57	**45.39**	44.14	**44.88**	43.82

"Ref" denotes using the estimated (maximum likelihood) phylogenetic tree from reference MSA; "Ran" denotes using the randomly generated tree; "Ori" denotes the aligner original version; "Ada" denotes using the adaptive guide tree construction method. Better results are shown in bold.

Reference

1. Ye, Y., Cheung, D.W., Wang, Y., Yiu, S.-M., Zhan, Q., Lam, T.-W., Ting, H.-F.: GLProbs: Aligning Multiple Sequences Adaptively. In: Proceedings of the International Conference on Bioinformatics, Computational Biology and Biomedical Informatics, pp. 152–160 (2013)

A Markov Clustering Based Link Clustering Method for Overlapping Module Identification in Yeast Protein-Protein Interaction Networks[*]

Yan Wang[1,2], Guishen Wang[1], Di Meng[1], Lan Huang[1,**],
Enrico Blanzieri[2,**], and Juan Cui[3]

[1] College of Computer Science and Technology, Key Laboratory of Symbolic
Computation and Knowledge Engineering of Ministry of Education
Jilin University, Changchun, China
[2] Department of Information and Communication Technology
University of Trento, Povo, Italy
[3] Department of Computer Science and Engineering
University of Nebraska at Lincoln, Lincoln, NE, USA
huanglan@jlu.edu.cn, blanzier@disi.unitn.it

Abstract. Previous studies have shown that many overlapping components among the modular structures in protein-protein interaction (PPI) networks reflect common functional components shared by different biological processes. In this paper, we proposed a Markov clustering based Link Clustering (MLC) method to identify the overlapping modular structures in PPI networks. MLC method calculates the extended link similarity that represents the relevance between links, the interactions between proteins, and derives a similarity matrix. It then uses markov clustering to partition the link similarity matrix and obtains overlapping modules in the original network automatically without much parameters and threshold constraints. Our experimental validation on two benchmark networks with known reference classes and the Yeast PPI network, respectively, show that MLC outperforms the original Link Clustering and the classical Clique Percolation Method with higher EQ/ENMI/DR evaluation and better GO enrichment performance. It is particularly interesting that, on Yeast PPI network, MLC also identifies new functional modules in which genes do not show significant correlation among their expressions. Overall, the MLC method has demonstrated promising potentials in identifying the core biological modules or important pathways in different organisms through studying the interplay between functional processes.

Keywords: Overlapping Module, Protein-protein Interaction, Markov Clustering, Link Clustering.

[*] This work is supported by the Natural Science Foundation of China (61175023), Jilin Innovation Team Project (20121805), the Ph.D. Program Foundation of MOE of China (20120061120106), and the Science-Technology Development Project of Jilin Province of China (20130522111JH, 20130522114JH, 20140101180JC).
[**] Corresponding authors.

M. Basu, Y. Pan, and J. Wang (Eds.): ISBRA 2014, LNBI 8492, p. 385, 2014.
© Springer International Publishing Switzerland 2014

Protein Function Prediction: A Global Prediction Method with Multiple Data Sources

Jun Meng[1], Xin Zhang[1], and Yushi Luan[2,*]

[1] School of Computer Science and Technology,
Dalian University of Technology, Dalian, China
[2] School of Life Science and Biotechnology,
Dalian University of Technology, Dalian, China

Multiple types of genomics and proteomics data are being available by heterogeneous high-throughput experiments. As each data source captures only one aspect about proteins' properties, it's necessary and wise to integrate these heterogeneous high-throughput data sources which bring a more complete picture about protein functions. The MS-KNN method [1] shows three data sources for protein function prediction: protein-protein interaction (PPI), gene expression and sequence similarity. Yu [2] proposed a method called TMEC to capture the relationships between pairs of proteins, between pairs of functions, and between proteins and functions. However, many methods predict functions without fully considering the properties of each data source. In order to use these data effectively, we choose appropriate methods for different data source to construct networks and merge those networks.

In this paper, we choose three Yeast data sources for protein function prediction, PPI, gene expression and protein sequence, which roles for function annotation were introduced by MS-KNN. The data sources are downloaded from the Biological General Repository for Interaction Datasets (BioGRID), the *Saccharomyces* Genome Database (SGD) and the MIPS Comprehensive Yeast Genome Database (CYGD), respectively. To calculate the weights between pairs of proteins in a PPI network, edge clustering coefficient [3] is a suitable measure which can evaluate the importance of edges in PPI and describe how close the two proteins are. For gene expression data, Pearson correlation coefficient is a frequently used coefficient to express the degree of linear relationship between two sets of gene expression value. For protein sequence, in consideration of poor similarity between proteins, we extract protein sequence' PseAAC [4] features and calculate the inner product distance between two proteins by the features, instead of sequence homologous similarity-based method. As the efficiency of methods for predicting protein functions from networks depend on the number of non-zero interactions, we sparse dense networks. Therefore, we retain k-nearest neighbors for each protein and set the rest to zero for gene expression and protein sequence network. Then, a naïve Bayesian fashion [5] is used to combine the networks. Finally, a global propagation algorithm [6] is designed on the combined

---

* Corresponding author.

M. Basu, Y. Pan, and J. Wang (Eds.): ISBRA 2014, LNBI 8492, pp. 386–387, 2014.

network, which takes the known function annotations for protein as the sources of 'function flow'. The method considers the global and local network topology.

The experimental results show that the proposed global propagation algorithm by iterating the combined network method has superior over MS-KNN and TMEC with high accuracy of protein function prediction.

References

1. Lan, L., Djuric, N., Guo, Y.H., Vucetic, S.: MS-kNN: protein function prediction by integrating multiple data sources. BMC Bioinformatics 14, S8 (2011)
2. Yu, G., Rangwala, H., Domeniconi, C., Zhang, G.J., et al.: Protein Function Prediction using Multi-label Ensemble Classification. IEEE ACM T. Comput. Bi. 10, 1 (2013)
3. Wang, J., Li, M., Wang, H., Pan, Y.: Identification of essential proteins based on edge clustering coefficient. IEEE ACM T. Comput. Bi. 9, 1070–1080 (2012)
4. Chou, K.C.: Prediction of protein cellular attributes using pseudo-amino acid composition. Proteins Struct. Funct. Genet. 43, 246–255 (2001)
5. Von Mering, C., Jensen, L.J., Snel, B., et al.: String: known and predicted protein-protein associations, integrated and transferred across organisms. Nucleic Acids Res. 33, D433–D437 (2005)
6. Vanunu, O., Magger, O., Ruppin, E., et al.: Associating genes and protein complexes with disease via network propagation. PLoS Comput. Biol. 6, e1000641 (2010)

A microRNA-Gene Network in Ovarian Cancer from Genome-Wide QTL Analysis

Andrew Quitadamo, Frederick Lin, Lu Tian, and Xinghua Shi

Department of Bioinformatics and Genomics
College of Computing and Informatics
University of North Carolina at Charlotte
Charlotte NC 28223, USA
{aquitada,flin8,ltian,x.shi}@uncc.edu

Ovarian cancer is the most deadly reproductive cancer in women. A better understanding of the biological mechanisms of ovarian cancer is needed for earlier diagnosis and more effective treatment. Differential microRNA(miRNA) expression and miRNA/mRNA dysregulation have been associated with ovarian cancer. Whole-genome miRNA and mRNA sequencing provides a new prospective to study these aberrations for their associations with ovarian cancer.

In this study, we perform a genome-wide QTL analysis between miRNA and gene expression in ovarian cancer, using data from The Cancer Genome Atlas (TCGA) [1]. The results from such QTL analysis provided a network new of the relationship between miRNA and gene expression. We found that all of the identified miRNAs were reported previously to be associated with different diseases, and particularly, the majority of these miRNAs were shown to be associated with ovarian cancer. Our results replicated several cancer genes [2], and provided a list of candidate cancer genes as well. In summary, we showed that our integrative analysis would help understand the molecular mechanism of disease manifestation and progression, and eventually result in better prognosis, diagnosis and treatment of ovarian cancer.

Keywords: Cancer genomics, ovarian cancer, microRNAs, RNA sequencing, Quantitative Trait Loci (QTL) analysis.

References

1. Cancer Genome Atlas Research Network: Integrated genomic analyses of ovarian carcinoma. Nature 474(7353), 609–615 (2011)
2. Atlas of Genetics and Cytogenetics in Oncology and Haematology,
   http://atlasgeneticsoncology.org/Genes/Geneliste.html

M. Basu, Y. Pan, and J. Wang (Eds.): ISBRA 2014, LNBI 8492, p. 388, 2014.
© Springer International Publishing Switzerland 2014

K-Profiles Nonlinear Clustering

Kai Wang[1] and Tianwei Yu[2,*]

[1] Department of Mathematics and Computer Science,
Emory University, Atlanta, Georgia, USA
[2] Department of Biostatistics and Bioinformatics, Emory University, Atlanta, Georgia, USA

Modern technologies such as microarray, deep sequencing, liquid chromatography–mass spectrometry (LC-MS) etc make it possible to measure the expression levels of thousands of genes/proteins simultaneously to unravel important biological processes. Detecting nonlinear relationships are most useful in the context of exploratory knowledge discovery from large biological datasets, when data structure itself is not yet well understood. Nonlinear relations, which were mostly unutilized in contrast to linear correlations, are prevalent in high-throughput data. In many cases, it can model biological relationships more precisely and reflect critical patterns in the biological systems.

Clustering is usually taken as the first step towards elucidating hidden patterns and understanding the mass of data. However, no single clustering algorithm tops all performance charts due to its built-in biases on datasets [1]. Well-defined relationship/distance measurement and cluster profiles play crucial roles in the process. Using the general dependency measure, Distance based on Conditional Ordered List (DCOL) that we introduced before [2], we designed the nonlinear K-profiles clustering method, which can be seen as the nonlinear counterpart of the K-means clustering algorithm with statistical testing incorporated to remove prevalent noise in biological data. It not only outperformed our previous General Dependency based Hierarchical Clustering (GDHC) algorithm and the traditional K-means algorithm in our simulation studies, but also showed much improved computational efficiency in contrast with GDHC. Real data analysis showed its capability to detect novel nonlinear patterns in high-throughput data. It will be discussed in detail in this talk.

References

1. D'Haeseleer, P.: How does gene expression clustering work? Nat. Biotechnol. 23(12), 1499–1501 (2005)
2. Yu, T., Peng, H., Sun, W.: Incorporating Nonlinear Relationships in Microarray Missing Value Imputation. IEEE/ACM Trans. Comput. Biol. Bioinform. 8(3), 723–731 (2011)

---

* Corresponding author.
1518 Clifton Rd NE, Rm 334, Atlanta, GA 30322.

M. Basu, Y. Pan, and J. Wang (Eds.): ISBRA 2014, LNBI 8492, p. 389, 2014.
© Springer International Publishing Switzerland 2014

Estrogen Induced RNA Polymerase II Stalling in Breast Cancer Cell Line MCF7

Zhi Han[1,2], Lu Tian[3], Jie Zhang[4], Tim Huang[5], Raghu Machiraju[6], and Kun Huang[2,*]

[1] College of Software, Nankai University, Tianjin, China
[2] Department of Biomedical Informatics, The Ohio State University, Columbus, Ohio, USA
[3] Department of Health Policy and Research – Biostatistics, Stanford University, USA
[4] OSU Biomedical Informatics Shared Resource, The Ohio State University, USA
[5] Department of Genetic Medicine, University of Texas Health Science Center, USA
[6] Department of Computer Science and Engineering, The Ohio State University, USA

RNA polymerase II (PolII) stalling is an important phenomenon in gene regulation. This is an important cellular process in response to stress [1]. For the genes with PolII stalling, while the PolII molecules accumulate at their promoter regions, their transcription processes are paused. Here we investigate PolII stalling induced by estrogen in the breast cancer cell line MCF7 by integrating data from ChIP-seq and microarray technologies. We take a rule based approach to identify genes with PolII stalling after 17β-estrodial (E2) treatment in MCF7 cells. E2 treatment activates estrogen receptor which plays important roles in the majority of breast cancers [2]. We use ChIP-seq data for PolII and gene expression microarray data for MCF7 before and after treatment of E2.

Our method includes several main steps: First, we identify genes with enriched PolII binding segment using a signal processing based algorithm we have previously developed. This algorithm identifies both long and short enriched regions from ChIP-seq data [3]. Next, we select genes whose PolII enrichment levels increase at the promoter regions but decrease over the gene bodies after E2 treatment. Finally, we zoom into genes whose expression levels decreased significantly ($p < 0.05$ and mean fold change > 1.5) after E2 treatment. We also apply similar rules to identify genes released from PolII stalling after E2 treatment.

Our method identified 92 genes which satisfy our criteria and demonstrate PolII stalling induced by E2 in MCF7 cell line while only 3 genes show PolII stalling released by E2. Functional analysis of identified genes shows that E2 induced PolII stalling is highly relevant to cancer development pathways. This suggests that E2 treatment potentially can cause a stress response of the breast cancer cells which leads promotion or disruption of cancer related biological functions.

References

[1] Baugh, L.R., et al.: RNA Pol II accumulates at promoters of growth genes during developmental arrest. Science 324, 92–94 (2009)
[2] Fox, E.M., et al.: ERbeta in breast cancer–onlooker, passive player, or active protector? Steroids 73, 1039–1051 (2008)
[3] Han, Z., et al.: A signal processing approach for enriched region detection in RNA polymerase II ChIP-seq data. BMC Bioinformatics 13(suppl. 2), S2 (2012)

---

[*] Corresponding author.

M. Basu, Y. Pan, and J. Wang (Eds.): ISBRA 2014, LNBI 8492, p. 390, 2014.
© Springer International Publishing Switzerland 2014

A Knowledge-Driven Approach in Constructing a Large-Scale Drug-Side Effect Relationship Knowledge Base for Computational Drug Discovery

Rong Xu[1] and QuanQiu Wang[2]

[1] Case Western Reserve University, Cleveland OH 44106, USA
[2] ThinTek, LLC, Palo Alto CA 94306

Introduction. It has been increasingly recognized that similar side effects of seemingly unrelated drugs can be caused by their common off-targets and that drugs with similar side effects are likely to share molecular targets [1]. Therefore, systems approaches to studying side effect relationships among drugs and integration of this drug phenotypic data with drug-related genetic, genomic, proteomic, and chemical data will facilitate drug target discovery and drug repositioning [2]. The availability of a comprehensive drug-side effect (SE) relationship knowledge base is critical for these tasks. Current drug phenotype-driven systems approaches rely exclusively on drug-SE associations extracted from FDA drug labels. However, there exists a large amount of additional drug-SE relationship knowledge in the large body of published biomedical literature. In this study, we present a novel knowledge-driven (KD) approach to automatically extract a large number of drug-SE pairs from 21 million published biomedical abstracts. We systematically analyzed extracted drug-SE pairs in combination with drug-related gene targets, metabolism, pathways, gene expression and chemical structure data. We show that these extracted drug-SE pairs have great potential in drug discovery.

Methods. Our study is based on the two key observations: (1) multiple side effects for a drug are often reported in the same sentences or abstracts; and (2) if a sentence contains a known drug-SE pair, then this sentence is likely to be SE-relevant. Other pairs in this SE-related sentence are likely to be drug-SE pairs. In this study, we used all known drug-SE pairs derived from FDA drug labels as prior knowledge to find SE-related MEDLINE sentences and abstracts, from which many additional drug-SE pairs that have not included in FDA drug labels are then extracted. We compared the KD approach to a support vector machine (SVM)-based approach. The entire experimental process consists of the following steps: (1) Build a local MEDLINE search engine; (2) Develop, evaluate and compare the KD approach to a SVM-based approach; (3) Extract drug-SE pairs from MEDLINE; and (4) Systematically analyze the correlation between drug-associated side effects and drug gene targets, metabolism genes, chemical similarity, and disease indications. For the text corpus, we used 21,354,075 syntactically parsed MEDLINE records (119,085,682 sentences).

Results. First, we used known drug-SE associations derived from FDA drug labels as prior knowledge to automatically find SE-related sentences and abstracts. We then extracted a total of 49,575 drug-SE pairs from MEDLINE sentences and 180,454

M. Basu, Y. Pan, and J. Wang (Eds.): ISBRA 2014, LNBI 8492, pp. 391–392, 2014.

pairs from abstracts. On average, the KD approach has achieved a precision of 0.335, a recall of 0.509, and an F1 of 0.392, which is significantly better than a SVM-based machine learning approach (precision: 0.135, recall: 0.900, F1: 0.233) with a 73.0% increase in F1 score. Through integrative analysis, we demonstrate that the higher-level phenotypic drug-SE relationships reflect lower-level genetic, genomic, and chemical drug mechanisms. In addition, we show that the extracted drug-SE pairs can be directly used in drug repositioning.

Conclusions. In summary, we automatically constructed a large-scale drug phenotype relationship knowledge, which in combination with other genetic, genomic and chemical data resources, can have great potential in computational drug discovery.

References

1. Campillos, M., Kuhn, M., Gavin, A.C., Jensen, L.J., Bork, P.: Drug target identification using side-effect similarity. Science 321(5886), 263–266 (2008)
2. Hurle, M.R., Yang, L., Xie, Q., Rajpal, D.K., Sanseau, P., Agarwal, P.: Computational drug repositioning: From data to therapeutics. Clinical Pharmacology and Therapeutics (2013)

Systems Biology Approach
to Understand Seed Composition

Ling Li[1,*], Wenxu Zhou[2], Manhoi Hur[1], Joon-Yong Lee[1], Nick Ransom[1],
Cumhur Yusuf Demirkale[3], Zhihong Song[2], Dan Nettleton[3], Mark Westgate[4],
Vidya Iyer[5], Jackie Shanks[5], Eve Syrkin Wurtele[1], and Basil J. Nikolau[2,*]

[1] Department of Genetics, Development and Cell Biology
[2] Department of Biochemistry,
Biophysics and Molecular Biology
[3] Department of Statistics
[4] Department of Agronomy
[5] Department of Chemical and Biological Engineering,
Iowa State University, Ames, Iowa 50011, USA
{liling,dimmas}@iastate.edu

Abstract. As the propagule that ensures the dissemination of plants, seeds also
support human activity as one of the major products of agriculture. The
biochemical storage reserves that are deposited within the seed during its development chemically fall into three general categories, proteins, oils and carbohydrates. The seed reserves are biosynthesized by the programmed expression of a
metabolic network during seed development. In most commercial lines of
soybean grown in the Midwestern states of the US, seeds are composed of 40%
protein, 20% oil, 15% soluble carbohydrates, and 15% fiber (http://www.asa-
europe.org/SoyInfo/composition_e.htm). There is considerable knowledge concerning the basic biochemical processes by which imported carbon and nitrogen
is converted to the final products, protein, oil and carbohydrate. However, there
is a great deal to be learned concerning the molecular, biochemical and genetic
mechanisms that regulate this complex metabolic network. Recent developments in genomics have provided the catalogue of genes that would be required
for this process. We have taken advantage of combined metabolomics and transcriptomics technologies to identify the global developmental and biochemical
transcriptomics network, and ultimately determine structure and composition of
the mature seed. Also, we have coupled this with bioinformatics and metabolic
flux analyses to gain insights as to the biochemical programs that determine
soybean seed development. For this purpose, we have developed Plant &
Microbial Metabolomics Resource (PMR, http://www.metnetdb.org/pmr), a
platform to empower the use of metabolomics data in the development of hypotheses concerning the organization and regulation of metabolic networks, and
MetNet systems biology platform (http://www.metnetdb.org) for plant 'omics, a
web-based framework which enables interactive visualization of metabolic and
regulatory networks. This combination of genetic resources, high-throughput
experimental data and bioinformatic analyses has revealed sets of specific
genes, genetic perturbations and mechanisms, and metabolic changes that are
associated seed composition during soybean seed development.

Keywords: *Glycine max*, Evans, seed development, gene expression, metabo-
lomic change, seed composition, PMR, MetNet.

---

* Corresponding authors.

M. Basu, Y. Pan, and J. Wang (Eds.): ISBRA 2014, LNBI 8492, p. 393, 2014.
© Springer International Publishing Switzerland 2014

Prediction of the Cooperative *cis*-regulatory Elements for Broadly Expressed Neuronal Genes in *Caenorhabditis Elegans*

Chen Xu and Zhengchang Su

Department of Bioinformatics and Genomics, University of North Carolina at Charlotte, 351 Bioinformatics Building, 9201 University City Blvd, Charlotte, NC 28223, USA

How cell types are derived by gene regulatory programs is a fundamental problem in developmental biology. The nervous system of the *Caenorhabditis elegans* (*C. elegans*) provides an excellent model to gain a good understanding of this problem. It has been shown that common structural features shared by diverse types of neurons in *C. elegans*, such as axons, dendrites and synapses, etc. are determined by a set of genes broadly expressed in neuronal cells, known as pan-neuronal genes [1]. However, so far very little is known about the transcriptional regulatory mechanisms of these genes. In this research, we found that two of our identified putative motifs tend to concur in the vast majority of upstream intergenic regions of the pan-neuronal genes [2], and hence are likely to be a *cis*-regulatory module (CRM). Interestingly, this module is also widely distributed in the whole *C. elegans* genome and is highly conserved between *C. elegans* and *Caenorhabditis briggase* (*C. briggase*). In addition to the general control by a common CRM, the pan-neuronal genes may rely on other *cis*-regulatory motifs in order to further specify their functions. We found that some identified motifs were harbored by different subsets of pan-neuronal genes which could be significantly related to different functions respectively. These results suggest that the pan-neuronal gene features are defined by the collective usage of the two regulatory mechanisms. Our computational results should provide some hints of *cis*-regulatory mechanisms of the pan-neuronal genes and a useful guide to experimental validations.

References

[1] Hobert, O., Carrera, I., Stefanakis, N.: The molecular and gene regulatory signature of a neuron. Trends in Neurosciences 33, 435–445 (2010)
[2] Ruvinsky, I., Ohler, U., Burge, C.B., Ruvkun, G.: Detection of broadly expressed neuronal genes in C. Elegans. Developmental Biology 302, 617–626 (2010)

M. Basu, Y. Pan, and J. Wang (Eds.): ISBRA 2014, LNBI 8492, p. 394, 2014.
© Springer International Publishing Switzerland 2014

Improving the Mapping of the Smith-Waterman Sequence Database Search Algorithm onto CUDA GPUs[*]

Chao-Chin Wu[1], Liang-Tsung Huang[2], Lien-Fu Lai[1], and Yun-Ju Li[1]

[1] Department of Computer Science and Information Engineering
National Changhua University of Education, Changhua 500, Taiwan
{ccwu,lflai}@cc.ncue.edu.tw, icecloud6666@gmail.com
[2] Department of Biotechnology, Mingdao University, Changhua 523, Taiwan
larry@mdu.edu.tw

Sequence alignment is one of the most important methodologies in the field of computational biology. The most widely used sequence alignment algorithm may be the Smith-Waterman algorithm because of its high sensitivity of sequence alignment even though it has higher time complexity of algorithm [1]. To enable the Smith-Waterman algorithm produce exact results in a reasonably shorter time, much research has been focusing on using various high-performance architectures to accelerate the processing speed of the algorithm. In particular, it becomes a recent trend to use the emerging accelerators and many-core architectures to run the Smith-Waterman algorithm.

Modern general-purpose (Graphics Processing Units) GPUs are not only powerful graphics engines, but also highly parallel programmable processors [2]. Today's GPUs use hundreds of parallel processor cores executing tens of thousands of parallel threads to rapidly solve large problems, now available in many PCs, laptops, workstations, and supercomputers. Because of the availability and the popularity, GPUs have been used to implement the Smith-Waterman algorithm, where CUDASW++ is the leading reseach that provides the fast, publicly available, solution to the exact Smith-Waterman algorithm on commodity hardware. CUDASW++ 3.0 is the latest version, which couples CPU and GPU SIMD instructions and carries out concurrent CPU and GPU computations [3].

This paper focuses on how to improve CUDASW++, especially for short query sequences. We observe that the shared memory in each streaming multiprocessor is not fully utilized in CUDASW++. Therefore, the execution flow of the Smith-Waterman algorithm is rearranged to fully utilize the shared memory for reducing the amount of slow global memory access. We have added our approach to CUDASW++ 2.0 and run experiments on nVIDIA Tesla C1060 and C2050. Experimental results demonstrate that our approach outperforms CUDASW++.

References

1. Smith, T.F., Waterman, M.S.: Identification of Common Molecular Subsequences. Journal of Molecular Biology 147, 195–197 (1981)
2. CUDA GPUs, https://developer.nvidia.com/cuda-gpus
3. Liu, Y., Wirawan, A., Schmidt, B.: CUDASW++ 3.0: Accelerating Smith-Waterman Protein Database Search by Coupling CPU and GPU SIMD Instructions. BMC Bioinformatics 14, 117 (2013)

---

[*] This work is supported by the contract, NSC102-2221-E-018-024.

M. Basu, Y. Pan, and J. Wang (Eds.): ISBRA 2014, LNBI 8492, p. 395, 2014.
© Springer International Publishing Switzerland 2014

Isomorphism and Similarity
for 2-Generation Pedigrees

Haitao Jiang[1], Guohui Lin[2], Weitian Tong[2], Daming Zhu[1], and Binhai Zhu[3]

[1] School of Computer Science and Technology,
Shandong University, Jinan, Shandong, China
{htjiang,dmzhu}@sdu.edu.cn
[2] Department of Computing Science, University of Alberta,
Edmonton, Alberta T2G 2E6, Canada
{weitian,guohui}@ualberta.ca
[3] Department of Computer Science, Montana State University,
Bozeman, MT 59717, USA
bhz@cs.montana.edu

In this paper, we follow the work by Kirkpatrick *et al.* [4] to consider the isomorphism and similarity problems for the simplest (unlabeled) pedigrees — 2-generation pedigrees, where the isomorphism and similarity problems are both studied. We show that the isomorphism problem is GI-hard (GI — Graph Isomorphism) even for 2-generation pedigrees. If the 2-generation pedigrees are monogamous (i.e., each individual at level-1 can mate with exactly one partner) then the isomorphism testing problem can be solved in polynomial time.

Subsequently, we relax the similarity measure for two general 2-generation pedigrees by using the minimum number of isomorphic $\langle i, j \rangle$-families which they can be decomposed into. Here, an $\langle i, j \rangle$-family is a sub-family of a couple with i female children and j male children. It turns out that this can be formulated as a Minimum Common Integer Pair Partition (MCIPP) problem, generalizing the NP-complete Minimum Common Integer Partition (MCIP) problem [1]. We then exploit a new property of the optimal solution for MCIPP, and show that MCIPP is Fixed-Parameter Tractable [2,3].

Acknowledgments. This research is partially supported by NSF of China under grant 60928006, 61070019 and 61202014, by NSF of Shandong Province under grant ZR2012FQ008, and by NSERC of Canada.

References

1. Chen, X., Liu, L., Liu, Z., Jiang, T.: On the minimum common integer partition problem. ACM Trans. on Algorithms 5(1) (2008)
2. Downey, R., Fellows, M.: Parameterized Complexity. Springer (1999)
3. Flum, J., Grohe, M.: Parameterized Complexity Theory. Springer (2006)
4. Kirkpatrick, B., Reshef, Y., Finucane, H., Jiang, H., Zhu, B., Karp, R.: Comparing pedigree graphs. J. of Computational Biology 19(9), 998–1014 (2012)

M. Basu, Y. Pan, and J. Wang (Eds.): ISBRA 2014, LNBI 8492, p. 396, 2014.
© Springer International Publishing Switzerland 2014

VFP: A Visual Tool for Predicting Gene-Fusion Base on Analyzing Single-end RNA-Sequence

Ye Yang[1,2] and Juan Liu[1,*]

[1] School of Computer, Wuhan University, Wuhan, Hubei, China
[2] Military Economy Academy, Wuhan, Hubei, China
liujuanjp@163.com

Gene fusion is a key factor in sarcomas, lymphomas, leukemias and so on. In recent years, some fusion detection algorithms were published to search the fusion by the data produced on next generation sequencing platform, Such as TopHat-Fusion[1], FusionHunter[2], FusionSeq[3]. But these algorithms have some common defects, such as the limitation of the operating system, the confusion of the parameter setting and so on. In order to help biologist to quickly discover the target of the treatment, we have developed VFP to predict gene-fusion from single-end RNA-sequencing reads. VFP employs seed index strategy and octal encoding operations for sequence alignments and uses several rules to score and filter the potential fusion genes. We tested VFP by a simulated dataset and two real datasets and found that VFP can detect known and novel fusions in lymphoma and melanoma datasets.

There are many extensions and modifications of VFP, some of them will be mentioned in the talk.

References

1. Kim D, Salzberg SL: TopHat-Fusion: an algorithm for discovery of novel fusion transcripts. Genome Biology. 12(2011),R72.
2. Li Y, et al: FusionHunter: identifying fusion transcripts in cancer using paired-end RNA-seq. Bioinformatics. 27(2011), 1708–1710 .
3. Sboner A. et al: FusionSeq: a modular framework for finding gene fusions by analyzing paired-end RNA-sequencing data. Genome Biology. 11(2010),R104.

---

* Corresponding author.

M. Basu, Y. Pan, and J. Wang (Eds.): ISBRA 2014, LNBI 8492, p. 397, 2014.

A Novel Method for Identifying Essential Proteins from Active PPI Networks

Qianghua Xiao[1,2], Xiaoqing Peng[1], Fangxiang Wu[1,3], and Min Li[1]

[1] School of Information Science and Engineering,
Central South University, Changsha 410083, China
[2] School of Mathematics and Physics,
University of South, HengYang 421001, China
[3] Division of Biomedical Engineering,
University of Saskatchewan, Saskatoon, SK, S7N 5A9, Canada

Essential proteins are vital for cellular survival and development. Identifying essential proteins is very important for helping us understand the way in which a cell works. Rapid increase of available protein-protein interaction (PPI) data has made it possible to detect protein essentiality at the network level. A series of centrality measures have been proposed to discover essential proteins based on the PPI networks. However, the PPI data obtained from large scale, high-throughput experiments generally contain false positives. It is insufficient to use original PPI data to identify essential proteins.

In this paper, we firstly adopt a dynamic model-based method to filter noisy data from time-course gene expression profiles. Second, a threshold of each protein is calculated from a threshold function of σ, the protein is active at a time point if its expression level is higher than the threshold. Two proteins are regarded as co-expression if they are all active at the same time point. Finally, an active PPI network is constructed by combining gene expression data with PPI data.

The classical centrality measures, like Degree centrality(DC), Local Average Connectivity Centrality (LAC), Edge Clustering Coefficient (NC), Betweenness Centrality (BC), Closeness Centrality (CC), and Subgraph Centrality (SC), are methods which can be applied to identify essential proteins based on network topology. These centrality measures are redefined and performed to identify essential proteins on the active PIN. The experimental results on yeast network show that the performance of centrality measures to identify essential proteins are considerably improved based on the active PPI network, compared with original PPI network, in terms of the number of identified essential proteins in top %k percentage and a jackknife methodology. At the same time, the results also indicate that most essential proteins are active.

Acknowledgement. This work is supported in part by the National Natural Science Foundation of China under Grant No.61370024, No.61232001, and No.61379108, the Program for New Century Excellent Talents in University (NCET-12-0547), Science and Technology Plan Projects of Science and Technology Bureau of Hengyang City (grant 2013KJ29).

M. Basu, Y. Pan, and J. Wang (Eds.): ISBRA 2014, LNBI 8492, p. 398, 2014.

RAUR: Re-alignment of Unmapped Reads with Base Quality Score

Xiaoqing Peng[1], Zhen Zhang[1], Qianghua Xiao[1,2], and Min Li[1]

[1] School of Information Science and Engineering,
Central South University, Changsha 410083, China
[2] School of Mathematics and Physics, University of South,
HengYang 421001, China

In recent years, with the emergencies of next-generation genome sequencing technologies, many software tools have been developed to efficiently and accurately align short reads to the reference genome. However, for most alignment tools, the edit distances or the allowed mismatches are limited, thus some reads cannot be mapped if their mismatches in any hits exceed the allowable differences.

Some trimmed-like strategies appear in some alignment programs and try to handle the problem. For example, Bowtie2 and BWA-MEM in BWA can perform local read alignment for long reads by maximizing the alignment score. However, the false positive sites are also introduced, since the maximum alignment score can't make sure that high quality bases are involved.

In this article, we propose a method (RAUR) to re-align the unmapped reads. A trimming strategy used in RAUR is to figure out the longest and most confident and informative segment of a read based on base quality score. RAUR can be applied on any alignment tool if there are reads which can't be aligned by it. It adopts an iterative progress to trim the unmapped reads until the reads can be confidently mapped or can't be mapped in any progress. To evaluate the performance of RAUR, we apply RAUR on the simulated reads and real reads of human genome with different lengths by comparing with BWA, Bowtie2, and SOAP2 with different settings. From the *precision* and *alignment rate*, we can find out that RAUR can improve the *alignment rate*s greatly, especially for long reads, while the *precision*s are still comparative with the original alignments. RAUR proposes a new insight for re-aligning unmapped reads, which can contribute to the downstream analysis.

Acknowledgement. This work is supported in part by the National Natural Science Foundation of China under grant nos. 61232001, 61379108, and 61370172, Hunan Provincial Innovation Foundation For Postgraduate (CX2013B070), and Science and Technology Plan Projects of Science and Technology Bureau of Hengyang City (grant 2013KJ29).

PIGS: Improved Estimates
of Identity-by-Descent Probabilities
by Probabilistic IBD Graph Sampling

Danny S. Park[1], Yael Baran[2], Farhad Hormozdiari[3], and Noah Zaitlen[1]

[1] University of California San Francisco, San Francisco CA 94143, USA
[2] Tel Aviv University, Tel Aviv, Israel
[3] University of California Los Angeles, Los Angeles CA 90095, USA

Identifying segments of the genome that are identical-by-descent (IBD) between individuals is a fundamental concept in genetics. IBD data are used in numerous applications including demographic inference, heritability estimation, and disease loci mapping. Therefore, the identification of IBD segments from genome-wide genotyping studies, and more recently sequencing studies, has important implications for studies of complex human traits.

Current methods for detecting IBD fall into two categories: multiway or pairwise. Multiway methods detect IBD over multiple haplotypes simultaneously and leverage the clique structure of true IBD. Although powerful, these approaches are generally computationally expensive since the number of potential IBD relationships at a locus is $O(2^{h(h-1)/2})$, where h is the number of haplotypes. As a result, many state-of-the-art methods estimate the probability of IBD between pairs of haplotypes independently [1]. The result is a much more efficient method but at the expense of loss of power when detecting smaller IBD segments (<1 centimorgans) [2].

We develop a hybrid approach (PIGS), which combines the computational efficiency of pairwise methods and the power of multiway methods. It leverages the IBD clique structure to simultaneously compute the probability of IBD conditional on all pairwise estimates. We show over extensive simulations, that PIGS yields a substantial increase in the number of identified small IBD segments. We observed a 95% increase in the total number of identified IBD segments of 0.5 centimorgans and a 40% increase in identified IBD segments across all sizes.

Given the substantial improvement in the number of identified IBD segments from our method, we expect that the approach will greatly facilitate the discovery of new loci from IBD-based disease association studies.

References

1. Browning, B.L., Browning, S.R.: Improving the accuracy and efficiency of identity-by-descent detection in population data. Genetics 194(2), 459–471 (2013)
2. He, D.: IBD-Groupon: an efficient method for detecting group-wise identity-by-descent regions simultaneously in multiple individuals based on pairwise IBD relationships. Bioinformatics 29(13), i162–i170 (2013)

M. Basu, Y. Pan, and J. Wang (Eds.): ISBRA 2014, LNBI 8492, p. 400, 2014.
© Springer International Publishing Switzerland 2014

Clustering PPI Data through Improved Synchronization-Based Hierarchical Clustering Method

Xiujuan Lei[1,2,*], Chao Ying[3], Fang-Xiang Wu[4], and Jin Xu[5]

[1] School of Computer Science, Shaanxi Normal University, Xi'an, Shaanxi 710062, China
[2] School of Electronics Engineering and Computer Science,
Peking University, Beijing, 100871, China
`xjlei@snnu.edu.cn`
[3] School of Computer Science, Shaanxi Normal University, Xi'an, Shaanxi 710062, China
`nbsschao@163.com`
[4] Division of Biomedical Engineering, University of Saskatchewan,
Saskatoon, SK S7N 5A9, Canada
`faw341@mail.usask.ca`
[5] School of Electronics Engineering and Computer Science,
Peking University, Beijing, 100871, China
`jxu@pku.edu.cn`

Abstract. Clustering algorithm is the main method to identify function module of protein-protein interaction (PPI) network, but traditional methods have advantages and corresponding drawbacks. The synchronization-based hierarchical clustering (SHC) algorithm was improved in this paper. Firstly, the PPI data was preprocessed via spectral clustering (SC) method which transforms the high-dimensional similarity matrix into a low dimension matrix. Then the SHC algorithm is used to perform clustering. In SHC algorithm, hierarchical clustering are achieved by enlarging the local neighborhood distance of objects synchronizing continuously, while the hierarchical search is very difficult to find the optimal local neighborhood distance of synchronizing and the efficiency is not high. So the glowworm swarm optimization (GSO) algorithm was adopted to determine the optimal threshold of the local neighborhood distance of synchronization automatically. The algorithm is tested on the PPI dataset. The results show that the improve algorithm is better than the traditional algorithms in *precision*, *recall* and *f-measure*.

Keywords: Protein-Protein Interaction network, glowworm swarm optimization algorithms, synchronization-based hierarchical clustering, spectral clustering algorithm.

1 Introduction

Protein-protein interaction (PPI) network is an important research field in the bioinformatics. Identifying the function of protein complex is critical for understanding

---

* Corresponding author.

M. Basu, Y. Pan, and J. Wang (Eds.): ISBRA 2014, LNBI 8492, pp. 401–403, 2014.
© Springer International Publishing Switzerland 2014

disease mechanisms, diagnosis and therapy. Recently there are a large number of clustering methods applied to discover modules in PPI network, but they suffered some degree of shortcomings. The cluster number of spectral clustering(SC)[1] must be predefined and the clustering algorithm is sensitive to noise data. Synchronization-based hierarchical clustering(SHC)[2] is very difficult to find the optimal local neighborhood distance of synchronizing and the efficiency is not high. In order to overcome these defects, we preprocess the PPI data via transforming the high-dimensional similarity matrix into a low dimension matrix inspired by spectral clustering, then the SHC algorithm is used to perform clustering and glowworm swarm optimization algorithm(GSO)[3] is used to find the optimal threshold of local neighborhood distance of synchronizing.

2   Methods and Results

The pretreatment process of spectral clustering need to construct a similarity matrix A, follow as:

$$A_{ij} = \begin{cases} w\dfrac{|N_i \cap N_j| + 1}{\min(|N_i|,|N_j|)} + (1-w)\dfrac{\sum_{k\in I_{i,j}} w(i,k) \cdot \sum_{k\in I_{i,j}} w(j,k)}{\sum_{s\in N_i} w(i,s) \cdot \sum_{t\in N_j} w(j,t)}, & i \neq j \\ 0, & i = j \end{cases} \tag{1}$$

Then constructing Laplacian matrix L on the basis of matrix A. Matrix X consist of matrix L's eigenvector the first three eigenvalue corresponding, normalize matrix X. Clustering the processed data make use of synchronization-based hierarchical clustering and replace hierarchical search by GSO algorithm, which to find the optimal threshold of local neighborhood distance in SHC.

Dynamic synchronous model applied in this paper follows as:

$$x_i(t+1) = x_i(t) + \frac{1}{|N_\varepsilon(x(t))|} \sum_{y\in N_\varepsilon(x(t))} \sin(y_i(t) - x_i(t)) \tag{2}$$

The objective function(*fval*) of GSO algorithm follows as:

$$fval = \sum_{i=1}^{x} \left\{ (2 \cdot m_{H_i} / (n_{H_i} \cdot (n_{H_i} - 1)))^\rho \cdot \left(\sum_{u,v\in H_i, w_{u,v}\in W} w_{u,v} \Big/ \sum_{v\in H_i, w_{v,k}\in W} w_{v,k} \right)^{1-\rho} \right\} \tag{3}$$

The maximum value of *fval* corresponds to the optimal local neighborhood distance of synchronization.

Hierarchical clustering method based on dynamic synchronous model proposed the concept of neighborhood closures, reducing the running time of synchronization clustering algorithm. Meanwhile the efficiency and accuracy of the algorithm is improved by using GSO algorithm to determine the optimal choice thresholds of local neighborhood distance of synchronizing. The *recall* and *precision* values of the new algorithm are improved and the anti-noise ability is better compared with SC and SHC algorithms, but the time complexity is relatively higher and still need to be decreased.

Acknowledgment. This paper is supported by the National Natural Science Foundation of China (61100164, 61173190), Scientific Research Start-up Foundation for Returned Scholars, Ministry of Education of China ([2012]1707) and the Fundamental Research Funds for the Central Universities, Shaanxi Normal University (GK201402035, GK201302025).

References

[1] Ng, A.Y., Jordan, M.I., Weiss, Y.: On spectral clustering: analysis and an algorithm. In: Advances in Neural Information Processing Systems, vol. 14, pp. 849–856. MIT Press, Cambridge (2001)

[2] Huang, J., Kang, J., Qi, J., Sun, H.: A hierarchical clustering method based on a dynamic synchronization model. Science China: Information Science 43(5), 599–610 (2013)

[3] Krishnanand, K.N., Ghose, D.: Glowworm swarm optimization: a new method for optimizing multi-modal functions. International Journal of Computational Intelligence Studies 1(1), 93–119 (2009)

Order Decay in Transcription Regulation in Type 1 Diabetes[*]

Shouguo Gao[1], Shuang Jia[2,3], Martin J. Hessner[2,3], and Xujing Wang[1,**]

[1] Bioinformatics and Systems Biology Core, Systems Biology Center,
National Heart, Lung and Blood Institute, NIH, Bethesda, MD 20892, USA
[2] The Max McGee National Research Center for Juvenile Diabetes,
Department of Pediatrics at the Medical College of Wisconsin and the Children's Research
Institute of the Children's Hospital of Wisconsin, 8701 Watertown Plank Road,
Milwaukee, Wisconsin, 53226, USA
[3] The Human and Molecular Genetics Center, The Medical College of Wisconsin,
8701 Watertown Plank Road, Milwaukee, Wisconsin, 53226, USA
xujing.wang@nih.gov

In this study, using type 1 diabetes (T1D) as a model system, we investigate two outstanding challenges in the development of molecular signatures of complex traits: the incorporation of gene interaction structure in signature definition, and the integration of multiple Omics data types to refine signature. The T1D data consists of our previously published transcription profiles in control *peripheral blood mononuclear cells* (*PBMC*) induced by sera of 142 human subjects from unrelated healthy controls (uHC), and 3 T1D family cohorts: recent onset (RO-T1D), and healthy siblings of probands that are at high (HRS) or low (LRS) genetic risk for T1D. First both weighted and non-weighted co-expression networks were separately constructed in each cohort and were compared. Several network measures, including edge weight and degree distribution, Shannon's entropy, the λ-coefficient, and h-index, were determined. We found that overall the co-expression networks induced by the RO-T1D cohort are significantly weaker, exhibiting a broad spectrum loss of order and control. More specifically, all T1D family cohorts induced more active and orderly transcription coordination among the innate immunity genes, consistent with our previous report of them sharing a heightened innate inflammatory state. On the other hand, higher coordination of the adaptive immunity genes was only induced by the LRS cohort, potentially explaining their low risk for disease. All the network measures also pointed to the same story, and additionally the importance of the innate immunity genes in determining the transcriptome state of the T1D family cohorts. Next, we integrated the protein-protein interaction (PPI) and the transcriptomic co-expression networks, and focused specifically on the smallest functional units of PPI, the protein complexes (PC). A PC is considered active in a cohort, if its co-expression network is percolated. We found that the RO-T1D cohort activated a significant less number of PC than the others. Overall, whether it is co-expression or protein interaction networks, the four cohorts show striking differences and can be clearly discriminated based on network structural measures. In contrast, gene expression levels alone, without the consideration of underlying interaction networks, could barely differentiate the cohorts. In summary these findings demonstrate the advantage network based metrics in defining molecular signatures.

---

[*] The rights of this work are transferred to the extent transferable according to title 17 U.S.C. 105.
[**] Corresponding author.

M. Basu, Y. Pan, and J. Wang (Eds.): ISBRA 2014, LNBI 8492, p. 404, 2014.
© Springer International Publishing Switzerland 2014

Simulated Regression Algorithm for Transcriptome Quantification

Adrian Caciula[1], Olga Glebova[1], Alexander Artyomenko[1], Serghei Mangul[2],
James Lindsay[3], Ion I. Măndoiu[3], and Alex Zelikovsky[1]

[1] Georgia State University, Atlanta GA, 30303, USA
[2] University of California, Los Angeles CA, 90095, USA
[3] University of Connecticut, Storrs CT, 06269, USA

RNA-Seq is a cost-efficient high-coverage powerful technology for transcriptome analysis. We propose a novel algorithm for transcriptome quantification from RNA-seq data ($SimReg$) which uses regression to find transcript frequencies for which the simulated read counts match the observed read counts.

$SimReg$ first aligns the reads to existing transcript library and then counts the equivalent reads, i.e., reads aligned to the same set of transcripts. The bipartite graph with vertices corresponding to transcripts and read classes is split into connected components which can be treated independently. For each component we simulate high coverage reads and estimate $D_{\mathcal{R},\mathcal{T}} = \{d_{r,t}\}$, where $d_{r,t}$ is the portion of reads from transcript $t \in \mathcal{T}$ belonging to read class $r \in \mathcal{R}$.

Initial transcript frequencies are estimated by minimizing the squared deviation between observed read class frequency $O_{\mathcal{R}} = \{o_r\}$ and expected read class frequency $E_{\mathcal{R}} = D_{\mathcal{R},\mathcal{T}} \times F_{\mathcal{T}}$, where $F_{\mathcal{T}} = \{f_t\}$ are the portions of reads emitted by transcripts. The squared deviation is minimized by the following quadratic program: $\sum_{r \in \mathcal{R}} \left(\sum_{t \in \mathcal{T}} d_{r,t} f_t - o_r \right)^2 \to \min | \sum_{t \in \mathcal{T}} f_t = 1$ and $f_t \geq 0$.

Next $SimReg$ repeatedly updates the frequency estimates by (1) simulating reads according to current estimates $F_{\mathcal{T}}$, (2) finding deviation between simulated and observed reads, $\Delta_{\mathcal{R}} = S_{\mathcal{R}} - O_{\mathcal{R}}$, (3) obtaining corrected read frequencies $C_{\mathcal{R}} = O_{\mathcal{R}} - \Delta_{\mathcal{R}}/2$, and (4) updating estimated transcript frequencies $F_{\mathcal{T}}$ based on corrected read class frequencies $C_{\mathcal{R}}$.

We tested $SimReg$ on several test cases using simulated human RNA-Seq data. Experiments on synthetic RNA-seq datasets show that the proposed method improves transcriptome quantification accuracy compared to previous methods. The results show better correlation compared with currently best method $RSEM$ [1].

Reference

1. Li, B., Dewey, C.: Rsem: accurate transcript quantification from rna-seq data with or without a reference genome. BMC Bioinformatics 12(1), 323 (2011)

M. Basu, Y. Pan, and J. Wang (Eds.): ISBRA 2014, LNBI 8492, p. 405, 2014.
© Springer International Publishing Switzerland 2014

Author Index